POWER SYSTEMS SIGNAL PROCESSING FOR SMART GRIDS

POWER SYSTEMS SIGNAL PROCESSING FOR SMART GRIDS

Paulo Fernando Ribeiro
Technological University of Eindhoven, The Netherlands

Carlos Augusto Duque
Federal University of Juiz de Fora, Brazil

Paulo Márcio da Silveira
Federal University of Itajubá, Brazil

Augusto Santiago Cerqueira
Federal University of Juiz de Fora, Brazil

This edition first published 2014
© 2014 John Wiley and Sons Ltd

Registered office
John Wiley & Sons Ltd, The Atrium, Southern Gate, Chichester, West Sussex, PO19 8SQ, United Kingdom

For details of our global editorial offices, for customer services and for information about how to apply for permission to reuse the copyright material in this book please see our website at www.wiley.com.

Library of Congress Cataloging-in-Publication Data

Ribeiro, Paulo F.
  Power systems signal processing for smart grids / Paulo F. Ribeiro, Paulo Marcio da Silveira, Carlos Augusto Duque, Augusto Santiago Cerqueira.
     1 online resource.
   Includes bibliographical references and index.
  Description based on print version record and CIP data provided by publisher; resource not viewed.
   ISBN 978-1-118-63921-4 (MobiPocket) – ISBN 978-1-118-63923-8 (ePub) – ISBN 978-1-118-63926-9 (Adobe PDF) – ISBN 978-1-119-99150-2 (hardback) 1. Electric power systems. 2. Signal processing–Digital techniques. 3. Smart power grids. I. Title.
  TK1005
  621.31′7–dc23

                                                                          2013023846

A catalogue record for this book is available from the British Library.

ISBN: 978-1-119-99150-2

Set in 10/12 pt Times by Thomson Digital, Noida, India

1 2014

Contents

About the Authors

Paulo Fernando Ribeiro achieved a PhD in Electrical Engineering from the University of Manchester and has worked in academia, industrial management, electric companies and research institutes in the fields of power systems, power electronics and power quality engineering, transmission system planning, strategic studies for power utilities, transmission and distribution system modeling, space power systems, power electronics for renewable generation, flexible AC transmission systems, signal processing applied to power systems, superconducting magnetic energy storage systems and smart grids. His professional experience includes teaching at US, European and Brazilian universities, and he has held research positions with the Center for Advanced Power Systems at Florida State University, EPRI and NASA. He is a Distinguished Lecturer and Fellow of the IEEE and IET and has written over 200 peer-reviewed papers, chapters and technical books. He is an active member of IEC, CIGRE and IEEE technical committees, including the chair of the IEEE Task Force on Probabilistic and Time-Varying Aspects of Harmonics and membership of the IEC 77A Working Group 9 (Power Quality Measurement Methods) and the CIGRE C4.112 (Guidelines for Power Quality Monitoring: Measurement Locations, Processing and Presentation of Data).

Carlos Augusto Duque achieved a BS degree in Electrical Engineering from the Federal University of Juiz de Fora, Brazil in 1986, and a MSc and PhD degree from the Catholic University of Rio de Janeiro in 1990 and 1997, respectively, in Electrical Engineering. Since 1989 he has been a Professor in the Electrical Engineering Faculty at Federal University of Juiz de Fora (UFJF), Brazil. During 2007 and 2008 he joined the Center for Advanced Power Systems (CAPS) at Florida State University as a visiting researcher. His major research works are in the area of signal processing for power systems including the development of a power quality co-processor, the time-varying harmonic analyzer and signal processing for synchophasor estimation. He is currently the head of the Research Group of Signal Processing Applied to Power Systems, UFJF and associated researcher of the Brazil National Institute of Energy. He has written over 120 peer-reviewed papers and chapters of technical books, and is the author of several patents.

Paulo Márcio da Silveira achieved a DSc degree in Electrical Engineering from the Federal University of Santa Catarina, Brazil in 2001. He has industrial design, academic and research experience in power system equipment, substation, protection and power quality issues, operation of power systems studies and development of protective devices and power quality monitoring algorithms for power utilities applications. He has conducted research on transmission and distribution system modeling, monitoring, measurement and signal processing for fault identification, fault location, protective relays, power quality and energy metering. He has worked as a consultant on power quality and power system protection, conducting research for different Brazilian utilities through the Brazilian Electricity Regulatory Agency (ANEEL). Dr Silveira was a visiting researcher at the Center for Advanced Power System at the Florida State University in Tallahassee, US in 2007, when he worked with real-time digital simulations. He is an associate professor at the Itajubá Federal University (UNIFEI) in Brazil, where currently he is also the coordinator of a post-graduate course on Power System Protection, the coordinator of the Electrical Compatibility for Smart Grid Study Center (CERIn), and the head of the Electrical and Energy System Institute of the UNIFEI.

Augusto Santiago Cerqueira achieved a DSc degree in Electrical Engineering at the Federal University of Rio de Janeiro, Brazil in 2002. In 2004, he began his academic and research activities at the Federal University of Juiz de Fora (UFJF), where he is currently an associate professor. His academic and research activities mainly involve electronic instrumentation, digital signal processing, computational intelligence for power systems and experimental high-energy physics. He has participated in and coordinated research projects related to power quality issues, applying signal processing and computational intelligence techniques for power quality monitoring and diagnosis. He is coordinator of the UFJF group at the Large Hadron Collider at CERN (European Organization for Nuclear Research), which conduct research into experimental high-energy physics instrumentation, signal processing and computational intelligence mainly for signal detection and estimation.

Preface

This book has grown out of a cooperation between friends who have a common interest, expertise and passion for power systems (PS) and signal processing (SP). It has evolved as a consequence of SP projects applied to power quality (PQ) and power systems in general.

The rapid growth of computational power associated with the cross-fertilization of applications and use of SP for analysis and diagnosis of system performance has led to unprecedented development of new methods, theories and models.

The authors have come to appreciate the potential for much wider applications of SP, prompted in particular by the modernization of electric power systems via the current and comprehensive developments associated with the implementation of smart grid (SG) technologies.

The increasing complexity of the electric grid requires intensive and comprehensive signal monitoring followed by the necessary signal processing for characterizing, identifying, diagnosing and protecting and also for a more accurate investigation of the nature of certain phenomena and events. SP can also be used for predicting and anticipating system behavior.

For electrical engineering SP is a vital tool for clarifying, separating, decomposing and revealing different aspects and dimensions of the complex physical reality of the operation of electrical systems, in which different phenomena are usually intricately and intrinsically aggregated and not trivially resolved.

SP can be qualified by the analytical aspects of the electrical systems, and can help to expose and characterize the diversity, unity, meaning and intrinsic purpose of electrical parameters, system phenomena and events.

As the electric grid becomes more complex, modeling and simulation become less capable of capturing the influence of the multitude of independent and intertwined components within the network. SP deals with the actual system and not with modeling abstraction or reduction (although it may be used in connection with simulations), so may clarify aspects of the whole through a multiplicity of analytical tools. Consequently, SP allows the engineer to detect and measure the behavior and true nature of the electric grid.

Today, the vast majority of analog signals are converted to digital signals. In the context of electrical systems, this conversion is carried out by numerous secondary smart digital devices that perform the tasks of controlling, metering, protecting, supervising or communicating with other components of the system. Moreover, the quality of such smart devices is enhanced by their ability to perform digital signal processing (DSP).

The term DSP is used to describe the mathematics, algorithms and techniques used to manipulate signals after they have been converted into a convenient digital form in order to

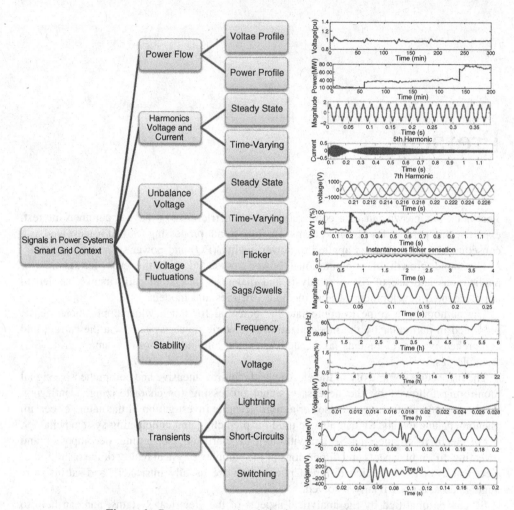

Figure 1 Power systems signals in the context of smart grids.

address a wide variety of needs such as the enhancement of visual images, recognition and generation of speech and compression of data for storage and transmission [1].

The aim of this book is to further promote the use of DSP within power systems, and to expand its application in the context of smart grids. Various techniques are presented, discussed and applied to typical and expected system conditions. Figure 1 illustrates a sample of the gamma of waveforms of typical power systems signals in a context of traditional and smart-grid power system environments.

Chapter 1 describes the motivation for the use of signal processing in different applications of power systems in the context of the smart grids of the future. A wide variety of digital measurements and data analysis techniques required to deliver diagnostic solutions and correlations is provided.

Chapter 2 provides a comprehensive list of power system events and phenomena in terms of time-varying voltage and current signals, characterizing these in terms of magnitude, phase and waveform. It will become apparent that many signals can be represented by a mathematical expression (e.g. exponential DC, faults, waveform distortions).

Chapter 3 describes the different aspects as related to voltage transformers, current transformers, analog filters and analog to digital converters. These components are sources of noise and errors, and impose speed constraints. Due to the lack of information about acquisition systems for electric power signals, this chapter addresses a few of the important demands that are generally neglected in common signal processing literature.

Chapter 4 covers discrete transforms essential in the analysis and synthesis of power systems signal processing. The chapter describes the discrete-time Fourier transform (DTFT), discrete Fourier transform (DFT) and z-transform, as well as a summary of the continuous transforms. Although these transforms are widely treated in several textbooks, the focus of the authors is on specific and common power systems applications.

Chapter 5 covers basic aspects of power system signal processing. These include digital signal operators (delay, adders, multipliers), digital signal operations (modulation, filtering, correlation and convolution), finite impulse response filters and infinite impulse response filters. Several power systems applications are used to illustrate these concepts.

Chapter 6 covers the multirate and sampling frequency alterations, a common time-variant method used in power systems to change the sampling frequency or to analyze a signal. Such an example is using filter banks or wavelet transform. (Filter banks and wavelet transform are covered in Chapter 9, but the digital principles for the implementation of these structures are presented in Chapter 6.) Offline and real-time frequency alterations for power systems application are also discussed.

In Chapter 7 the focus is on algorithms that are capable of estimating parameters such as phasor, frequency, RMS (root mean square), harmonics and transients (decaying exponential) for real-time and offline applications. The basic concepts of estimation theory are presented, including the Cramer–Rao lower bond (CRLB), the MVU estimator, BLUE and LSE estimators. The smart-grid environment is one of higher-complexity electrical signals, which need to be properly and accurately measured.

Chapter 8 covers the basic concepts of spectrum analysis and parametric and non-parametric spectrum estimations. Common errors in parametric estimation are covered, including aliasing, scalloping loss and spectrum leakage. Among the parametric methods discussed are the Prony, Pisarenko, MUSIC and ESPRIT methods.

Chapter 9 introduces a unified view of time-frequency decomposition based on filter banks and wavelet transforms for power system applications. The short-time Fourier transform (STFT) is presented, and the basic principle of filter banks theory and its connection with wavelets is discussed. The basic theory of the wavelet and relevant signal processing techniques are described. Guidance on how to choose the mother wavelet for power system applications is provided.

Chapter 10 covers pattern recognition as an essential enabling tool for the operation and control of the upcoming electric smart-grid environment. The chapter highlights the main aspects and necessary steps required for providing necessary tools to operate the grid of the future.

Chapter 11 presents the basic aspects of detection theory using the Bayesian framework and discusses the deterministic signal detection for white Gaussian noise.

Chapter 12 discusses the application of wavelet analysis to determine fluctuation patterns in generation and load profiles. This is achieved by the filtering of its wavelet components based

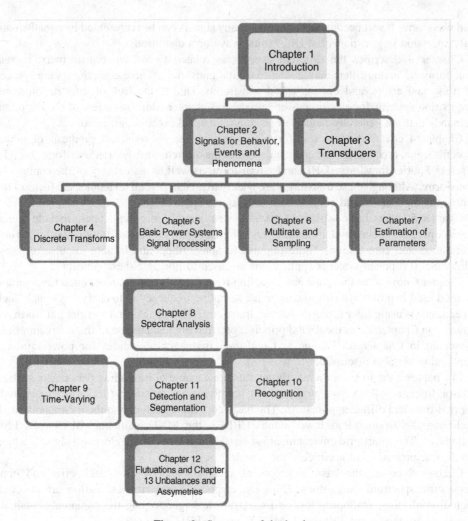

Figure 2   Structure of the book.

on their RMS values, from which it is possible to identify the most-relevant scaling factors. The procedure reveals fluctuation patterns which cannot be visualized via frequency decomposition methods.

Chapter 13 describes an application in which the evaluation of unbalances and asymmetries in power systems can be facilitated by the use of a time-varying decomposition method based on SW-DFT. The time-varying harmonics and their positive-, negative- and zero-sequence components are calculated for each frequency.

Figure 2 depicts the structure of the book.

Finally, some philosophical considerations with regards to the utilization and reception of this book (or any other book) is adapted below from the writings of British author C. S. Lewis:

'A scientific or engineering work such as this can be either *received* or *used*. When we receive it, we exercise our senses and imagination and various other powers according to a pattern suggested by the authors. When we *use* it we treat it as an assistance for our own activities.... *Using* is inferior to *receiving* because, in science and engineering, *using* merely facilitates, relieves or palliates our research/applications; it does not add to it.' [2]

The authors hope that the reader will both use and receive this book as a valuable and thought-provoking guide and tool.

References

1. Smith, S.W. (1997) *The Scientist and Engineer's Guide to Digital Signal Processing*, California Technical Publishing.
2. Lewis, C.S. (1961) *An Experiment in Criticism*, Cambridge University Press.

Accompanying Websites

To accompany this book, two websites have been set up containing MATLAB® files for additional waveforms of typical non-linear loads; these can be signal-processed by different techniques for further understanding. Two MATLAB®-based time-varying harmonic decomposition techniques are also available on site for waveform processing.

Please visit http://www.ufjf.br/pscope-eng/digital-signal-processing-to-smart-grids/
Password: dspsgrid
Or http://www.wiley.com/go/signal_processing

Readers are welcome to send additional waveforms for signals and MATLAB® scripts to be included in the database to Professor Paulo Fernando Ribeiro at pfribeiro@ieee.org.

Acknowledgments

The authors would like to thank PhD students Tulio Carvalho, Mauro Prates, Leandro Manso, Ballard Asare-Bediako, Vladimir Ćuk and Pedro Machado for their valuable support and for comments, suggestions and assistance in preparing simulations, illustrations and experiments used in this text. Thanks are also due to Dr Jan Meyer from the University of Dresden for his suggestions and contributions to Chapter 3, Dr Jasper Frunt for his contributions to Chapter 12 and Tulio Carvalho and Totis Karaliolios for their contributions to Chapter 13. Thanks also to Mrs Adriana S. Ribeiro for her proofreading of all chapters and helpful editorial suggestions.

The authors are especially grateful to The INERGE - Brazilian Institute of Electric Energy Science and Technology, Brazil, for the sponsoring Prof. Paulo Ribeiro as a visiting research professor during the preparation of this manuscript. The authors are also grateful to the Federal University of Juiz de Fora, Federal University of Itajubá, Technical University of Eindhoven, Netherlands, CNPq, and FAPEMIG, Brazil.

The authors would like to thank their wives and families for their support during the last couple of years of persistent and unrelenting production process, in which new ideas, concepts and experiments have been developed, updated and refined.

1

Introduction

1.1 Introduction

A power system is one of the most complex systems that have been made by man. It is an interconnected system consisting of generation units, substations, transmission, distribution lines and loads (consumers). Additionally, these encompass a vast array of other equipment such as synchronous machines, power transformers, instrument transformers, capacitor banks, power electronic devices, induction motors and so on. In this context the smart grid has contributed even further to this complex situation, of which a better understanding is required. Given these conditions, signal processing is becoming an essential assessment tool to enable the engineer and researcher to understand, plan, design and operate the complex and smart electronic grid of the future.

Signal processing is used in many different applications and is becoming an important class of tools for electric power system analysis. This is partly due to a readily available vast arsenal of digital measurements that are needed for the understanding, correlation, diagnosis and development of key solutions to this complex context of smart grids.

Measurements retrieved from numerous locations can be used for data analysis and can be applied to a variety of issues such as:

- voltage control
- power quality and reliability
- power system and equipment diagnostics
- power system control
- power system protection.

This book focuses on electrical signals associated with power system analysis in terms of characterization and diagnostics, or where signal-processing techniques can be useful such as for the analysis of possible concerns about individual loads and/or state of the system.

A large variety of equipment can be used to capture and characterize system variations. These include monitors, digital fault recorders, digital relays, various power system controllers and other intelligent electronic devices (IEDs). Furthermore, power system conditions and events require signal processing techniques for the analysis of its recorded signals. This book

Power Systems Signal Processing for Smart Grids, First Edition. Paulo Fernando Ribeiro, Carlos Augusto Duque, Paulo Márcio da Silveira and Augusto Santiago Cerqueira.
© 2014 John Wiley & Sons, Ltd. Published 2014 by John Wiley & Sons, Ltd.
Companion Website: http://www.wiley.com/go/signal_processing/

promotes attentiveness to issues in the signal processing community. It will provide an overview of these techniques for the understanding and promotion of solutions to its concerns.

1.2 The Future Grid

The future of the developed, developing and emerging countries in a global economy will rely even more on the availability and transport of electrical energy. It is believed that in the near future the global consumption of electrical energy will grow to unprecedented levels. Additionally, security and sustainability have become major priorities both for industry and society.

The deployment of sustainable/renewable energy sources is crucial for a healthy relationship between man and his environment. These changes are driven by a number of developments in society, where the transition to a more sustainable society is a priority. Moreover, the availability of various new technologies and the deregulation of the electric industry may have an additional impact on future developments.

The sustainable and low-carbon imprinting of a society and problematical energy storage requires an integrated power grid which will play a central role in the achievement of energy-efficiency targets and savings. However, the large-scale incorporation of renewable energy production and novel forms of consumption will substantially increase the complexity of its electricity distribution system. The urgency requirement of this complex smart energy grid is evident from the extensive research and development in this area.

An overall picture of this new complex infrastructure is shown in Figure 1.1, where the smart grid of the future can be seen as a merging of the power system and control information technologies.

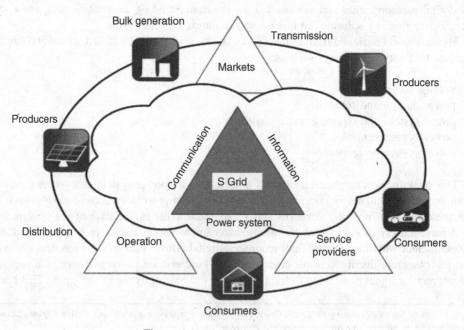

Figure 1.1 The grid of the future.

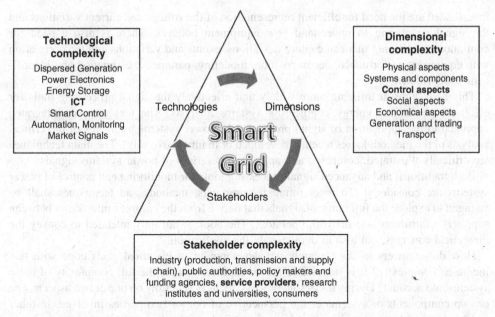

Figure 1.2 The complexity of the smart grid: technologies, stakeholders, dimensions.

The complexity of a smart grid (illustrated in Figure 1.2) might be classified as:

- dimensional complexity
- technological complexity
- stakeholder complexity.

The science and art of designing technological systems within a complex societal environment is a challenging job. In order to produce systems that are synchronized with all the different normative moments of each complexity, new projects must take into account the abovementioned evolving reality. In philosophical terms, a simultaneous realization of different laws and norms is required, where dimensional, technological and stakeholder issues with conflicting objectives and interests need to be accommodated in a well-integrated manner.

In this context, signal processing emerges as one of the most important and effective tools for investigating the operation of such a system.

1.3 Motivation and Objectives

In topics such as power quality, research has traditionally been motivated by the need to supply acceptable voltage quality to end-user loads where voltage, current and frequency deviations in the power system are normal concerns of a systems operator.

The characterization of the incompatibilities caused by these deviations requires an understanding of the phenomena themselves. Listed among the possible aspects to be investigated are the need for efficient representation of the voltage and current variations and the signal processing to understand how equipment behaves. There is also a need for continuous monitoring that can capture deviations, events and variations and the correlation with equipment performance, decomposition, modeling, parametric estimation and identification algorithms.

This book aims at utilizing more widely and effectively the signal processing tools for electrical power and energy engineering systems analysis. The text uses an integrated approach to the application of signal processing in power systems by means of the critical analysis of the methodologies recently developed or in innovative ways. The main techniques are critically illustrated, compared and applied to a variety of power systems signals.

Both traditional and advanced signal processing tools for monitoring and control of power systems are considered. To meet future requirements, methods and techniques shall be engaged to explore the full range of signals that derive from the complex interaction between suppliers, consumers and network operators. The book is not only intended to convey the theoretical concepts, but also to demonstrate the application.

How do engineers in the research and development of electrical grids cope with this increased complexity? It is impossible for an engineer to take the full complexity of these systems into account? During the design process, focus is generally on one or two aspects, one or two components or systems or the perspective of one or two parties involved. In other words, the complex system is reduced to a simplified, neatly arranged subsystem in order to design a new component, to study its performance and to optimize its stability. Through the years this has proven to be a very practical approach as long as the system does not experience major changes, allowing engineering judgment to be used in the simplification process. Unfortunately, a direct consequence of this is that it is not the whole system that is considered: only a reduced system.

In research and development, reduction is unavoidable. Engineers and researchers therefore have to be aware that they study and design in the context of reduced realities. As a consequence, they have to question themselves continuously whether they are missing any relevant dimensions. In practice, engineers cannot easily handle all the technical and non-technical dimensions of an electrical system due to the enormous complexity of smart grids and the requirements of all parties involved, including the requirements of governments and powerful stakeholders. As a consequence it is easy to miss relevant dimensions, to overlook important interactions between technical systems, to neglect the interests of certain parties and to lose a great amount of information. The interaction between multitudes of participants produces very complex signals that must be monitored and processed in order to determine the state of and developments around devices and systems, as depicted in Figure 1.3.

1.4 Signal Processing Framework

The condition of the grid can be fully assessed through the measurement and analysis of signals at different points in the system. Figure 1.4 illustrates the basic concept of signals and parameters that can be processed and derived in steps. First, three-phase signals are decomposed into time-varying harmonics and these are then processed by symmetrical components. The result provides the engineer with a unique tool to visualize the nature of time-varying imbalances and asymmetries in power systems.

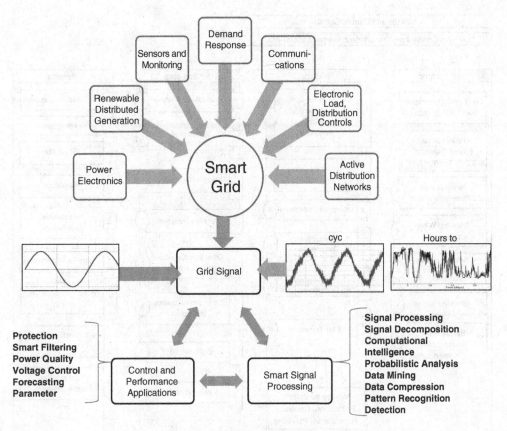

Figure 1.3 Signals, technologies and interactions.

Figure 1.5 further summarizes the signal processing that includes the measurement, monitoring and processing sequences from acquisition, analysis, detection, extraction and classification of the waveforms which might carry useful information for identification of system events, phenomena and load characteristics.

As new signal processing tools are developed to deal with the smart grid developments, it useful to remember that the development of signal processing began in the late 1970s. Figure 1.6 shows the progression of these developments starting with the Fourier series and progressing to time-frequency decompositions, analyzers and advanced signal processing for smart grids. Figure 1.7 shows a summary of these signal processing aspects in the context of smart girds, emphasizing applications, techniques and specifications.

In Figure 1.8 a comprehensive approach to the use of signal processing is illustrated. Here it can be seen that voltage and current signals at a specific point (even in a remote location) can be used to determine impedance, power factors, power flow, stability and so on, where such information can be used by the system operator for more efficient control of the electric grid.

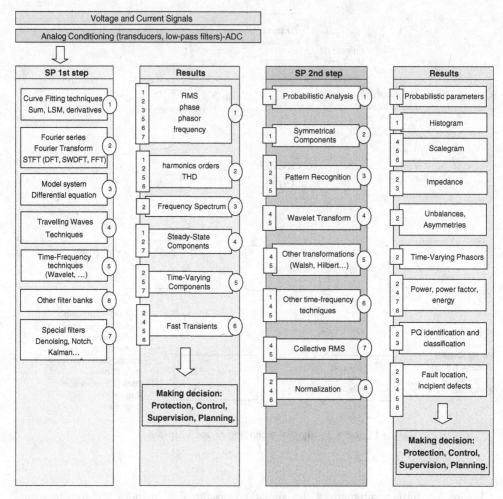

Figure 1.4 Basic concept of signals and parameters that can be processed and derived.

Finally, Figure 1.9 illustrates the perspective of a system, highlighting where signal processing can take place at different points within the network and providing crucial information to system operators.

Finally, the use of a phasor measurement unit (PMU), wide area networks (WANs), home area networks (HANs) and local area networks (LANs), together with developments in information and communications technology (ICT), can be integrated with power quality and energy measurements. Signal processing techniques can then be utilized to facilitate the control, protection and diagnosis of performance of the complex transmission and distribution of the micro cyber-physical smart grid of the future (see Figure 1.10).

Excellent literature has been published [1–7] describing the types of measurements and their technical specifications for power quality and other power systems operation performance requirements.

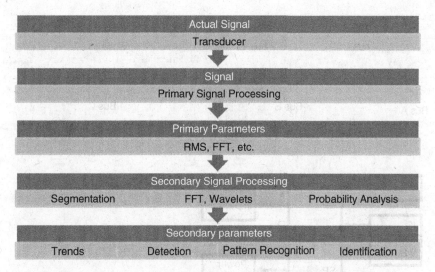

Figure 1.5 Measurement, monitoring and signal processing sequence.

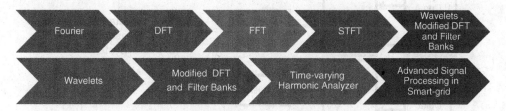

Figure 1.6 Summary of signal processing development.

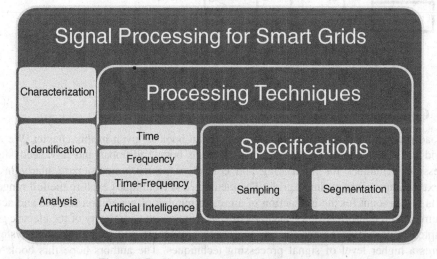

Figure 1.7 Signal processing techniques process.

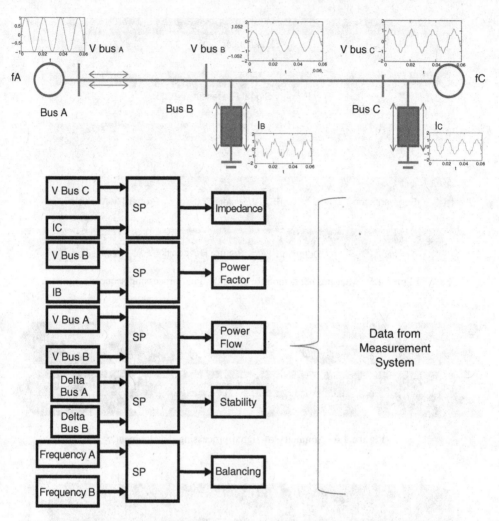

Figure 1.8 A comprehensive system-wide signal processing analysis.

1.5 Conclusions

A broad perspective of the material covered by the book is given in this chapter. We also expand on how to apply in an integrated fashion both traditional and advanced signal processing techniques for monitoring and control of power systems, particularly in the context of future complex smart grids. The methods and techniques explore the full range of signals that account for the interaction of a greater number of generation sources and active consumers with non-linear time-varying loads. The increased complexity of the electric grid, prompted by the development and implementation of smart grid technology and systems, requires a higher level of signal processing techniques. The authors hope this book will increase this awareness and assist with the visualization of solutions and applications.

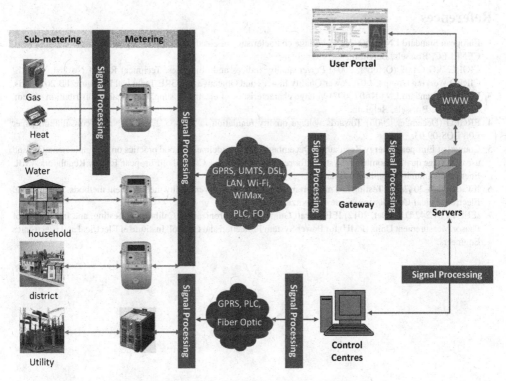

Figure 1.9 System perspective of signal processing.

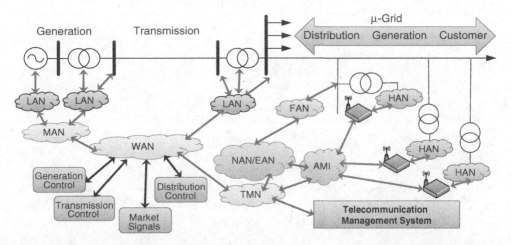

Figure 1.10 The complex transmission and distribution of the micro cyber-physical smart grid of the future.

References

1. European Standard EN50160 (1999) Voltage characteristics of electricity supplied by public distribution system, CENELEC, Brussels, Belgium.
2. CIGRE WG C4.07 (October, 2004) Power quality indices and objectives. Technical Report No 261. CIGRE/CIRED Working Group C4.07, Power Quality Indices and Objectives, CIGRE Technical Brochure TB 261, Paris.
3. European Standard EN50160 (2007) Voltage characteristics of electricity supplied by public distribution system, CENELEC, Brussels, Belgium.
4. ERGEG (December, 2006) Towards voltage quality regulation in Europe. ERGEG public consultation paper E06-EQS-09-03.
5. Council of European Energy Regulators (December, 2012) Guidelines of good practice on the implementation and use of voltage quality monitoring systems for regulatory purposes. Council of European Energy Regulators ASBL Energy Community Regulatory Board.
6. IEC 61000-4-30 (2003) Testing and measurement techniques – power quality measurement methods. International Electrotechnical Commission, Geneva, Switzerland.
7. IEEE PC37.242/D11 (Oct, 2012) IEEE Draft Guide for Synchronization, Calibration, Testing, and Installation of Phasor Measurement Units (PMU) for Power System Protection and Control. Institute of Electrical and Electronics Engineers.

2

Power Systems and Signal Processing

2.1 Introduction

A key aspect of signal processing in power systems is determining which parameters should be measured and to what accuracy, as well as which signal processing methods provide the best characterization and analysis of the signals to be investigated. For example, in many types of studies only the voltage measurements are necessary for an adequate evaluation. However, there are many reasons to measure the current, frequency and active and reactive power of a power system.

The study and application of digital signal processing techniques for the control, protection, supervision and monitoring of smart grids requires an understanding of the electrical system behavior under both normal and unusual or uncharacteristic situations. For any reading the basic sinusoidal signal (voltage and current) may be modified for different reasons, and as such will present distinguishing features in its waveforms.

In this chapter we describe the main phenomena in power systems of time-varying and/or steady-state conditions, in terms of voltage and current. The aim is to characterize each of those taking into consideration its magnitude, phase and waveforms. Furthermore, evidence will be provided showing that many of these signals may be represented by a mathematical expression, such as exponential DCs, faults, harmonics and others.

This is not however the case for completely chaotic signals, such as ferroresonance, sub-synchronous oscillations and voltage fluctuations. The processes that generate such events involve highly non-linear elements such as arc resistances, steel core and so on.

Taking the above into account it can be said that many of the phenomena mentioned can be recreated in simulation models, but others can only be represented by their specific measurements.

Finally, this chapter will demonstrate the importance of knowledge of the phenomena that occur in power systems in order that the correct tools for signal processing can be properly applied. This is especially true in the context of increased system complexity due to the advent of smart grids.

Power Systems Signal Processing for Smart Grids, First Edition. Paulo Fernando Ribeiro, Carlos Augusto Duque, Paulo Márcio da Silveira and Augusto Santiago Cerqueira.
© 2014 John Wiley & Sons, Ltd. Published 2014 by John Wiley & Sons, Ltd.
Companion Website: http://www.wiley.com/go/signal_processing/

The following sections describe the different types of signals in electrical systems under different conditions. Due to the wide range of possible waveforms, only the most common waveforms are mentioned here. The representation of the electrical waveforms of the electric grid can be compared to the representation of the electrocardiogram (ECG) representing the electric function of a human heart, giving insight into its health and function. In the same manner, an evaluation of the electrical signals of a power grid can give the electrical engineer the ability to diagnose and predict possible malfunctions of the electric system.

2.2 Dynamic Overvoltage

2.2.1 Sustained Overvoltage

Sustained overvoltage means an increased voltage of an industrial frequency (50–60 Hz) above the rated values. This overvoltage can appear in different regions of the power system such as a generator output or a load terminal. Figure 2.1 is an illustration of a sustained overvoltage.

In general, the excess of reactive power is the primary cause of the overvoltage in an electrical power system. Over a given time period, the reactive power consumed by inductive loads is no longer consumed due to an abnormal occurrence. The immediate effect of this excess is the increase in voltage in different parts and components of the system.

Specifically, in a transmission line the overvoltage can occur in its receiving terminal, either by load rejection or during an energization, when the terminal is opened. This effect is known as the Ferranti effect and is due to the voltage drop across the line impedance and by the absorption of a capacitance-charging current. This can happen when the line is energized with an open-ended terminal impedance and results in a voltage rise at the terminal. Both the

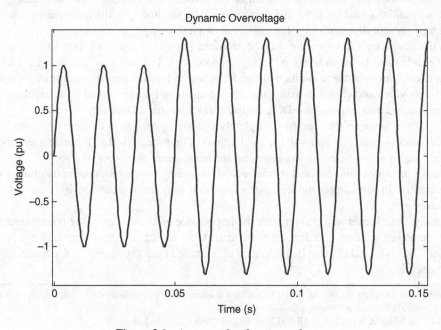

Figure 2.1 An example of an overvoltage.

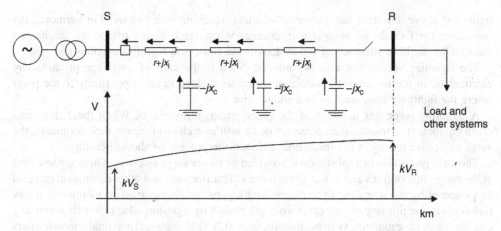

Figure 2.2 Voltage profile in an open-ended transmission line.

inductance and capacitance are therefore responsible for the production of this phenomenon. This will be more pronounced the longer the line and the higher the service voltage. In reality, the resistance, inductance and capacitance values of a long transmission line must be considered as distributed parameters. Figure 2.2 illustrates the effect of reactive power flowing (not active power) in the direction of the voltage source when the voltage at the receiving terminal (R) is higher than its sending terminal (S).

Depending on its intensity and duration, a sustained overvoltage causes the deterioration of the insulation characteristics of the power equipment.

For transformers and shunt reactors, for example, the overvoltage can result in:

- excessive current due to a saturation of the core; such a current will be distorted with harmonics and consequently cause unwanted interference in the rest of the system;
- local damage due to overheating, since the magnetic field during the saturation is sustained at a high level; and
- premature aging (loss of insulation characteristics).

Adequate surge protection is therefore necessary to disconnect the equipment and/or the transmission lines.

2.2.2 Lightning Surge

Lightning can be a source of significant voltage surges. It can hit anywhere in an electric system, and affects the equipment and connected loads. This is true for both high-voltage (HV) and low-voltage (LV) devices.

Electric charges build up in thunderclouds to such an extent that they can break through the atmospheric insulation. This may result in an electric discharge from cloud to ground, and such a current can reach 20–200 kA.

If a lightning discharge occurs directly or in the vicinity of a transmission line, it will cause the movement of electrical charges (outbreaks). These discharges travel close to the speed of light and move through the power conductors reaching substations. Furthermore, the substation itself might be subject to lightning. When these high currents are discharged through the earthing structure of the system, these can cause significant voltage surges.

The lightning surges can also be induced. Such is the case of a voltage produced by electrostatic or electromagnetic induction on cables located in close proximity to the point where the lightning hits, such as in a shield wire.

A lightning surge has duration of the order micro–milliseconds. When these transients reach equipment (a transformer, a reactor or an insulator chain) through their terminals, the surge can cause cracks in the insulation and start the process of short-circuiting.

The main protection in a substation is provided by power surge arresters. These are installed at the entry point of lines and power transformers. Transformers and other equipment can also be protected by the so-called overvoltage 'spark gaps'. It is important to mention that any transients in the primary circuit (high voltage) caused by lightning also affect its secondary circuits through capacitor voltage transformers (CVTs), voltage (potential) transformers (VTs), current transformers (CT) or through electrostatic and/or electromagnetic induction.

The standard lightning current has a fast rise followed by a slow decay. The typical waveform has a rise of 1.2 μs and decay of 50 μs. Such a waveform can be generated in a HV laboratory in order to test equipment for lightning currents. The waveform, shown in Figure 2.3, can be described by the equation:

$$v(t) = V_0 \left(e^{-t/t_b} - e^{t/t_a} \right) \tag{2.1}$$

where V_0 is initial voltage, t is time and t_a and t_b are time to reach 30% and 90% of the peak value, respectively. In this case, $t_a = 71$ μs and $t_b = 0.2$ μs.

Not all instruments are equipped to receive or capture a lightning strike on a voltage signal, due to the short duration of the phenomenon. Figure 2.4 depicts an example of such a voltage signal.

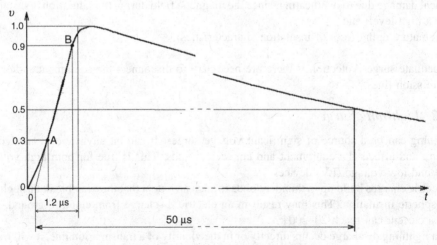

Figure 2.3   Standard lightning voltage.

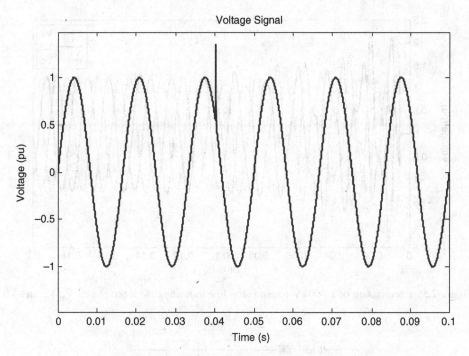

Figure 2.4   Voltage spike from lightning surge.

2.2.3 Switching Surges

Switching in power systems (close and re-closing operations) causes switching surges in high-voltage systems and their auxiliary control circuits.

The switching surge occurs in many different forms and has many different sources. Normally these are associated with a change in the operating state of the system, which means a switching involving trapped energy and it's release.

Transients in power circuits are caused by the transition from one state to another as, for example, when a circuit breaker opens. In this transitory period the energy accumulated in the electromagnetic fields is redistributed causing transients, as shown in Figure 2.5.

The exchange of energy between electric and magnetic fields occurs not only in the fundamental frequency (50 or 60 Hz); there will also be oscillations between the related fields at other frequencies. These will depend on the involved inductances and capacitances of the circuit.

The radian frequency of the transient component is given by:

$$\omega_S = \sqrt{\frac{1}{LC} - \left(\frac{r}{2L}\right)^2} \tag{2.2}$$

where ω_S is the natural angular frequency (rad/s); r is damping resistance; L is inductance; and C is capacitance. If $r = 0$, $\omega_s = \omega_0 = \sqrt{\frac{1}{LC}}$.

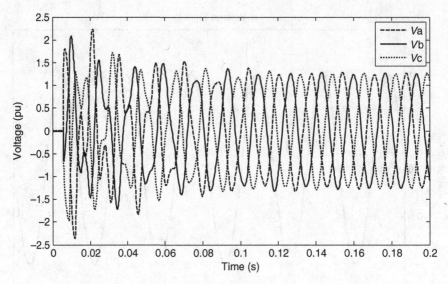

Figure 2.5 Energization of a 400 kV transmission line indicating the three phases V_a, V_b, and V_c.

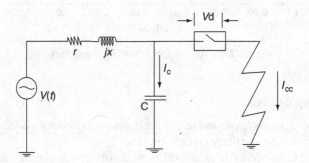

Figure 2.6 Cleaning a short circuit.

For the circuit shown in Figure 2.6 it can be demonstrated that the terminal voltage at the opened circuit-breaker (V_d) overlaps the fundamental signal. This is approximately defined:

$$V_d = V_c = V_M \left[1 - e^{-t/\tau} \cos(\omega_0 t)\right] \tag{2.3}$$

where V_c is the terminal voltage at the shunt capacitance and V_M is the fundamental amplitude of the source,

$$V(t) = -V_M \cos(\omega_0 t) \tag{2.4}$$

and τ is the time constant of the circuit:

$$\tau = \frac{2L}{r}. \tag{2.5}$$

The high-frequency damped transient term $V_M = e^{-t/\tau}\cos(\omega_0 t)$ is normally referred to as transient recovery voltage (TRV). Its maximum value ($V_{d,max}$) depends on r (damping

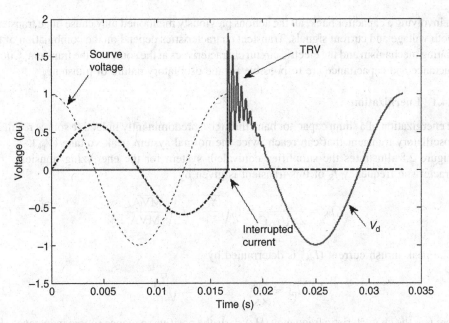

Figure 2.7 A transient recovery voltage at the breaker terminals. When $r \neq 0$.

resistance) and on frequency (ω_0), and its value is within the range:

$$V_M < V_{d,max} < 2V_M. \tag{2.6}$$

If r is considered equal to zero, we have a non-damped transient with frequency f_S, defined:

$$f_S = \frac{1}{2\pi}\sqrt{\frac{1}{LC}}. \tag{2.7}$$

In this case, $V_{d,max} = 2V_M$.

The characteristic of the transient recovery voltage V_d is shown in Figure 2.7.

Circuit breakers are designed with special devices to minimize the magnitude and the impact of these transients. If the rate of rising of the TRV (RRRV) in kV s^{-1} exceeds the rate of recovery characteristic of the dielectric strength or the voltage-withstanding capability between the contacts, the breaker will be unable to hold off the voltage and a re-strike will occur. Re-strike is defined as the breaker conduction of a current half-cycle after the successful interruption at a zero-crossing current. Sometimes the circuit breakers are specified for this special switching of critical circuits.

2.2.4 Switching of Capacitor Banks

It is common practice to install a shunt capacitor in order to improve the power factor and the voltage profile at all voltage levels of an electric power system. These shunt capacitor banks are switched on and off as necessary. The switching operations include energizing, de-energizing, fault clearing, reclosing and energization of a capacitor bank when another bank is already in operation. The latter operation is known as 'back-to-back switching'.

If involving a capacitor bank, all the actions previously mentioned may cause huge transients in both voltage and current signals. Transient characteristics depend on the combination of the initiating mechanism and the electric circuit characteristics at the source of the transient. Circuit inductance and capacitance are responsible for the oscillatory nature of transients.

2.2.4.1 Energization

The energization of a shunt capacitor bank through a predominantly inductive source results in an oscillatory transient that can reach twice the normal system peak voltage (V_{pk}).

Figure 2.8 illustrates the simplified equivalent system for the energizing transient. The characteristic frequency f_S of this transient is given by:

$$f_S = \frac{1}{2\pi\sqrt{L_S C}} \approx f_0 \sqrt{\frac{X_C}{X_S}} \approx f_0 \sqrt{\frac{\text{MVA}_{sc}}{\text{MVA}_r}}. \tag{2.8}$$

The peak inrush current (I_{pk}) is determined by:

$$I_{pk} = \frac{V_{pk}}{Z_S} \quad \therefore \quad Z_S = \sqrt{\frac{L_S}{C}} \tag{2.9}$$

where f_S is the characteristic frequency (Hz); L_S is the positive sequence source inductance (H); C is the capacitance of the capacitor bank (F); f_0 is the system frequency (50 or 60 Hz); X_S is the positive sequence source impedance (Ω); X_C is the capacitive reactance of capacitor bank (Ω); MVA_{sc} is the three-phase short circuit capacity (MVA); MVA_r is the three-phase capacitor bank rating; V_{pk} is the peak line-to-ground bus voltage (V); and Z_S is the surge impedance (Ω).

Because the capacitor voltage cannot change instantaneously, energizing a capacitor bank results in an immediate system voltage drop toward zero. This is followed by an oscillating transient voltage superimposed on the fundamental power frequency waveform. The peak voltage magnitude depends on the point-on-wave (i.e. the instant of energization), and can reach twice the normal system voltage (V_{pk} in per-unit) under worst-case scenarios.

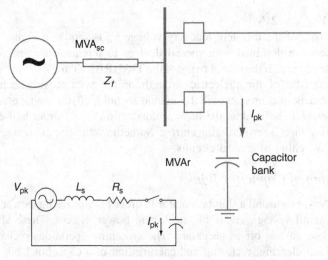

Figure 2.8   System representation of a capacitor bank.

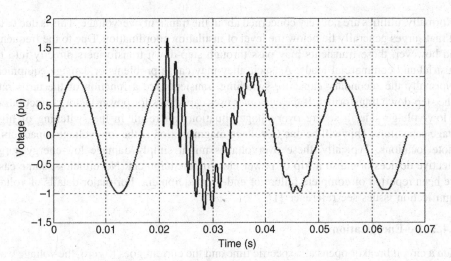

Figure 2.9 Voltage transient during capacitor energizing.

Nevertheless, in a real system the transient magnitude is less than the theoretical 2.0 per-unit energization when considering the system losses, loads and other damped elements. Typically the real magnitude levels range from 1.2 to 1.8 per-unit and the transient frequencies generally fall in the range of 300–1000 Hz. Figure 2.9 illustrates the voltage oscillation during the energization of a capacitor bank, obtained by simulation. Figure 2.10 illustrates the waveforms from a digital fault recorder (DFR). Note that the frequency of the voltage transient is the same as for the capacitor inrush current.

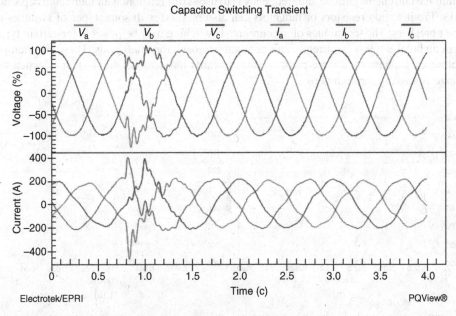

Figure 2.10 Real oscillography of a capacitor bank energization (source www.pqview.com).

Normally, utilities are not very concerned about the transient overvoltage. This is due to the fact that surges generally lie below the level of insulation coordination. Due to the frequency band however, these transients may pass through step-down transformers directly into the industrial and commercial loads. As such, they may cause problems or damage equipment.

Normally the secondary capacitor-switching transients are a function of the turns ratio of the step-down transformer. If the customer uses capacitors for power factor corrections on the low-voltage side, a severe overvoltage situation can result in the switching of high-voltage capacitors. This will be due to voltage magnification at the low-voltage capacitors in remote locations. Typically, these overvoltages might simply damage low-energy surge-protective devices or cause a trip of power electronic equipment. Nevertheless, some cases have been reported of complete failure of end-user equipment. For major details of voltage magnification issues see reference [1].

2.2.4.2 De-Energization

When a circuit breaker opens at a specific time and the current goes to zero, the voltage wave at the open end (where the capacitor bank is installed) goes into a DC mode and the charge remains. This is due to the presence of the shunt capacitance at the end of the circuit, as shown in Figure 2.11a. It is very important to understand the transient recovery voltage (TRV) across the circuit breaker. Figure 2.11b represents these waveforms. Note that the TRV can reach 2 pu. If the TRV magnitude exceeds the allowed ratings of the breaker, a re-strike may occur.

2.2.4.3 Back-to-Back Switching of Capacitor Banks

This event is normally associated with the high-intensity inrush of currents, with high frequencies overlapping the fundamental component. Nowadays, the most common practice to limit the current magnitude and frequency is to use series reactors with individual capacitor banks. Pre-insertion resistors or inductors can also be used with some types of switches or circuit breakers. The techniques of synchronized switching may be possible for certain types of circuit breakers; this is currently an important approach for smart grids. These are attempts to close each phase of a three-phase system towards its ideal condition, that is, when the voltage is passing through zero.

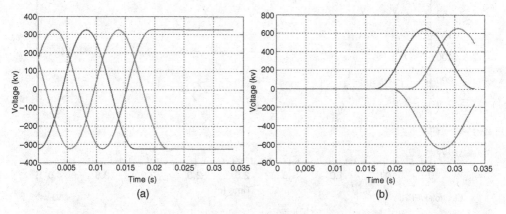

Figure 2.11 Capacitor bank de-energization: (a) voltages across the capacitor; (b) TRV.

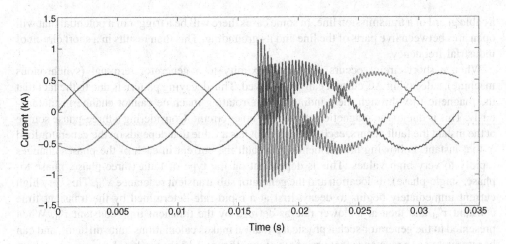

Figure 2.12   Capacitor inrush current during a back-to-back energization.

The frequency and magnitude of the inrush current during back-to-back switching depends on the size of the existing capacitor in service, the impedance of the discharging loop and the instantaneous voltage at the terminals of the capacitor bank at the time of the energization (point-on-wave).

Normally, the impedance of the loop is much lower than the system impedance. This causes a higher inrush current if compared to the current during the energization of an isolated bank. Figure 2.12 illustrates the inrush current during back-to-back energization, characterized by the presence of high-frequency components.

Although the high frequency lasts only a few milliseconds, this may exceed the momentary transient-frequency capability of the switching device. It can also cause fuse blows, false operation of protective relays and excessive errors in current transformers located at the feeder cables of grounded-wye capacitor banks.

2.3   Fault Current and DC Component

Transmission and distribution lines are the components of a power system most exposed to environmental conditions. Rain, wind, lightning, fire, objects carried by the wind, birds and airplanes are among events that may affect the operation of a distribution system or a transmission line. When there is a breakdown of the insulation between conductors, this can be a path to an abnormal flow of current. The current tends to flow through the area of low resistance, bypassing the rest of the circuit. In such an event, the current can be very large or almost negligible, depending on configuration of the system and mainly of the grounding. In any case, this is known as a short circuit or simply a 'fault'.

In power plants and substations, short-circuits occur involving buses, connections, switchgear, transformers, reactors, capacitor banks, power electronic circuits and other equipment.

Among the abovementioned events, the most common cause of a short circuit in a transmission line is lightning. A lightning strike can directly reach an energized cable or a cable guard of a transmission line. In some cases there will be a trigger of a potential that will open arcs between live parts of the line and its grounding. This then results in a short circuit of industrial frequency.

When a short circuit occurs in close proximity to a generator terminal (synchronous machine), a decaying AC current can be observed. This decaying pattern is due to the fact that the magnetic flux through the windings of the rotating machine cannot change instantaneously. This is due to the magnetic nature of its inner circuit. Considering a three-phase source at the instant the fault occurs, each phase current has a value that depends on the generator load at that instant. Following the incidence of a fault the current in each of the phases changes rapidly to very high values. This is dependent on the type of fault (three-phase, phase-to-phase, single-phase), its location and the generator sub-transient reactance x''_d. This very high current immediately begins to decay, first at a rapid rate determined by the transient time constant τ''_d and then at a slower rate as defined by the transient time constant τ'_d. When presented to the generator such a physical situation makes calculations quite difficult, and can be interpreted as a reactance that varies over time. Figure 2.13 shows this phenomenon at three non-discrete levels of current as established in one phase of a three-phase generator. The sub-transient period (2 or 3 cycles) is characterized by x''_d and τ''_d. The transient period (several cycles) is characterized by x'_d and τ'_d, and the current in steady-state or synchronous period. This is imposed by the direct axis reactance (x_d). It is important to note that if a short circuit occurs far from a generator, this AC decaying does not exist practically.

Furthermore, all short circuits in an AC circuit are also associated with the appearance of a DC component. Figure 2.14 illustrates the concept involved in a switching inductive or a capacitive circuit. For the circuit of Figure 2.14 consider the voltage source E:

$$E = E_M \sin(\omega t + \psi); \tag{2.10}$$

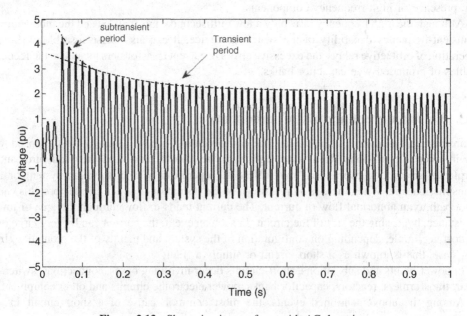

Figure 2.13 Short-circuit waveform with AC decaying.

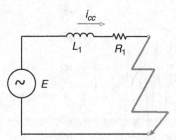

Figure 2.14 Equivalent circuit.

the impedance to the point of fault:

$$Z_1 = \sqrt{x_1^2 + R_1^2};$$ (2.11)

and the electrical angle of the circuit:

$$\theta = \arctan\frac{x_1}{R_1}.$$ (2.12)

The differential equation for this circuit is:

$$R_1 i_{cc} + L_1 \frac{d}{dt} i_{cc} = E \sin(\omega t + \psi)$$ (2.13)

where the angle ψ characterizes the instant in which the circuit is closed or short circuited. The solution of this equation shows that the short-circuit current i_{cc} has two terms:

$$i_{cc} = I \sin(\omega t + \delta) - I \sin\delta\, e^{-t/T_1}.$$ (2.14)

The first term of the equation is a sinusoidal term, where $\delta = \psi - \theta$. The second is a non-periodic term that decays exponentially over time and is dependent on the primary time constant T_1, defined:

$$T_1 = L_1/R_1.$$ (2.15)

This second term is normally referred as the DC component or DC offset of the fault current. The intensity of the DC component depends on the point-on-wave. Equation (2.14) shows that depending on the time instant in which the switching occurs, the value of the DC component magnitude can be higher or lower (even zero, if $\psi = \theta$). On the other hand the duration of the DC component depends on the time constant T_1. In a distribution system the normal value for L_1/R_1 is lower than the L_1/R_1 relation in a transmission system. As a consequence, the DC component disappears very quickly as shown in Figure 2.15b, compared to a DC component in transmission lines occurrences as in Figure 2.15a.

Figure 2.16 shows a real waveform of a current during a phase-to-ground fault. Note the presence of the DC offset that causes an asymmetry in relation to the time axis. In some cases the transient overvoltage at high frequencies may be marked with the DC component, considering the voltage drop on the impedance of the source.

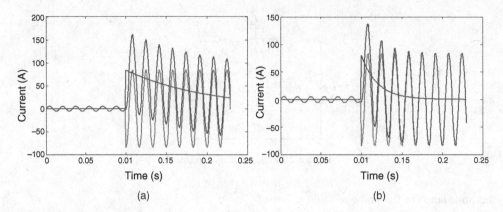

Figure 2.15   Short-circuit waveform with $\psi = 90°$ and (a) $T_1 = 0.1$; and (b) $T_1 = 0.02$.

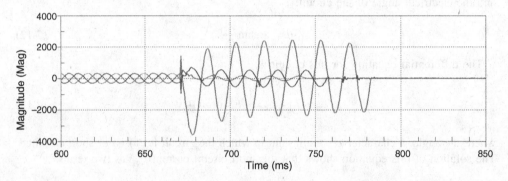

Figure 2.16   A phase-to-ground fault with DC offset.

The three-phase waveforms during a fault close to a generator can be more complex. Since the DC component is mixed with the AC current decay, the short-circuit current can be approximated [2]:

$$i_{cc(t)} = \sqrt{2}\left\{ \left[\left(I_k'' - I_k'\right)e^{\frac{1}{\tau_d''}} + \left(I_k' - I_k\right)e^{\frac{1}{\tau_d'}} + I_k \right]\sin(\omega t - \delta) + I_k'' e^{\frac{1}{T_1}}\sin\delta \right\} \qquad (2.16)$$

where I_k'', I_k' and I_k are the RMS currents in transient and sub-transient periods and in a steady state, respectively. Figure 2.17 illustrates the typical case of a three-phase fault close to a generator.

Finally, it is important to remember that there are different types of faults. They may occur as a result of the abovementioned situations, involving grounding or not:

- single-phase fault (phase-to-ground);
- phase-to-phase fault;
- phase-to-phase-to-ground; and
- three-phase fault.

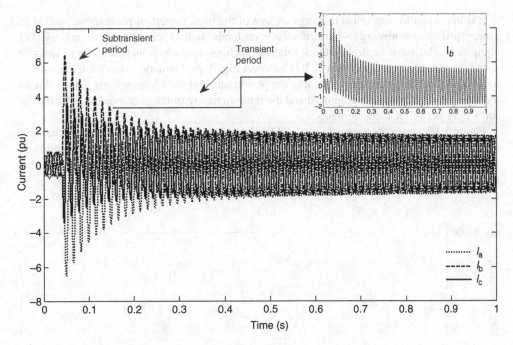

Figure 2.17   Three-phase short-circuit waveforms.

It is observed that the incidence of a single-phase fault is always higher than other disturbances. The statistics published in literature show that around 85–90% of all short-circuits in transmission lines are single-phased.

2.4   Voltage Sags and Voltage Swells

Voltage sags (dips) are defined as an attenuation voltage signal with a duration of approximately 0.5 cycles up to 1 min. The typical magnitude is 0.1–0.9 pu. This phenomenon is often associated with short circuits; however, it can also be caused by switching transients of heavy loads such as large motors. The standards [3,4] define undervoltage as a similar phenomenon; its duration is longer than 1 min however.

The phenomenon 'voltage sag' can be completely defined by measuring its magnitude (RMS value) and its duration. The magnitude depends on the fault conditions: location, resistance and the electrical system conditions such as as topology, source impedance ratio (SIR), short-circuit level, grounding system and so on. The duration of the voltage sag is directly related to the timing of the power system protection. This includes fault detection, fault magnitude measurements, trip command by the relay or the time of fault clearing by the circuit-breaker.

In a three-phase system the voltage magnitude and duration is however also dependent on the kind of fault and the connection of the transformer(s) windings between the fault location and the point of the interest. This is normally the point of common coupling (PCC), where the sensitive load is installed. Such an example is that of a step-down Dy transformer. A single phase-to-ground fault on the primary side (D) will result in voltage sags on the secondary side (y) in two phases.

It is important to emphasize that sags are one of the most important phenomena studied in power quality, considering its harmful effects on loads such as information technology (IT) equipment, electronic control systems, rotating machines (mainly induction motors), variable speed drivers, contactors and relays. It is however highly problematic to establish the indices to be used as a reference when assessing the performance of the whole system. This is due to the number of points to be monitored and the random nature of the phenomenon, which calls for a long period of measurements.

Figure 2.18 represents a record from a DFR showing (a) a phase-to-ground fault current and (b) the respective voltage sag of this phase in a 230 kV system. In this case, the propagation of the voltage sag through a Dy transformer is also measured in two phases on 69 kV sub-transmission system (c).

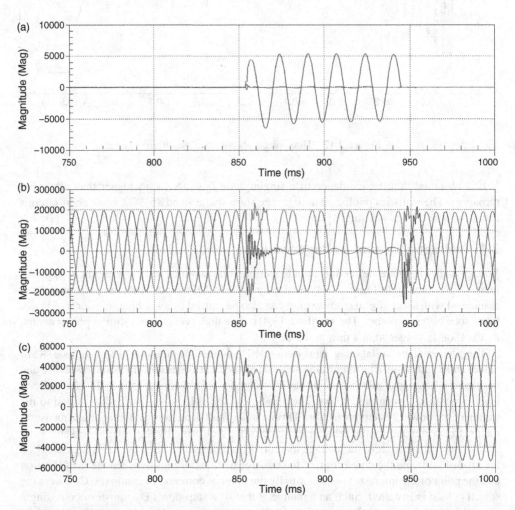

Figure 2.18 Voltage sags caused by short circuit: (a) phase-to-ground fault current in 230 kV system; (b) the voltage sag measured at the 230 kV busbar; and (c) the voltage sage measured at the 69 kV system.

Voltage swells are defined as an increase of the voltage signal with a duration of 0.5 cycles up to 1 minute, of typical magnitude 1.1–1.8 pu. This phenomenon can also be associated with short circuits. It can cause malfunctioning or damage of electronic equipment, controllers, computers and so on. In isolated systems or power sources (generators, transformers) with neutral earth through high impedance, normally the three-phase voltage profile during a phase-to-ground fault will present an increase of the voltages in at least two phases. The magnitude of the overvoltage ($t > 1$ min) or swell (0.5 cycle $< t < 1$ min) depends on the earthing level or on the relation between zero sequence impedance and positive sequence impedance (Z_0/Z_1).

Equation (2.17) gives an idea of the voltage increase during a phase-to-ground fault:

$$e = \frac{1}{2}\sqrt{3\left(\frac{X_0/X_1}{(X_0/X_1) + 2}\right)^2 + 1} \tag{2.17}$$

where e is the factor of increase of the single-phase voltage, defined:

$$V_{\text{phase-to-ground}} = eV_{\text{phase-phase}}. \tag{2.18}$$

Consider a system with a high-impedance earthing and $X_0/X_1 = 10$. In this case, during a solid phase-to-ground fault with voltage going to zero the sound lines (single-phase voltage) will be 0.88 of the phase-phase voltage. This means an increase of 1.53 pu or 53% on the single-phase voltages of the system. In the case where $X_0/X_1 = \infty$ (theoretical isolated system), this results in $V_{\text{phase-to-ground}} = V_{\text{phase-phase}}$ since $e = 1$ (i.e. an increase of 73% or 1.73 pu on the sound phase voltage). On the other hand, in a solid grounding system where $X_0/X_1 = 1$, there will be no swells (or overvoltage).

A swell occurs less frequently compared to voltage sags, as the great majority of transmission and distribution systems are not isolated. A complementary study of short-duration voltage variations (sags and swells), including interruptions, can be read in [5,6].

2.5 Voltage Fluctuations

Voltage fluctuations can be described as random or cyclical voltage waveforms. These can be observed when a load requires abrupt variations of current, especially when reactive components are present. The characteristics of voltage fluctuations depend on the load type and the power system capacity. Basically, there are two important parameters for voltage fluctuations: (1) the frequency; and (2) the magnitude. The best-known load that causes this voltage variation is an arc furnace, where the voltage waveform varies in magnitude due to the fluctuating nature or intermittent operation of connected loads. Typically the magnitudes of these variations do not exceed 10% of the rated voltage.

This variation in magnitude is usually much lower than the sensitivity threshold of most equipment. Consequently, operational problems are experienced only on rare occasions. The main disturbing effect of these voltage fluctuations are changes in the illumination intensity of light sources, commonly called scintillation or flicker. 'Flicker' is the term used to refer to the subjective sensation that is experienced by the human eye when changes occur in the illumination intensity of some kinds of lamps, mainly incandescent. Technically, a voltage fluctuation is an electromagnetic phenomenon. Flicker is an undesirable result of the voltage fluctuation manifested in the luminosity emitted by incandescent lamps. Unfortunately,

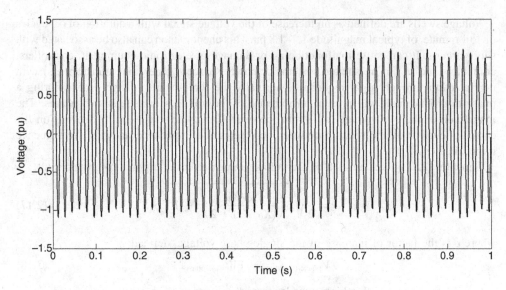

Figure 2.19 A typical voltage fluctuation.

these two terms are often used synonymously in the literature. Figure 2.19 illustrates an example of a fluctuating voltage waveform.

The flicker signal is defined by its RMS magnitude expressed as a percentage of the fundamental; IEC 61000-4-15 [7] describes the methodology as well as the specifications in terms of instruments for flicker measurement. An example of RMS voltage fluctuation caused by arc furnace operation is depicted in Figure 2.20.

It is important to mention that, over the last 10 years, worldwide renewable energy capacity for many technologies has increased at rates of 10–60% annually. Wind power and many other

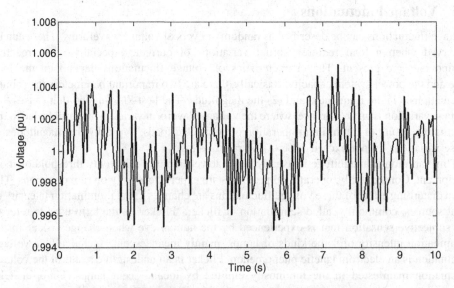

Figure 2.20 RMS voltage fluctuation.

renewable technologies such as photovoltaic cells have experienced accelerated growth. Typically, 10–100 MW are produced in many different areas and connected to the local grid. Many challenges arise from this new scene in terms of operation, especially considering that wind and photovoltaic (PV) power systems are often sources of voltage fluctuations. A reasonable system to control voltage fluctuations should therefore exist.

2.6 Voltage and Current Imbalance

Synchronous generators are three-phase voltage sources in power systems. The voltages should be of the same magnitude in each of the three terminals, positioned at 120° in relation to each other, for a practically balanced and symmetrical system. In practice however it is impossible to obtain a 100% balanced system since the many loads are primarily single-phase loads, especially when in a low- or medium-voltage network. As a consequence, currents that flow through these systems are no longer balanced.

Special three-phase loads can also cause voltage unbalances such as arc furnaces. These may have different impedances in the path of the high current during the melting process [8]. Unbalanced voltages may appear for various other reasons, for example, the self-impedance and mutual impedance between different phases in a non-transposed transmission line.

One of the main tools to assess the imbalance of current and voltage signals is the method of symmetrical components. Unbalanced currents flow through the system and can be decomposed into positive, negative and zero sequences, depending on the electric network and the transformers connections.

These currents will produce additional losses of power (energy). Furthermore, they may cause undesirable heating at some points (wiring connections), and different voltage drops in each phase of a three-phase network. These cause unbalanced voltages at the point of common coupling. In turn, the unbalanced voltages may seriously affect three-phase loads. This is particularly true for motors and asynchronous rectifiers. For example, a negative sequence voltage will induce double-frequency currents at the rotor of a machine, producing additional heating and pulsating torques.

Figure 2.21 shows the three-phase voltages measured in a PCC coupled to a weak system (low level of short circuit).

Finally, it is important to mention that most international standards present limits and indices of voltage imbalance; these are outwith the scope of this book, however. For more details of unbalanced and asymmetrical voltages and currents see [5].

2.7 Harmonics and Interharmonics

Harmonics have always been present in electric power systems. However, due to the widespread use of power electronics during the last decades, these have increased in magnitude. As such, current and voltage harmonic studies have become very important issues in all kind of installations.

The typical definition of a harmonic is 'a sinusoidal component of a periodic wave or quantity having a frequency that is an integral multiple of the fundamental frequency'. [9]. When a harmonic or harmonics are added in a periodic and sinusoidal signal its waveform becomes distorted, representing a non-ideal condition for an electrical grid as shown in Figure 2.22.

Harmonics can be treated by their numbers or orders, taking into account the fundamental frequency as reference. For a fundamental frequency of 60 Hz, the fifth harmonic means

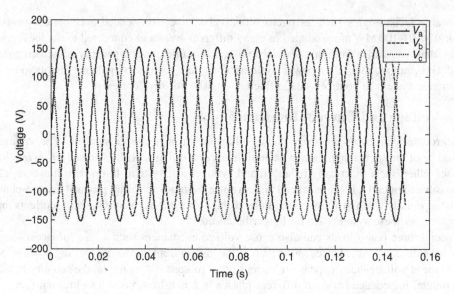

Figure 2.21 Unbalanced voltage (VT secondary) in a PCC.

5 × 60 Hz or 300 Hz and the eleventh means 11 × 60 or 660 Hz. Figure 2.23 illustrates some harmonics components.

A periodic distorted or non-sinusoidal waveform can be obtained by adding different sine and cosines, each one a frequency multiple of the fundamental. A rectangular waveform can be composed in principle by the sum of all odd-order harmonics (sine) whose amplitudes are equal to the inverse of their order. In practice this particular case of the rectangular waveform does not converge uniformly, even considering the sum of a large number of harmonic components. This is known as the *Gibbs phenomenon* [10].

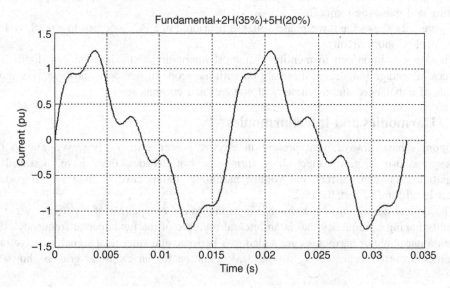

Figure 2.22 Distorted waveform by harmonics.

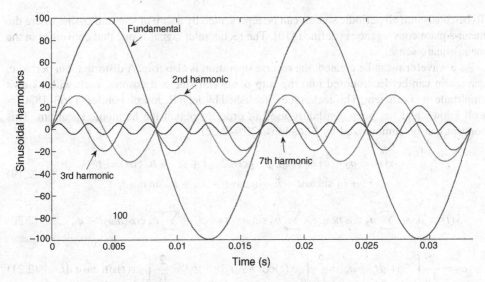

Figure 2.23   Examples of harmonic orders.

Figure 2.24 depicts the sum:

$$\sum_{n=1}^{41} \frac{1}{n}\sin(n2\pi f_0 t).$$

Note that the maximum value of the oscillation does not reduce, even if more terms are added to the series. The Gibbs phenomenon occurs for every signal that presents discontinuity.

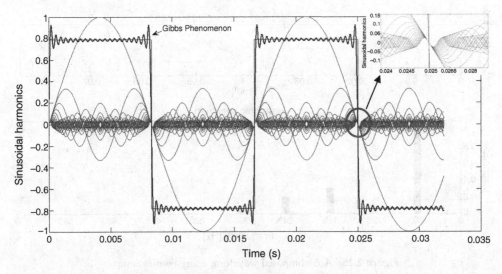

Figure 2.24   The composition of a function and the Gibbs phenomenon.

To be consistent, all periodic signals can be represented by a sum of sine and cosine where the mean-square convergence is defined [10]. The rectangular signal is one that converges in the mean-square sense.

As a waveform can be created, the reverse operation is also true. A distorted, but periodic, waveform can be decomposed into the sum of several sine and cosines, each with its own amplitude or coefficient. This technique established by Jean B. Joseph Fourier (1768–1830) is well known and has been constantly used in order to obtain the harmonic spectrum. This concept can be summarized by Equations (2.19–2.22):

$$x(t) = a_0 + a_1 \cos \omega_0 t + a_2 \cos 2\omega_0 t + \ldots + a_n \cos n\omega_0 t$$
$$+ b_1 \sin \omega_0 t + b_2 \sin 2\omega_0 t + \ldots + b_n \sin n\omega_0 t, \tag{2.19}$$

$$x(t) = a_0 + \sum_{n=1}^{\infty} a_n \cos n\omega_0 t + \sum_{n=1}^{\infty} b_n \sin n\omega_0 t = a_0 + \sum_{n=1}^{\infty} c_n \cos(n\omega_0 t - \phi_n), \tag{2.20}$$

$$a_0 = \frac{2}{T_0} \int_0^T x(t)dt; \quad a_n = \frac{2}{T_0} \int_0^T x(t)\cos n\omega_0 t \, dt; \quad b_n = \frac{2}{T_0} \int_0^T x(t)\sin n\omega_0 t \, dt, \tag{2.21}$$

$$c_n = \sqrt{a_n^2 + b_n^2} \quad \therefore \quad \phi_n = \arctan\left(-\frac{b_n}{a_n}\right). \tag{2.22}$$

The harmonic spectrum (as shown in Figure 2.25) is obtained by this technique, referred to as the Fourier series. The theoretical treatment of the Fourier series will be presented in Chapter 4.

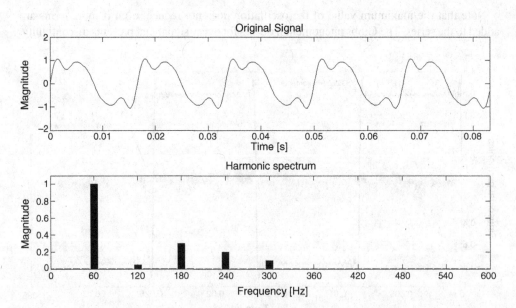

Figure 2.25 A decomposed waveform using Fourier series.

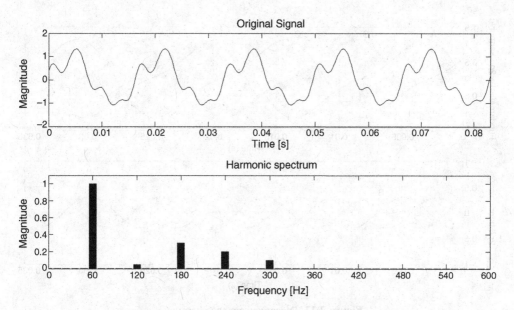

Figure 2.26 The influence of the harmonics phase.

The magnitude of a certain harmonic order depends on the type of the non-linear load (or loads), as well as on the actual conditions of the electrical system. As shown in Equation (2.22) the phase of each harmonic (phase spectrum) can also be obtained. It should be noted that the phase of a harmonic changes the final waveform, as shown in Figure 2.26, although the harmonic content remains the same. The phase of a harmonic in relation to its fundamental frequency can assume any value between 0 and 2π.

In a three-phase system the harmonics flowing through the phases of a system in a balanced way can be decomposed in terms of its symmetrical components. In other words, each order has a sequence that can be positive, negative or zero. Figures 2.27–2.29 illustrate this point, presenting an example of each sequence.

Table 2.1 illustrates how harmonic sequence corresponds to the RMS value of the harmonic h: all triple (or multiple) harmonics are of zero sequence. These are also called homopolar harmonics. Among the sources of harmonics in power systems, three groups of equipment can be distinguished: (1) with a magnetic core (transformers, reactors, rotating machines); (2) special loads (arc furnaces, arc welders, discharge lamps); and (3) electronic and power electronic equipment.

Figures 2.30–2.33 introduce some waveforms, their sources (compact fluorescent lamp, television, notebook and six-pulse rectifier, respectively) and their harmonic spectrum. Table 2.2 summarizes the current waveform for several devices.

Although it is not the focus of this chapter, the harmonic levels in a grid must be supervised and monitored. For this, it is necessary to design instruments that quantify the overall harmonic content, both for current and voltage. Among several factors, one of the most

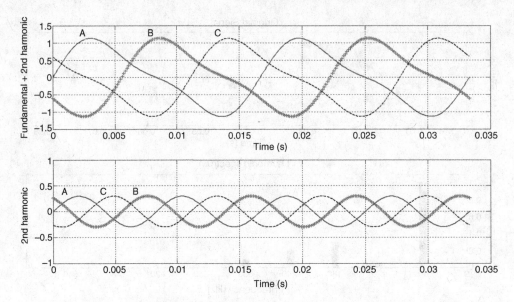

Figure 2.27 Negative sequence harmonic.

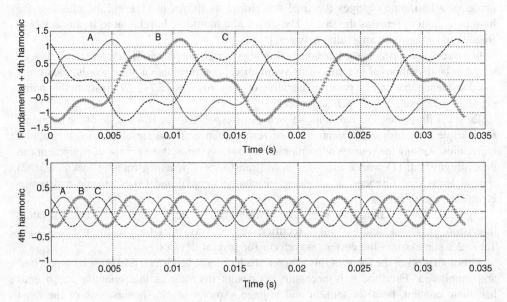

Figure 2.28 Positive sequence harmonic.

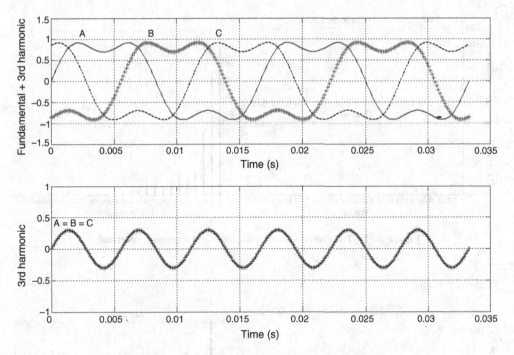

Figure 2.29   Zero sequence (or homopolar) harmonic.

Table 2.1   Harmonic sequence.

Order h	0	1	2	3	4	5	6	7	8	9	10	11	12	13	...
Sequence	0	+	−	0	+	−	0	+	−	0	+	−	0	+	...

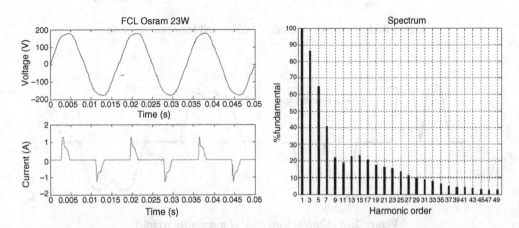

Figure 2.30   Compact fluorescent lamp: current waveform and harmonic spectrum.

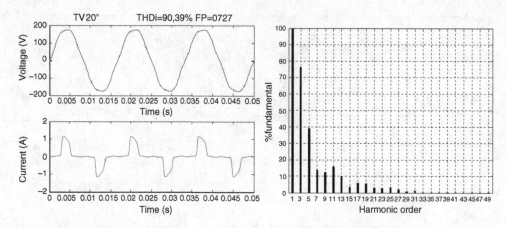

Figure 2.31 Television: current waveform and harmonic spectrum.

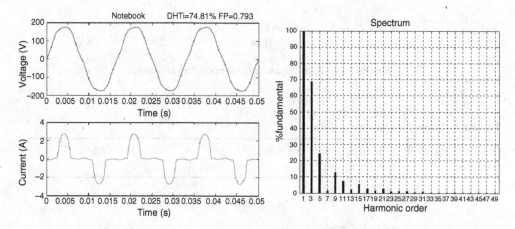

Figure 2.32 Notebook: current waveform and harmonic spectrum.

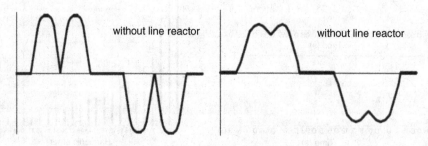

Figure 2.33 Current waveform of a six-pulse rectifier.

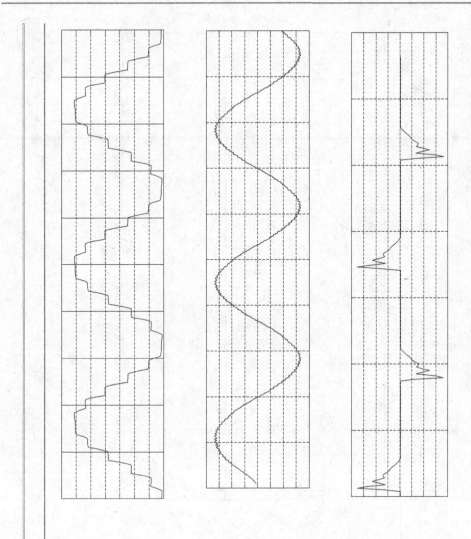

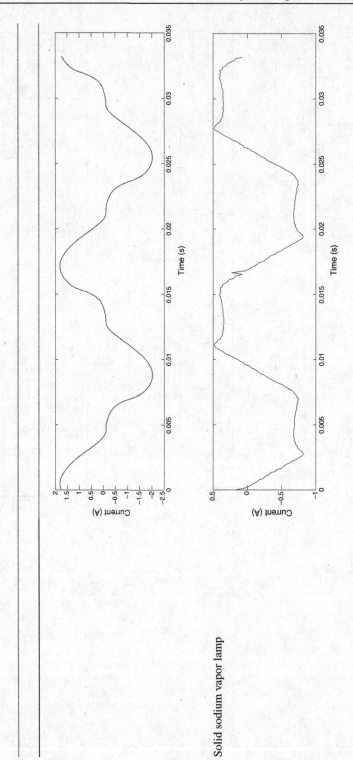

Solid sodium vapor lamp

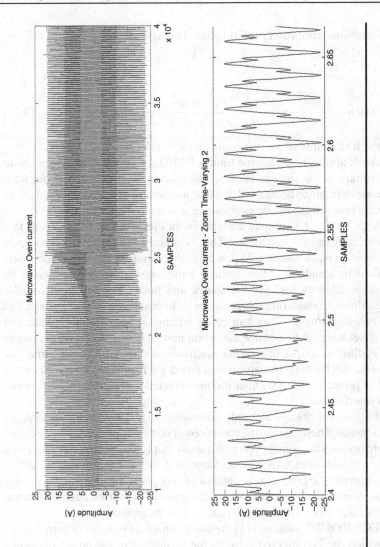

Microwave oven

Microwave zoom

common is the total harmonic distortion (THD) factor. This is used for voltage or current, according to:

$$\mathrm{THD}_f = \frac{\sqrt{h_2{}^2 + h_3{}^2 + h_4{}^2 + \ldots + h_n{}^2}}{h_1} = \frac{\sqrt{\sum_{n=2}^{N_h} h_n^2}}{h_1} \tag{2.23}$$

where h represents the RMS value of the harmonic and n is the harmonic order.

International Standards always require some limits of THD for voltage and current, as in relation to the rated voltage of the system. However, there is some controversy about what should be the best values or limits to be adopted, the best measurement protocol to be used by instrumentation and the best procedure for establishing a measurement campaign.

The presence of high levels of harmonic currents in an electrical network can in fact jeopardize the system components and loads. This is mainly if the point of common coupling (PCC) is linked to weak sources. In this case, the voltage THD can reach levels above those established by standards. These voltage harmonic distortions are caused by voltage drops on the impedances between the source and the PCC when the harmonic currents flow through them. Considering the presence of both non-sinusoidal voltage and current, the effects can be harmful. They can cause additional losses of the phase and neutral conductors, appearance of non-active distortion power, deterioration of the power factor of the plant, additional torque in rotating machines, additional losses of iron and copper in transformers, the burning of capacitors caused by resonance overvoltages or improper operations of protective relays. Other problems include measurements errors and telecommunications interference.

A distorted waveform is not always and only composed of harmonics that are integer multiples of the fundamental. These can also be composed of non-multiples, commonly called interharmonics. Interharmonics are signals with a frequency that is a non-integer multiple of a fundamental frequency. Studies of electrical events associated with interharmonics are still in progress, but there is currently a great deal of interest in this phenomenon. Interharmonics have recently become more significant since the many types of power electronic systems, cycle converters and similar have led to an increase in their magnitude.

According to the IEC 61000-2-1 standard [11] the interharmonics (voltages or currents) are further frequencies which can be observed between the harmonics of the power frequency voltage and current which are not an integer of the fundamental: 'they can appear as discrete frequencies or as a wide-band spectrum'.

By analogy to the order of a harmonic, the order of interharmonic is given by the ratio of the interharmonic frequency to the fundamental frequency (f_i/f_1). The frequency of 92 Hz represents an interharmonic of 1.533 'order'. Also according to the IEC recommendation, the order of an interharmonic is denoted m [12]. If this ratio is less than unity the frequency is also referred to as a subharmonic frequency. The term 'subharmonic' does not yet have any official definition; it is simply a particular case of an interharmonic and its frequency is less than the fundamental frequency. However, the term has been used in many references.

It is not straightforward to deal with interharmonics because their presence in a signal makes it a non-periodic signal within a given observation window, causing the spread or spillover to the next frequency domain. It can be observed in Figure 2.34 that considering the actual data window, the waveform is non-periodic. In fact, the waveform content is composed

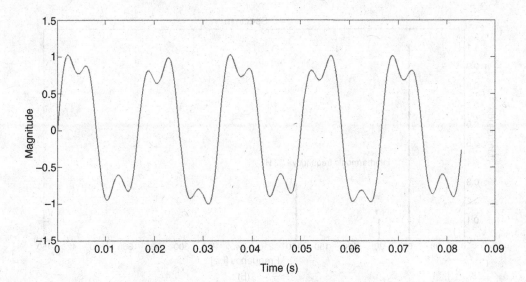

Figure 2.34   A waveform with interharmonic.

by 1 pu of the fundamental, 0.3 pu of the third harmonic and 0.11 pu of a non-integer multiple, in this case 92 Hz. As consequence, the conventional spectral analysis will cause false frequencies if an incorrect data window is utilized, as shown in Figure 2.35a.

The basic sources of interharmonics include variable-load electric drives, arcing loads, static converters (in particular with direct and indirect frequency converters or inverters) and ripple controls. Interharmonics can also be caused by oscillations in systems comprising series or parallel capacitors, during switching processes or where transformers are subject to saturation.

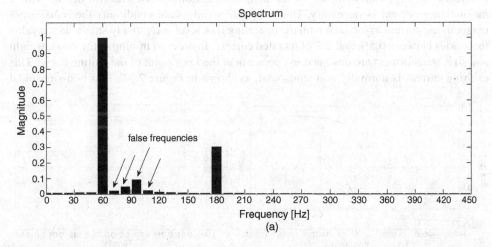

Figure 2.35   From the waveform with interharmonics: (a) a 'false' harmonic spectrum; and (b) the true components.

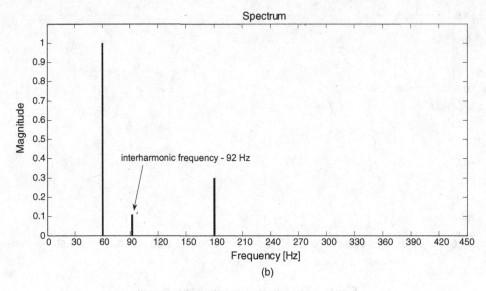

Figure 2.35   (*Continued*)

The most common effects of interharmonics are thermal effects, low-frequency oscillations in mechanical systems, disturbances in fluorescent lamps and electronic equipment operation. Others are interference with control and protection signals in power supply lines, overloading passive parallel filters for high-order harmonics, telecommunication interference and acoustic disturbances. More about harmonics and interharmonics see [22].

2.8   Inrush Current in Power Transformers

A power transformer works by transferring apparent power from one side to another with an efficiency of 95–98%. In order to do this, an electromagnetic transformation with a magnetizing current is necessary. Under normal steady-state conditions the transformer magnetizing current associated with the operating flux level is relatively small, with a value that varies between 0.5% and 2% of its rated current. In order to minimize the costs, weight and size, transformers are designed to operate near the knee point of the exiting curve. This exciting current is normally non-sinusoidal, as shown in Figure 2.36. This non-sinusoidal

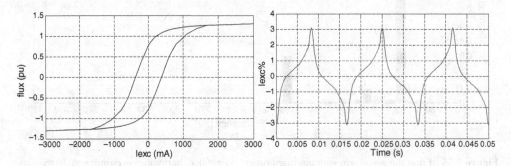

Figure 2.36   Hysteresis cycle and the exciting current.

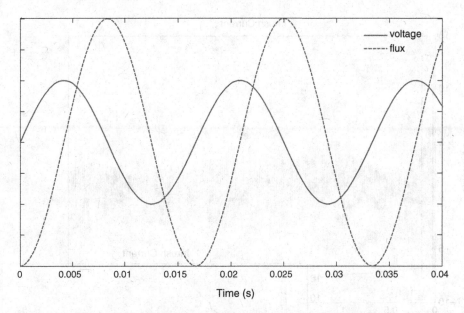

Figure 2.37   A graphical relationship between voltage and flux.

current is due to the magnetization curve and the hysteresis cycle, illustrated in the figure. In other words, on a minor scale, a transformer is a source of harmonic currents during its normal operation.

On the other hand, when a transformer is initially connected to a source of AC voltage, there may be a substantial surge of current through the primary winding of the transformer. This inrush current is required to establish its magnetic field.

It is known that the rate of change of the instantaneous flux in a transformer core is proportional to the instantaneous voltage drop across its primary winding. In a continuously operating transformer, flux and voltage are phase-shifted by 90° since the flux is the integral of the voltage:

$$\phi = \frac{1}{N_t} \int \sin \omega t \, dt = -\frac{1}{\omega N_t} \cos \omega t \qquad (2.24)$$

where N_t is the number of turns. Figure 2.37 illustrates the relationship between voltage and flux. If the winding inductances were linear, the current would have exactly the same waveform as the flux. Both would be lagging the voltage waveform by 90°.

Let us assume that the primary winding of a transformer is suddenly connected to an AC voltage source at the exact moment in time that the instantaneous voltage is at its maximum positive value. Under these circumstances, both core flux and coil current start from zero and build up to the same peak values experienced during continuous operation. There is theoretically no inrush current in this scenario; however, the inductance is not linear and saturation can be expected to occur, especially since transformers are usually working near the knee point. Taking the flux to twice its normal maximum will cause hard saturation, requiring a very large exciting current. However, this is not the worst-case scenario. If a transformer

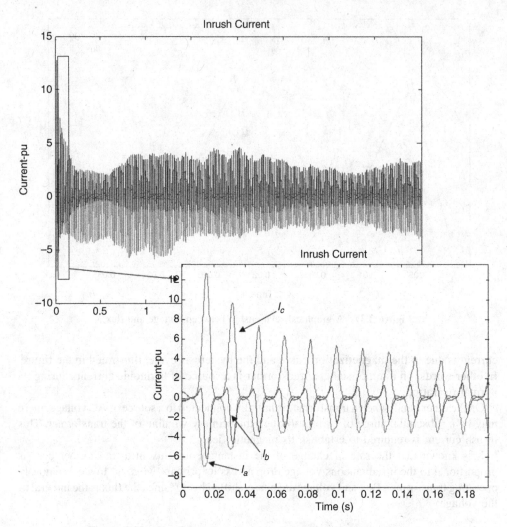

Figure 2.38 A real inrush current in a three-phase bank transformer.

is energized at the zero point of the voltage wave with a maximum residual flux in the same direction, the saturation may be greater and the inrush current will be established reaching 5–20 times the rated current. Moreover, the current waveform will be non-sinusoidal and fully offset from the time axis as sharp pulses, as depicted in Figure 2.38.

For three-phase transformers, as can be seen in Figure 2.38, each phase will have a different exciting current since the point-on-wave at which excitation begins is different for each phase voltage. Thus, even if one phase experiences non-saturation (non-residual flux, point-on-wave at its peak), the other two phases will necessarily have a distorted waveform with great magnitude. In brief, the magnitude of the inrush current strongly depends on the exact time of the connection. If the transformer has some residual magnetism in its core at the moment of the connection to a source, the inrush could be even more severe.

The decay of the inrush current is fast for the first few cycles, but after this it decays slowly and it will take several seconds for the current to reach normal levels. The time constant depends on: (1) the transformer size:10 cycles for small transformers to 1 min for a large one [13]; (2) the resistance of the system; (c) the variable inductance of the transformer. In some cases the current waveform will still be distorted for a relatively long period such as 30 min after the energization point [14]. Other elements that influence the inrush current are the core design, the type of three-phase connection and the sort of steel used for its construction.

It is important to emphasize that the inrush current is rich in harmonics. It contains all orders of harmonics. However, excluding the significant DC offset also present, the second- and the third-order harmonics are by far the greatest in magnitude. The second harmonic will be present in all inrush waveforms of all three phases. Its proportion will vary with the degree of saturation and the presence of DC offset in the core flux. The second harmonic magnitude has been reported as being about 20–60% of the transformer's rated current value. Higher harmonics are also present, but their proportion is much smaller than that of the second and third.

Other factors that control the magnitude and duration of an inrush current are the conditions surrounding the energization of a transformer or a bank transformer. These include: (1) initial energization, normally producing the maximum value; (2) recovery inrush from a fault next to the transformer, when the voltage returns to normal value after the action of the protection system; and (3) sympathetic inrush. The latter is when a current occurs in an energized transformer when a nearby transformer (paralleling a second transformer) is energized. This last case is due to the fact that the flowing inrush current finds a parallel path in the previously energized bank. The DC offset flowing may actually saturate the core of the previously energized bank, causing a kind of inrush as shown in Figure 2.39.

It is important to emphasize that the inrush current is a very interesting time-varying harmonics process, since each harmonic can be seen changing its magnitude and phase during the entire process. However, the main concern regarding an inrush current is the transformer protection. Under normal conditions, the inrush cannot activate the protection system. For a transformer to operate correctly however, the protective relay needs to distinguish between an internal fault and an inrush current. Nowadays, with more intelligent numerical relays, this task is relatively simple.

In order to minimize the magnitude and the distortion of the energized current, a new system of controls for smart grids must be developed so that accurate synchronization can lead to the closing instant of a transformer's circuit breakers.

2.9 Over-Excitation of Transformers

The magnetic flux inside a transformer core is directly proportional to the applied voltage and inversely proportional to the power frequency. Overvoltage and/or under-frequency conditions can cause over-excitation conditions that can saturate the transformer core.

This over-excitation causes heating and increases the exciting current, as well as noise and vibration in a transformer. A typical waveform can be seen in Figure 2.40. The most significant harmonics of this waveform are the third, with about 40% of the fundamental and the fifth with 20%. Normally the fifth harmonic is used to detect over-excitation, since the third can be filtered by a delta connection of the transformer or the delta connection of the CTs.

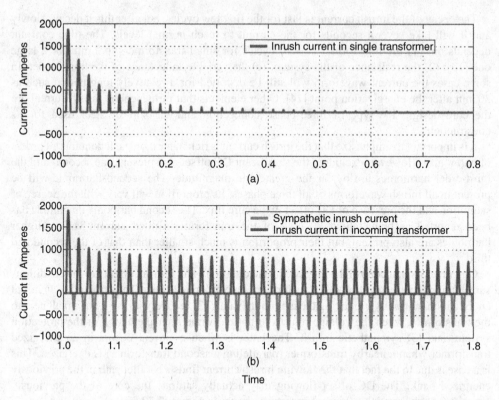

Figure 2.39 An example of a sympathetic inrush.

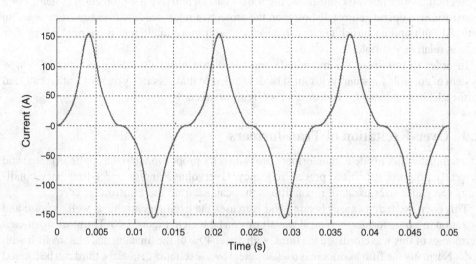

Figure 2.40 Exiting current of an overxcited transformer.

2.10 Transients in Instrument Transformers

Instrument transformers are used for measurement, control and protective applications together with equipment such as meters, relays and other devices of control. Their objective in electric systems is of great importance, considering that these devices are the means by which the values of high current and high voltage are reduced to values such as 1–5 A or 100–120 V, enabling them to be handled by the secondary instruments. This offers the advantage that measurements and protective equipment can be standardized on a few values of current and voltage.

Nowadays there is an entire range of conventional instrument transformers that cover low-voltage 600 V to extra-high voltage of 1000 kV, such as current transformers (CTs), voltage (potential) transformers (VTs) and capacitor voltage transformers (CVTs).

Instrument transformers are designed to present good accuracy in steady-state or during faults in the electrical systems for which they are specified. However, certain events can extrapolate the magnitude levels for which they are prepared. This can result in a distorted waveform of the secondary signal and can be caused by the performance of its own instrument transformers. These secondary transients may affect the performance of the protection and control systems, endangering the reliability of the entire system. Of the three devices mentioned above, CTs and CVTs are of most interest in terms of transients on their secondary terminals. The main issues are discussed in the following sections.

2.10.1 Current Transformer (CT) Saturation (Protection Services)

The performance of a CT can be assessed by considering its ability to accurately reproduce the primary current waveform at its secondary. Normally, two different aspects must be considered: symmetrical and asymmetrical fault current.

2.10.1.1 Symmetrical Fault Current

AC saturation occurs when the current magnitude is higher than expected during a symmetrical fault for which the CT was designed. Usually the accuracy of a CT (typically 5–10% for protective devices) must be confirmed with a standard burden connected in its secondary from the rated current up to 20 times the rated current. If the magnitude of a fault current exceeds 20 times the nominal current, the CT enters into a saturation regimen. It is worth mentioning that every CT has its magnetization curve. Its flux density β is defined by Equation (2.25):

$$\beta = k\frac{E}{N_2 f A} \tag{2.25}$$

where E is electromotive force; N_2 is the number of secondary turns; f is the system frequency; A is the cross-section of the core; and k is a manufacturing constant. The secondary rated voltage of a CT which will be available to the relay is calculated based on this factor of 20. Suppose a CT with rated current equal to 500-5A and a nominal standard burden of 25 VA or 1 Ω of impedance (VA $= 1 \times 5^2$). For this example, when the current reaches 20 times the rated current, the terminal voltage will be 100 V (20×5 A $\times 1$ $\Omega = 100$ V). This is actually very close to the knee point of the excitation curve. Any current magnitude higher than a factor of 20 greater will cause an increase in the excitation current, which may be above 10% of the nominal value; this is an unacceptable error. In addition to the error itself, such a current

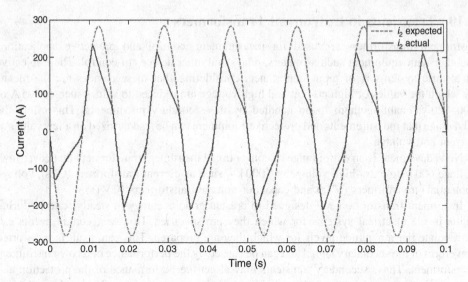

Figure 2.41 Secondary current of a CT during AC saturation (actual and expected).

waveform is distorted, consequently distorting the secondary current flowing to the secondary devices (relays, meters, electronic devices, etc.). Figure 2.41 shows a secondary current waveform of a CT 100-5A, C100 ($Z_{\text{burden}} = 0.5 + j0.866 \, \Omega$) with a symmetrical fault current equal $4000 \, A_{\text{rms}}$ ($40 \times 100 \, A$). Figure 2.42 is another example with a burden equal to $Z_{\text{burden}} = 1 + j0 \, \Omega$.

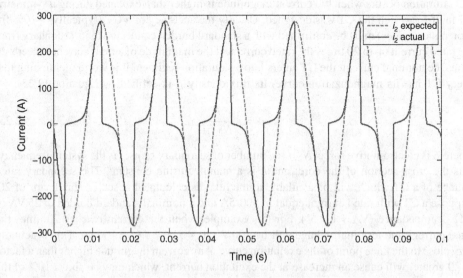

Figure 2.42 Secondary current of a CT during AC saturation (actual and expected).

2.10.1.2 Asymmetrical Fault Current

From the definition of short-circuit current (Equation (2.14)), we assume a short-circuit current that is completely asymmetrical i.e. $\delta = 90°$. This gives:

$$i_1 = i_{cc} = -I\left[\cos(\omega t) - e^{-t/T_1}\right] \qquad (2.26)$$

where T_1 is the time constant of the system, given by Equation (2.13), or $T_1 = X_1/\omega R_1$ where X_1 and R_1 are the reactance and the resistance of the primary system up to the fault point.

Considering a lossless CT with resistive burdens, the following expression for the flux inside the CT core can be derived:

$$\phi = R_{2T}I\left[\underbrace{\frac{e^{-\gamma t} - e^{-\alpha t}}{\alpha - \gamma} - \frac{e^{-\alpha t}\sin\varphi}{\sqrt{\alpha^2 + \omega^2}}}_{\phi\,DC} + \underbrace{\frac{\sin(\omega t + \varphi)}{\sqrt{\alpha^2 + \omega^2}}}_{\phi\,\text{alternated}}\right] \qquad (2.27)$$

where

$$\varphi = \arctan\frac{R_{2T}}{\omega L_m}; \qquad (2.28)$$

where L_m is the magnetizing inductance and R_{2T} is the total resistance of the secondary side (burden plus secondary coil). Further, $\gamma = 1/T_1$ and $\alpha = 1/T_2$, with T_2 being the time constant of the CT, given by

$$T_2 = \frac{L_m}{R_{2T}}. \qquad (2.29)$$

If we consider the case $L_m \to \infty$, Equation (2.27) reduces to

$$\phi = -R_{2T}IT_1\left[e^{-t/T} - 1\right] + \frac{R_{2T}I}{\omega}\sin(\omega t) \qquad (2.30)$$

which comprises another exponential term added to an alternated term and whose evolution with time is illustrated by Figure 2.43.

It is important to note that the relationship between the maximum exponential flux and alternated flux is given by

$$\frac{\phi_{\text{exp,max}}}{\phi_{\text{alt,max}}} = \omega T_1 = \omega\frac{L_1}{R_1} = \frac{x_1}{R_1}. \qquad (2.31)$$

Equation (2.31) highlights the importance of the X/R ratio of the primary system when analyzing CT transient performance e.g. if considering that this ratio will directly influence the saturation time.

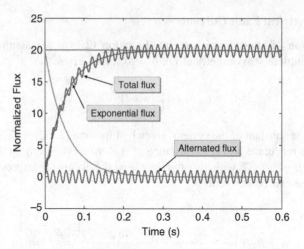

Figure 2.43 Theoretical flux evolution.

The previous consideration is of a theoretical nature since L_m may assume a high value but not infinity (∞). Thus, when $L_m \gg R_{2T}$ and hence $\varphi \cong 0$, Equation (2.27) can be simplified as:

$$\phi = \frac{R_{2T}I}{\alpha - \gamma}[e^{-\gamma t} - e^{-\alpha t}] + \frac{R_{2T}I}{\sqrt{\alpha^2 + \omega^2}}\sin(\omega t). \tag{2.32}$$

In this case, the core flux also consists of an AC and a DC component. The DC component is composed of two different exponentials: (1) γ, defined by the time constant of the primary system (i.e. related to the X_1/R_1 ratio) and (2) α, defined by the time constant of the CT.

It is well known that the DC component reduces proportionally with the reduction of the X/R ratio of the complete circuit, from generator to short-circuit point. In generators the ratio of sub-transient reactance to resistance may reach values as high as 70 [15]. On the other hand, in circuits far away from generators (e.g. utility power distribution systems and industrial power systems), the X/R ratio is lower and the DC component decays quickly.

Figures 2.44 and 2.45 depict the calculated flux, assuming no saturation. The first case is for a system with $T_1 = 0.053$ s ($X_1/R_1 = 20$, typical for power transmission systems). The second case is for a system with $T_1 = 0.013$ s ($X_1/R_1 = 5$, typical for power distribution systems). For both examples, $T_2 = 0.663$ s.

Each design CT core has a saturation level or factor (according to IEEE Std C37.110 [16] guidelines), denoted K_s. This depends mainly on core section and type of lamination material (stalloy, mumetal, etc.), and is calculated from the expression:

$$K_s = V_x/V_s \tag{2.33}$$

where V_x is the saturation voltage and V_s is the secondary voltage for a symmetrical fault.

Considering Figures 2.44 and Figure 2.45, if the core has a saturation level of 6 pu for the first case of Figure 2.44 the CT will saturate as soon as the first cycles occur. On the other hand, the CT for the second case (Figure 2.45) will not saturate. The saturation therefore also

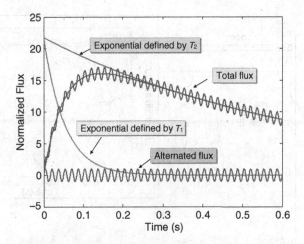

Figure 2.44 Flux evolution ($T_1 = 0.053$ s and $T_2 = 0.663$ s).

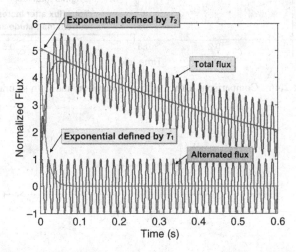

Figure 2.45 Flux evolution ($T_1 = 0.013$ s and $T_2 = 0.663$ s).

depends on two important external factors: the level of fault and the X/R ratio. Figure 2.46 shows in detail a comparison of a 20% increase in magnitude of the fault current and a change in X/R ratio from 5 to 7. For this case the CT saturates in approximately 0.02 s; the greater the fault current and the X/R ratio, the smaller the time to saturation.

The question now is: what happens to the secondary current when the core saturates? From the theoretical model of saturation depicted in Figure 2.47, we can attempt to answer this question. The model shows that when the core tends towards saturation, the primary current becomes equal to the magnetizing current i_0 and the secondary current i_2 tends to zero. Furthermore, when the flux falls below the saturation level the secondary current i_2 follows the primary current i_1, as illustrated in Figure 2.48. Observe that when the transient flux decreases

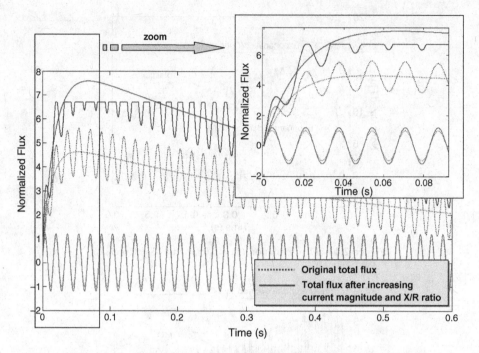

Figure 2.46 Comparison: a flux without saturation and with saturation.

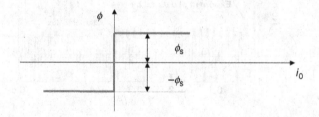

$i_1 = i_2$ (before the saturation) => $i_0 = 0$
$i_1 = i_0$ (during saturation) => $i_2 = 0$ ($e_2 = 0$)

Figure 2.47 A theoretical model of saturation curve.

below the saturation level (the bold line), the waveform of the secondary current is no longer distorted.

In practice, the model shown in Figure 2.48 does not occur. The current does not tend to zero instantaneously, but there is a smoothed change. The data presented in Figure 2.49 are the result of a simulation, representing the secondary current during a CT saturation considering CT 100-5A, C100 ($Z_{burden} = 0.5 + j0.866 \,\Omega$). It was submitted to an asymmetrical fault of 1000 A_{rms} with maximum DC offset. Note that the current in this case does not reach 20 times the rated current; instead, the saturation occurs due to the DC component of the fault.

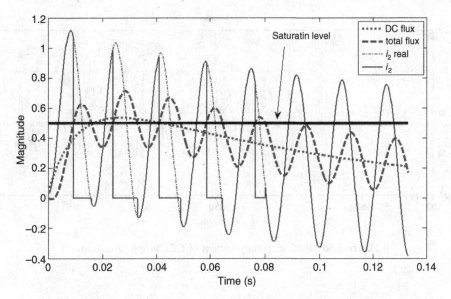

Figure 2.48 Theoretical model of secondary current of a saturated CT.

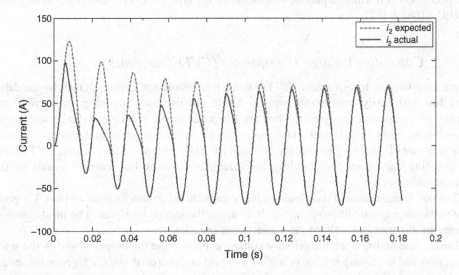

Figure 2.49 Secondary current of a CT during saturation caused by DC offset.

Figure 2.50 depicts a real waveform; during a three-phase fault two CTs go into saturation.

Finally, it is important to observe that CT saturation time is normally a function of current fault, core flux density, CT parameters, CT burden including wire impedance and the DC time constant. Sometimes the time to saturation is long enough for high-speed relaying devices to

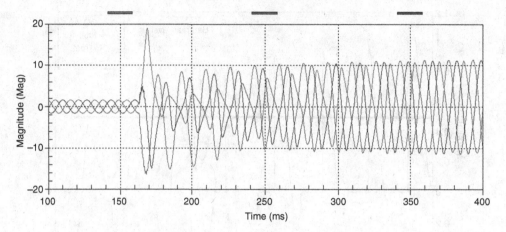

Figure 2.50   Real secondary current of CTs during saturation.

detect faults prior to its CT collapse. In many cases however, the CT saturation may cause improper operation of protective relays, jeopardizing the system performance.

Specific CT performance parameters, including CT type TPX, TPY and TPZ, can be found in IEC 61869-6 [17].

2.10.2 Capacitive Voltage Transformer (CVT) Transients

Capacitive voltage transformers (CVTs) convert transmission class voltages to standardized low and easily measurable values. These are used for metering, protection and control of a high-voltage system. They are the predominant source of voltage signals for impedance relays in HV and extra-high voltage (EHV) systems and provide a cost-efficient way of obtaining secondary voltages for EHV systems. Additionally, CVTs serve as coupling capacitors for coupling high-frequency power line carrier signals to the transmission line.

The non-linear nature of the elements that constitute the primary circuit of the CVT gives rise to electromagnetic phenomena that may affect the secondary signal. The most common events are the transient voltage drop and ferroresonance.

During line faults, when the primary voltage collapses and the energy stored in the stack capacitors and the tuning reactor of a CVT needs to be dissipated, the CVT generates severe transients that affect the performance of protective relays. The higher the system impedance ratio, the worse the CVT transient will be.

Figure 2.51 presents an example of CVT operation during a fault occurring at the zero crossing of the primary voltage. As seen from the figures, the CVT transients can last for up to two cycles and reach a magnitude of up to 40% of the nominal voltage. This small signal seems of no importance; however, it can cause great difficulties for a protective relay to distinguish quickly between faults at the reach point and faults within the protection zone. More details about this kind of problem can be found in reference [18].

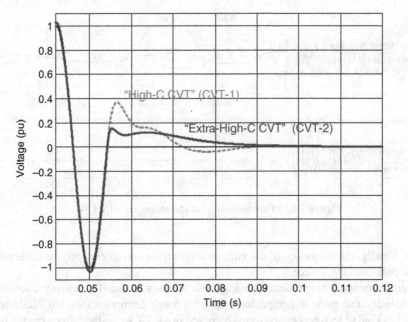

Figure 2.51 Transient voltage drop in the secondary of a CVT [18].

2.11 Ferroresonance

Ferroresonance is a very complex phenomenon. The book which covers it thoroughly has yet to be written and few papers on the topic have been published. According to reference [19], the following categories or classes of ferroresonant circuits have been reported:

1. transformer supplied accidently in one or two phases;
2. transformer energized through the grading capacitance of one or more open circuit breakers;
3. transformer connected to a series-compensated transmission line;
4. voltage transformer connected to an isolated neutral system;
5. capacitor voltage transformer;
6. transformer connected to a de-energized transmission line running in parallel with one or more energized lines;
7. transformer supplied through a long transmission line or a cable with a low short-circuit power.

Ferroresonance is a type of resonance. It can suddenly change from a steady-state response (sinusoidal frequency) to another with a chaotic behavioral response. It is characterized by a high-current fundamental-frequency state. However, it is also characterized for subharmonic, quasi-periodic and even chaotic waveforms and a random time duration in any circuit containing a non-linear inductor. The associated high overvoltage can cause dielectric and thermal problems to the transmission and distribution systems, as well as to secondary devices in CVTs or VTs. It may exhibit different modes of operation that are not experienced in linear

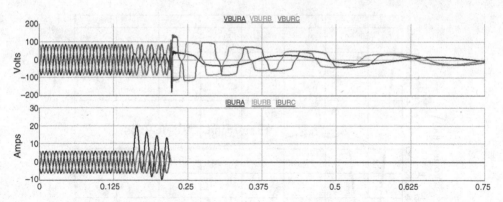

Figure 2.52 Ferroresonance at the secondary of a CVT.

systems. Finally, the frequency of the voltage and current waveforms may be different from the sinusoidal voltage source.

This phenomenon is not predictable using linear theories. Unusual solutions are necessary to detect, identify and prevent equipment damage through ferroresonance. To illustrate, Figure 2.52 is a kind of ferroresonance signal recorded in a CVT secondary after a circuit-breaker has cleared a fault. For a better understanding of the phenomenon, including ferroresonance modes, methods of identifying, modeling and preventing ferroresonance, see reference [20].

2.12 Frequency Variation

In normal operating conditions the system frequency experiences very small variations. In abnormal situations however, such as imbalance between generation and load, the frequency can experience large variations. All these events are supposed to be stabilized by the control and protection systems, leading the frequency back towards its tolerable limits of variation. In critical situations the protection and control systems can however fail. Under such circumstances the power system experiences losses of synchronism and stability. This phenomenon constitutes a vicious circle. Furthermore, because a large number of algorithms are used in control and protection, it is assumed that the frequency is constant. If there is a large deviation in frequency the algorithms cannot work properly. Figure 2.53 shows the frequency variation when the interlinked Brazilian system suffered a major system event. The plot shows the frequency in several sites spread along the Brazilian grid. The large variation of the frequency at certain sites after islanding two main regions can clearly be seen.

2.13 Other Kinds of Phenomena and their Signals

There are many others types of disturbances that can appear in a power system, such as: secondary arcs during a single-pole auto-reclosing power system oscillation or stable and unstable power swings; the loss of excitation in synchronous machines; the appearance of third harmonic voltage at generator neutral during stator winding phase-to-ground fault; and the non-zero current crossing during faults. Problems are also experienced with circuit break re-strike and re-ignition, reversal current, reversal voltage and sub-synchronous oscillations in compensated transmission lines.

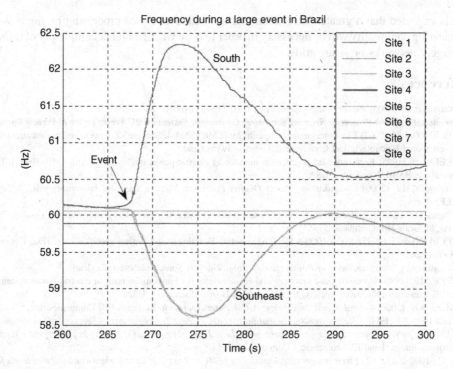

Figure 2.53 Frequency measured in several sites during a strong event in the interlinked Brazilian Power System.

Nowadays, all of the above and their signals (voltage and current) can be recorded using digital fault recorders (DFRs); numerical relays, phase measurement units (PMUs) and other intelligent electronic devices (IEDs) can also be used.

Analysis of system disturbance is very important since it can provide feedback regarding the integrity of the equipment and the overall system, answering the basic questions: What happened? Why did it happen? What is going to be done about it? For more about these questions and answers, see reference [21].

2.14 Conclusions

This chapter gives a comprehensive but concise overview of the most common power system phenomena (time-varying or not). These are discussed in terms of voltage, current signals and waveforms. The idea is to characterize many of them by considering the magnitude, phase and waveforms and to show that many signals can be represented by a mathematical expression (exponential DC, faults, harmonics in general, etc.). The chapter also highlights some of the challenges in applying signal processing techniques to electrical power signal analysis. Although voltage and current quality are one of the most common applications for these techniques, the approach and tools can be used by the entire power engineering community. This chapter gives an extensive list of other issues for which advanced signal processing tools may provide solutions.

It is expected that a greater integration and utilization of signal processing by the power engineering community will be necessary in order to cope with the higher complexity of future intelligent networks or smart grids.

References

1. Natarajan, R. (2005) *Power System Capacitors*, CRC Press, Boca Raton.
2. Reimert, D. (2006) *Protective Relaying for Power Generation Systems*, CRC Press, Taylor & Francis Group.
3. IEC61000-4-30 (2003) Electromagnetic compatibility (EMC), Part 4, Section 30: Power quality measurement. International Electrotechnical Commission, Geneva, Switzerland.
4. IEEE-1159 (1995) Recommended practice for monitoring electric power quality, IEEE Std. 1159–1995. IEEE Standards Board, Piscataway, USA.
5. Bollen, M.H.J. (2000) *Understanding Power Quality Problems: Voltage Sags and InterrruptionsWi*, Wiley–IEEE Press.
6. Dugan, R.C., McGranaghan, M.F., Santoso, S. and Beaty, H.W. (2003) *Electrical Power Systems Quality*, 2nd edn, McGraw-Hill Companies, Inc.
7. IEC61100-4-15 (1997) IEC 61000-4-15, Flickermeter: Functional and design specifications. IEC, Geneva, Switzerland.
8. Baggini, A. (2008) *Handbook of Power Quality*, John Wiley & Sons, Chichester, England.
9. IEEE-519. (1993) Recommended practices and requirements for harmonic control in electric power systems. IEEE Industry Applications Society, IEEE Standards Board, Piscataway, USA.
10. Mitra, S.K. (2006) *Digital Signal Processing: A Computer Approach*, McGraw–Hill Companies, Inc.
11. IEC61000-2-1 (1990) *Electromagnetic compatibility (EMC) - Part 2: Environment - Section 1: Description of electromagnetic environment for low-frequency conducted disturbances and signalling in public power supply systems.* International Electrotechnical Commission.
12. IEC 61000-2-2 (2002) *Electromagnetic compatibility (EMC) - Part 2-2: Environment-compatibility levels for low-frequency conducted disturbances and signalling in public low-voltage power supply systems.* International Electrotechnical Commission.
13. Blackburn, J.L. (2007) *Protective Relaying: Principles and Application*, 3rd edn, CRC Press.
14. Anderson, P.M. (1999) *Power System Protection*, IEEE Press.
15. GE Power. *Dimensioning of current transformers for protection application. Application note, GER3973.* GE Power.
16. IEEE Std. C37.110 (2007) *IEEE guide for the application of current transformers used for protective relaying purposes.* IEEE Power Engineering Society.
17. IEC61869-6 (2012) Current transformer for transient performance. Part 6. International Electrotechnical Commission.
18. Kasztenny, B., Sharples, D., Asaro, V. and Pozzuoli, M. (April 2000) Digital relays and capacitive voltage transformers: balancing speed and transient overreach. *3rd Annual Conference for Protective Relay Engineers, GE Power.*
19. Jacobson, D.A.N. (2003) *Examples of ferroresonance in a high voltage power system.* In Proceedings of Power Engineering Society General Meeting, IEEE Vol. 2, 13–17 July 2003.
20. Ang, S.P. (2010) *Ferroressonance simulation studies of transmission systems. Thesis.* University of Manchester.
21. Ibrahin, M.A. (2012) *Disturbance Analysis for Power Systems*, Wiley & Sons.
22. Acha, E. and Madrigal, M. (2001) *Power Systems Harmonics: Computer Modelling and Analysis*, Wiley & Sons.
23. Arrilaga, J. and Watson, N.R. (2003) *Power System Harmonics*, Wiley & Sons.

3

Transducers and Acquisition Systems

3.1 Introduction

Intelligent electronic devices (IEDs) such as protective relays, digital fault recorders (DFRs), energy meters, power quality monitors, signal analyzers or other secondary devices are required to have reasonably accurate reproduction under the conditions of their power system, whether normal or unusual operational conditions.

Between that power system and the core of a processing device a chain of elements however exists, as shown in Figure 3.1. The first components in connection with the power grid are instrument transformers. Instrument transformers are used for measurement, control and protective applications, together with a diverse array of equipment such as meters, relays and other devices of control. In electrical systems their role is of primary importance as they are the means of 'stepping down' the current or voltage of the system to measurable values, such as 1–5 A in the case of current transformer or 100–120 V in case of a voltage transformer. This offers the advantage that measurements and protective equipment can be standardized on only a few values of current and voltage.

Another important function of instrument transformers is to decouple (isolate) the primary circuit from the secondary. This means that there is no electrical connection between the primary and secondary circuits. The transfer of information between the primary and the secondary circuits is achieved through an electromagnetic transformation. There are basically three types of conventional instrument transformers: voltage transformer (VT), capacitive voltage transformer (CVT) and current transformer (CT).

Recent technology allows the replacement of a conventional instrument transformer by a non-conventional transducer, for example an optical current transformer (OCT), optical voltage transformer (OVT), Rogowski Coil and others.

The objective of this chapter is to present the basic concepts and types of transducers currently used as well as the recently introduced technologies.

Power Systems Signal Processing for Smart Grids, First Edition. Paulo Fernando Ribeiro, Carlos Augusto Duque, Paulo Márcio da Silveira and Augusto Santiago Cerqueira.
© 2014 John Wiley & Sons, Ltd. Published 2014 by John Wiley & Sons, Ltd.
Companion Website: http://www.wiley.com/go/signal_processing/

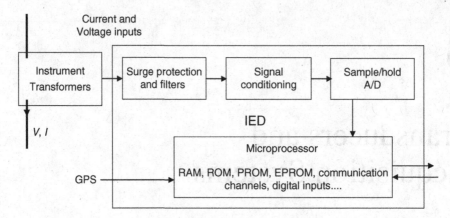

Figure 3.1   Block diagram of an IED.

3.2   Voltage Transformers (VTs)

As for any kind of transformer, VTs (formally potential transformers or PTs) obey the law of electromagnetic transformation as depicted in Figure 3.2. The nominal rate is given by $V_1/V_2 = N_1/N_2 = R_{VT}$.

This class of conventional transformers has both primary and secondary windings. The primary winding is connected directly to the power circuit, either between two phases or between one phase to ground depending on the rating of the transformer and on the requirements of the application. These are specially designed to accurately reflect the primary voltage signal of the power system, at a rated frequency. In a low voltage this is a secondary signal; these may therefore be used in any kind of low-voltage measurement or protective devices.

The flux density β in the magnetic circuit is given by Equation (2.23) (see Chapter 2), and the relation between flux density and current intensity H is not linear, as depicted in Figure 3.3. This curve represents the saturation curve of the transformer core.

The VTs can also be represented by their equivalent circuit as shown in Figure 3.4, where R_1, X_1, R_2, X_2, R_m and X_m are the resistance and the reactance of the primary circuit, secondary circuit and the fictitious magnetizing shunt element, respectively. When analyzing the equivalent circuit (Figure 3.4) the same characteristic curve that is shown in Figure 3.3 can be obtained in laboratory, relating the internal voltage E_2 and the magnetizing current I_0.

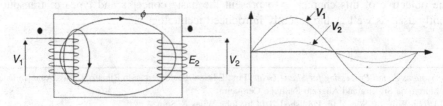

Figure 3.2   Voltage transformer.

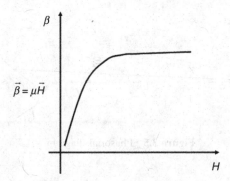

Figure 3.3 Magnetizing curve.

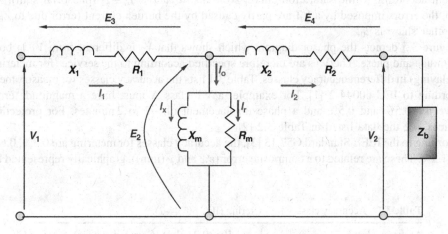

Figure 3.4 Equivalent circuit of VTs.

Equation (2.23) can also be written as

$$\beta = k\frac{E_2}{N_2 f A} = k\frac{[(R_2 + jX_2) + Z_B]I_2}{N_2 f A},\tag{3.1}$$

where Z_B is the burden impedance which represents the consumption of the secondary instruments.

It is important to emphasize that any VT introduces an error to the process, even in steady state. The main cause of this error is the voltage drop on the primary and on the secondary circuits, defined:

$$\frac{(K_n \cdot V_2 - V_1) \cdot 100}{V_1} = \frac{\Delta V}{V_1} \cdot 100\tag{3.2}$$

where as K_n is the VT nominal rate.

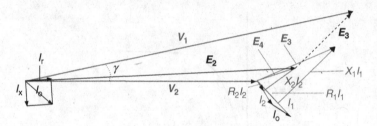

Figure 3.5 Phasorial diagram.

The secondary current I_2 and the magnetizing current I_0 are responsible for the voltage drops. In other words: the greater the burden, the greater the secondary current. This implies an increase of the internal voltage E_2, causing an increase of the magnetization current according to the saturation curve. Note also that $I_1 = I_0 + I_2$ (phasorial sum). As such, the errors imposed by a VT are partly caused by the burden current (error due to Z_B); the other share is I_0.

Figure 3.5 depicts the phasor diagram which shows that V_2 is different from V_1 in both magnitude and phase. The VTs are therefore specified according to the service (for metering or relaying) in different accuracy classes. Table 3.1 lists the accuracy classes for measurement according to IEC 60044-2 [1]. For example, a VT class A must have a magnitude error between -0.5% and 0.5% and a phase displacement of -2 to 2 minutes. For protection services, see the data listed in Table 3.2 [2].

According to the IEEE Standard C57.13 [3], the accuracy classes for metering are 0.3%, 0.6% and 1.2%. These are related to a composite error ($\varepsilon\%$ and γ (min)), graphically represented by

Table 3.1 Accuracy classes for metering (IEC 60044-2).

Accuracy class	$V = (0.9 - 1.1) \, V_{pn}$ (rated voltage)	
	$\varepsilon_V\%$	γ (min)
A	± 0.5	± 2
B	± 1.0	± 30
C	± 2.0	± 60

Table 3.2 Accuracy classes for relaying (IEC60044-2) (k depends on earthing/grounding mode; a value of 1.1 or 1.5 or 1.9 may be assumed)

ccuracy class	$V = (0.25 - 0.9) V_{pn}$		$V = (1.1 - k) \, V_{pn}$	
	ε (%)	γ (min)	ε (%)	γ (min)
E	± 3	± 120	± 3	± 120
F	± 5	± 250	± 10	± 300

Note: additional requirements about VTs were recently introduced by IEC 61869: Part 1–9 [2].

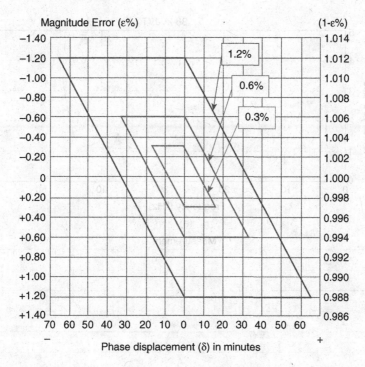

Figure 3.6 Parallegram of accuracy according to IEEE Standard C57.13 [3].

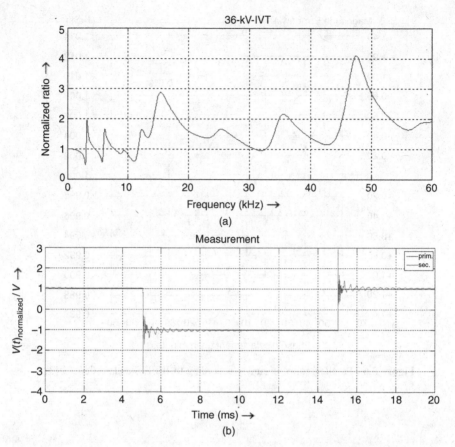

Figure 3.7 (a) Frequency response of a VT and (b) signal transformation of a square wave (measurements). Measurements taken at University of Dresden laboratory, Germany [4].

3.3 Capacitor Voltage Transformers

A high-voltage VT must have a large number of turns in the primary circuit as well as a lower primary current. For a VT in 138 kV the number of turns can be greater than 100 000 and the primary current lower than 2 mA. Thus, the design of a high-voltage VT is very complex considering the large number of turns of very thin wire, and can experience problems of fixation, isolation, resistance to breakage and so on. This makes such equipment very expensive.

Due to the issues above, it is common that in high- and extra-high-voltage systems (above 145 kV) a voltage transformer operating through a capacitive voltage divider is used. This equipment is referred to as a capacitive voltage transformer (CVT). In other words, a CVT is basically a capacitive voltage divider with a voltage transformer (inductive) connected to a point of medium voltage, normally 10–25 kV. Furthermore, a secondary divider is used at the low-voltage (100–120 V) level and is used for metering, protection and control as shown in Figure 3.8.

The capacitive divider represents an equivalent source of capacitive impedance and can therefore be compensated by a reactor (tuning inductance L) connected in series with the tapping point. An ideal reactor should allow for minimum error in the process.

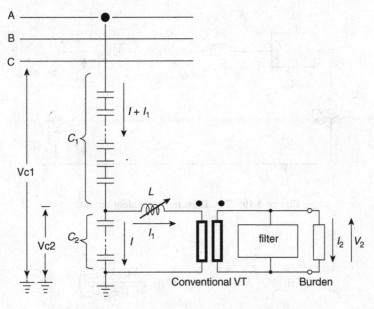

Figure 3.8 Capacitor voltage transformer.

The reactor L is normally adjustable in order to form a tuned circuit with a group of capacitors $C_1 + C_2$. This is achieved in such a fashion that the current load I_2 does not influence the accuracy of the secondary voltage V_2. Figure 3.9 shows the equivalent circuit of a CVT, emphasizing the compensation reactor, the VT equivalent circuit and the burden. An equivalent circuit when seen from the perspective of the burden terminals (i.e. with zero source impedance when the excitation circuit and the resistance of the primary and secondary winding of the VT are neglected) is shown in Figure 3.10, where

$$V_{\text{TH}} = \frac{XC_2}{XC_1 + XC_2} V_f \tag{3.3}$$

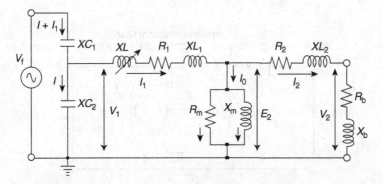

Figure 3.9 Capacitor voltage transformer equivalent model.

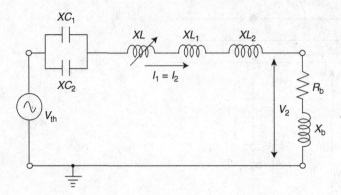

Figure 3.10 The Thèvenin equivalent circuit.

and

$$Z_{TH} = j\left(XL_1 + XL_2 + XL - \frac{XC_1 XC_2}{XC_1 + XC_2}\right). \tag{3.4}$$

Considering Figure 3.10, if the total inductive reactance is exactly equal to its capacitive reactance (C_1/C_2), the circuit will be tuned. This means that the resulting impedance will be equal to zero and no voltage drop is considered. This is true when

$$XL = \frac{XC_1 XC_2}{XC_1 + XC_2} - (XL_1 + XL_2). \tag{3.5}$$

The secondary terminal voltage is therefore equal to the Thèvenin voltage (i.e. $V_2 = V_{TH}$), and depends only on XC_1 and XC_2.

A CVT normally contains a particular ferroresonance suppression circuit, as shown in Figure 3.11. The ferroresonance suppression circuit does not however adversely affect the transient response. The analysis is similar to other types of ferroresonance circuits that were discussed in Chapter 2.

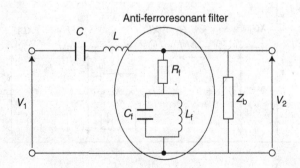

Figure 3.11 Anti-ferroresonant filter.

For information on the accuracy class and standard burden for metering and protective relays, consult the IEEE Standard 57.13 [3] and IEC 60044-5 [5]. It is important to remember that additional requirements about CVTs were recently introduced by IEC 61869, Part 1–9 [2]. The drawback of the CVT is that its accuracy is dependent on the harmonic content of the primary voltage. This may not be a problem since in HV and EHV the harmonic distortions are very small. In a future grid this may become more problematic, however.

Finally, it is important to mention that a capacitor voltage transformer serves as a coupling capacitor for power line carrier signals. These are normally of high frequency (30–500 kHz) and are conducted/dispersed through the transmission line (end-to-end) in order to have a unity protection via teleprotection schemes [6].

3.4 Current Transformers

The main purpose of a CT is to reduce the primary current from the power system to a measurable and standardized secondary current. A CT has the primary winding in series with the power circuit, at which it is necessary to make measurements, protection and monitoring.

CTs obey the same principle of electromagnetic transformation as for VTs. However, two special operating conditions exist where the primary current is absolutely independent of the transformer itself and works practically in short circuit, considering that the burden has very small impedance. Figure 3.12 illustrates the CT windings, and the transformation rate is given by $I_1/I_2 = N_2/N_1 = R_{CT}$.

Similarly to VTs, CTs can be represented by their equivalent circuit as shown in Figure 3.13; simplifications such as no primary impedance and discounting the secondary reactance need

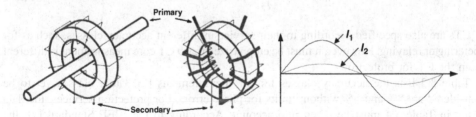

Figure 3.12 Current transformer.

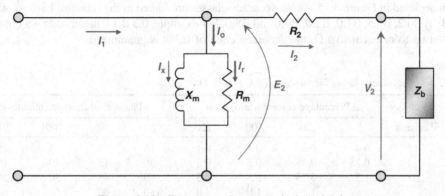

Figure 3.13 Current transformer equivalent circuit.

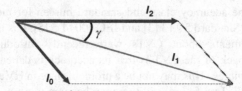

Figure 3.14 Phasorial diagram of currents.

to be taken into account. This second consideration is valid since the current flowing in a conductor through the hole in a toroidal core gives rise to a magnetic flux in the space surrounding the conductor including the core. If the core material is of high permeability, almost all of the created flux will be localized in the core material and the dispersion flux can be neglected.

Even in a steady state, a CT can introduce errors in the transformation. Its causes are related only to the magnetizing current I_0:

$$\dot{I}_1 = \dot{I}_2 + \dot{I}_0. \tag{3.6}$$

Figure 3.14 shows the phasor diagram and, in a hyperbolic way, I_2 is different from I_1 in magnitude and phase.

The magnitude of the error as related to the secondary current can be calculated as:

$$|\varepsilon|\% = \frac{|I_1 - I_2|}{I_2}100 = \frac{I_0}{I_2}100\%. \tag{3.7}$$

CTs are also specified according to their service in different accuracy classes, such as for metering or relaying. However, it must be emphasized that a CT core measurement is different from the CT for protection.

Table 3.3 lists the accuracy classes for CT measurements [7]. Other classes yet to be considered are 3% and 5% without limits for phase errors. For protection services, the data listed in Table 3.4 must be taken into account. According to the IEEE Standard [2], the accuracy classes for metering are 0.3%, 0.6% and 1.2%. These classes are related to a composite error ($\varepsilon\%$ and γ (min)), graphically represented by the accuracy classes parallelogram depicted in Figure 3.15. These accuracy classes are linked to the standard burdens such as B0.1, B0.2, B0.5, B1.0, B2.0, B4.0 and B8.0. For example, 0.3 B 1.0 means that if a burden is used up to or equal to 1 Ω, the accuracy class of 0.3% is guaranteed.

Table 3.3 Accuracy classes for metering (IEC 60044-1).

Accuracy class	±Percentage current (ratio) error				±Phase displacement (minutes)			
% rated current	5	20	100	120	5	20	100	200
0.1	0.4	0.2	0.1	0.1	15	8	5	5
0.2	0.75	0.35	0.2	0.2	30	15	10	10
0.5	1.5	0.75	0.5	0.5	90	45	30	30
1	3	1.5	1.0	1.0	180	90	60	60

Table 3.4 Accuracy classes for relaying (IEC 60044-1).

Class	Current error at rated primary current (%)	Phase displacement at rated current (minutes)	Composite error at rated accuracy limit primary current (%)
5P	±1	±60	5
10P	±3		10

Note: Additional requirements about CTs were recently introduced by IEC 61869: Part 1–9 [2].

The IEEE Standard C57.13 also describes how the CT for protection services must be specified. Firstly, all protection CTs must have a maximum relative error equal to 10%, up to 20 times the rated current. In other words, the CT should not be saturated within this limit. Considering this statement, the accuracy classes for a protection CT are designated by two symbols that effectively describe the performance for the permanent state: C and T.

Class C encompasses CTs where the leakage flux in the core has no appreciable effect on the transformation ratio within the limits of current (1–20 times I_{rated}) and has a specified standard burden. The ratio can be calculated through the excitation curves and equivalent circuits.

Class T includes CTs where the core leakage flux has an appreciable effect on the transformation ratio within the limits of current (1–20 times I_{rated}) and has a specified standard burden. The appreciable effect is defined as 1% of the difference between the current

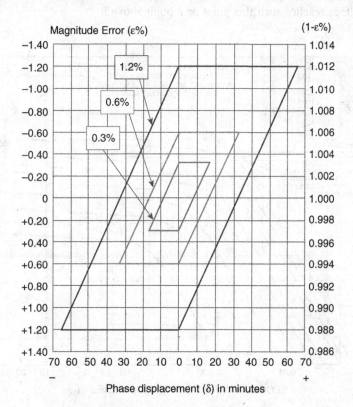

Figure 3.15 Accuracy classes parallelograms.

and calculated ratio corrections using the excitation curves. This transformation ratio should be tested.

Finally, the nominal secondary voltage must be specified. This voltage can be calculated when the secondary current is less than or equal to 20 times the rated current passing through the burden.

For example, consider a CT with the following data: $1200 - 5 - 5$ A; 0.3 B 1.0; C100. This describes a CT with a primary current rating equal to 1200 A and a secondary current rating equal to 5 A with two cores: one for measurement and the other for protection. The protection CT (core) is specified for a standard burden equal to 1.0 Ω. An example may be a short circuit: if the secondary current is equal to 100 A (20×5 A) the voltage at the burden terminals will be 100 V. This voltage is the approximated excitation curve knee-point.

It is important to mention that, for an error of 10%, the limit admitted for protection CTs of 5 A secondary is its own exciting current of 10 A, where the exciting voltage E_2 reaches a value above the knee-point curve; see the following equation and Figure 3.16 for further explanation:

$$\varepsilon_{20}\% = \frac{I_{0,20}}{20 \times 5} 100\% = I_{0,20}\%. \tag{3.8}$$

Chapter 2, Section 2.10, demonstrates what happens to the secondary current when the excitation voltage reaches such this point or a point above it.

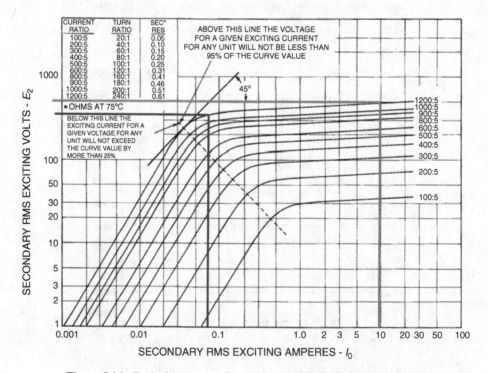

Figure 3.16 Excitation curves of a multi-ratio CT (IEEE Standard 57.13).

3.5 Non-Conventional Transducers

For decades, certain voltage and current transducers (the so-called unconventional transducers) were confined to the literature or else had very limited practical applications. Although they can be accurate, one of their major drawbacks is the poor ability to supply secondary power to feed instrumentation basically composed of voltage and current coils. With the advent of digital instrumentation the inconvenience of a high secondary burden was eliminated, increasing the future possibilities for the use of these efficient transducers.

The latest research in the field of voltage and current monitoring in high voltage has focused on obtaining, manufacturing and installing new transducers which are safer, more economical and technically advantageous. These new solutions proposed for measurement and protection of electrical systems promise to bring great benefits in both performance and applications in the near future, combined with lower costs.

These sensors are not exactly new; they operate on the basis of principles that have been known of since the beginning of the 20th century. However, it is only now that they are in great demand. This is mainly due to the proliferation and installation of intelligent electronic devices (microprocessor relays, meters, etc.) that require only the voltage and/or current with almost no secondary power.

To cater for these new technological tendencies, the most promising transduction devices for voltage and current are: (1) resistive voltage divider; (2) optical voltage transducer; (3) the Rogowski coil; and (4) optical current transducer.

3.5.1 Resistive Voltage Divider

The principle of the resistive voltage divider is depicted in Figure 3.17. It is used for the sensing of voltage and presents some advantages when compared to voltage transformers (VTs), such as: not saturable; linear; small; lightweight; and does not cause ferroresonance.

Due to its high linearity, one of the great benefits of a resistive voltage divider is the possibility of its use at several voltage levels. For example, the same sensor applied at 69 kV can also be used at 138 kV since it has sufficient isolation. The transducers do not need to be replaced. For this purpose several techniques can be employed such as derivation, using secondary voltage tapes along the resistive divider, the use of auxiliary resistive dividers (comparable to auxiliary VTs) or a simple adjustment of the dynamic range software of the

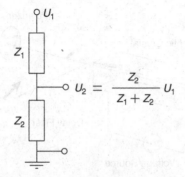

$$U_2 = \frac{Z_2}{Z_1 + Z_2} U_1$$

Figure 3.17 Voltage resistive divider.

analog/digital converters at the input of the instrumentation. As a result, this reduces the diversity of sensors needed for very different applications.

As opposed to conventional VTs, resistive voltage dividers do not cause ferroresonance and are not destroyed by this phenomenon. They can even be used in abnormal conditions of operation.

Resistive voltage dividers experience several major disadvantages however, listed below.

- *Losses by Joule effect.* Since the losses are proportional to the square of the voltage this effect becomes significant for very high voltage. High values of resistance should be used to minimize such losses.
- *Parasitic capacitances* play an important role in determining the accuracy of the divider. The higher the voltage level involved, the higher are the values of the resistors and the most significant are the values of these capacitances. As a consequence, there is a reduction in bandwidth of the voltage signal, which will limit the harmonic frequencies to the order of a few kiloHertz. This range is however sufficient for most power system applications, including harmonic analysis and measures of power quality.
- *No galvanic decoupling.* Due to the inherent coupling between primary and secondary current, special care must be taken to overcome this deficiency.
- *Low output capability* (small burden). Actually, this fact is guaranteed by the new generation of intelligent electronic devices.

To ensure high accuracy of the resistive dividers, the resistors must have the same coefficient of drift with temperature. It is possible to reach values of 0.2–0.5% of accuracy classes for the long-term stability, the effect of parasitic capacitances and drift with temperature.

3.5.2 *Optical Voltage Transducer*

This kind of transducer is better referred to as a voltage sensor through the piezo-optical effect. The operating principle of these sensors is based on the phenomenon of change in the physical size and shape of piezo-electric crystals when subjected to electric fields. These changes are detected by the rotation of polarized light through an optical fiber wrapped around a crystal. This phenomenon is known as the Pockel effect, illustrated by Figure 3.18. Since this effect is directly proportional to the electric field applied to the crystal, the applied voltage can be accurately measured.

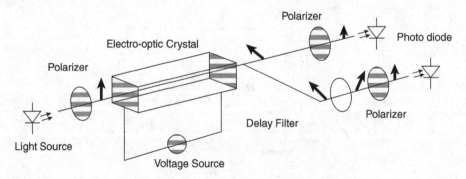

Figure 3.18 Optical voltage transducer.

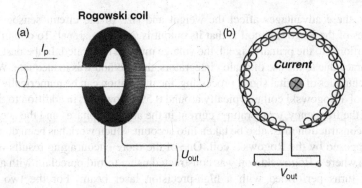

Figure 3.19 Rogowski coil.

Unlike a resistive voltage divider, an optical voltage transducer has the advantage of being galvanically decoupled. Other advantages include no losses by the Joule effect and a negligible effect of parasitic capacitances. However, disadvantages are the high cost associated with its production, complex technology and 90° of phase rotation on the output signal.

3.5.3 Rogowski Coil

The measurement principle of current through the Rogowski coil has been known since 1912. This coil consists of a winding, uniformly distributed in a core of non-magnetic material. The simplest possible arrangement consists of a toroid air core, where a great number of turns are wound around the coil and one turn comes back inside the toroid, as shown in Figure 3.19.

A Rogowski coil provides measurements that are galvanically decoupled. Furthermore, it operates over extremely large bands compared to the frequencies concerned in power systems. The frequency response of the coil can reach the order of megaHertz, as illustrated by Figure 3.20.

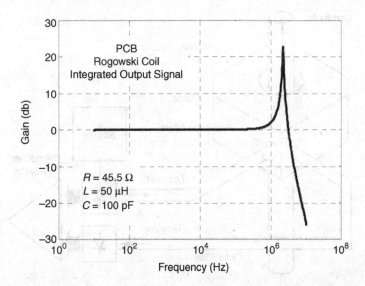

Figure 3.20 Frequency response of a Rogowski coil [8].

Additionally, these advantages affect the weight and size of the circuit sensor. One of the disadvantages of the Rogowski coil is that its output is a voltage signal. To obtain the current signal proportional to the primary signal, the voltage must be integrated. In the past, because of the low accuracy due to the use of analog integrators, this method was inadequate. With today's advanced techniques of digital signal processing, the integration can be numerically performed.

The error of a Rogowski coil is typically around 0.5%. However, in addition to the load, the influence of the frequency, temperature, current in the adjacent phases and the accuracy of its mechanical construction must also be taken into account. Much work has been done to reduce the errors imposed by the Rogowski coil. One of the more encouraging results is shown by Ramboz [9], where a Rogowski coil was constructed using toroid porcelain with a metallized surface and turns performed with a high-precision laser beam. For the two prototypes prepared, errors were found to be 0.05% and 0.26%. This may be considered excellent when compared to conventional coils.

3.5.4 Optical Current Transducer

The operation of an optical current transducer (OCT; see also magneto-optical current transducer or MOCT) is based on the Faraday effect, in which polarized light suffers a phase rotation in the presence of a magnetic field. Figure 3.21 illustrates the operating principle of an OCT. It is important to note that these devices are able to provide the necessary power to modern IEDs, since microprocessor devices only represent a small burden to the transducers.

The key element in the system is a Faraday-effect sensor composed of an element called a rotor, which is a block of a special glass or crystal. This rotor has an opening for the passage of the conductor in which current is being monitored. The fiber-optic cables are connected to the block through a set of lenses that guide and polarize the light beams. This block is passive and is the only component of the transducer that is installed at a high voltage level.

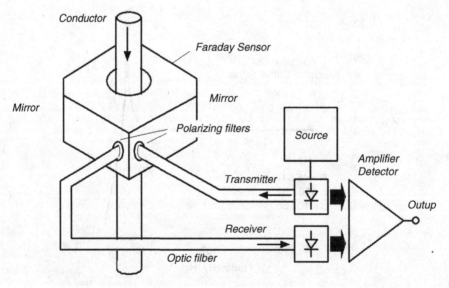

Figure 3.21 Optical current transducer.

A polarized light beam surrounds the rotor, which is also subjected to a magnetic field. A second polarizing filter captures the light beam. The interaction between light and magnetic field (proportional to the current) establishes a phase modulation to the light, captured by a photodetector diode that feeds an electronic amplifier. This amplifier produces a low voltage proportional to the instantaneous value of the current in the primary conductor.

It can be shown that, for an arbitrary closed path around a current conductor, the phase modulation depends only on the collected current, that is, the sensor does not respond to external fields.

A pair of fiber-optic cables is connected to each Faraday sensor. These cables carry the light between the sensor and the electronic module in the control room, sometimes hundreds of meters away. All these components are passive and stable over time. The active electronics system, that is, the light source and the signal processing circuit, is fully installed in the control room on a rack which easily accessible and not harmful to the environment.

The expression that relates the measured current and the angular variation is $\theta = 2\upsilon i$, where υ is the Verdet constant and i the instantaneous current. The selection of the crystal is always a compromise involving its optical characteristics, operating range and thermal stability. The combination of the sensor and electronic amplifier should maintain an accuracy of 0.2–0.5% in a range of 0.01–2 pu current levels. In this case, the transfer function approximates a linear function for angles of rotation in the range of ±25° since it uses a material of low υ to avoid amplification of noise. On the other hand, low values of υ can lead to large instantaneous errors; these are still much lower than those provided by conventional high-quality CTs, however.

Finally, since modern instrumentation represent negligible secondary loads (burdens), the application of unconventional transducers becomes increasingly attractive, either for cost/benefit or performance reasons. When the performance of conventional CTs is compared to special CTs and other unconventional current transducers, the first impressive characteristic is the non-saturation capability of the Rogowski coil and OCT. Although the secondary voltage provided by these devices is small (low burden), the linearity is maintained as can be seen in Figure 3.22. In other words, the secondary signal is a quasi-perfect copy of the primary signal for a great range of magnitude, in this case primary current.

It must be emphasized that this 'new' technology is increasingly used in electrical systems. There is no doubt that this is a path of no return. On the other hand, conventional VTs and CTs will still find their place in the real world for a long time.

3.6 Analog-to-Digital Conversion Processing

This chapter has so far been focused on instrument transformers. The next sections however focus on other issues related to signal processing, such as the analog-to-digital (A/D) conversion and the anti-aliasing filter. The electronic design and analysis are beyond the scope of this book; only the signal processing aspects are addressed here.

From an economical perspective, the price of an analog-to-digital converter has been continually decreasing as resolution and the conversion speed increase. However, an important point regarding IEDs is the choice of analog-to-digital converter (ADC). The engineer has to make decisions such as the converter rate, resolution and conversion time. Figure 3.23 represents the tradeoff between conversion rate (Hz) versus the number of bits. The right scale shows the percentage of active ADCs with a specific number of bits. For example, 34.88% of

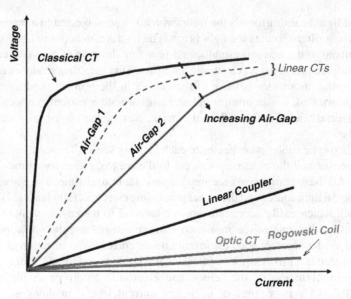

Figure 3.22 Voltage versus current characteristics of transducers (adapted from [8]).

the converters investigated had 12-bit resolution. The left scale shows the speed in gigaHertz of the ADC. It can be noticed that as the converter resolution increases, the speed decreases. This figure shows that for ADCs with the same number of bits, there is at least one that reaches the maximum frequency operation (right scale).

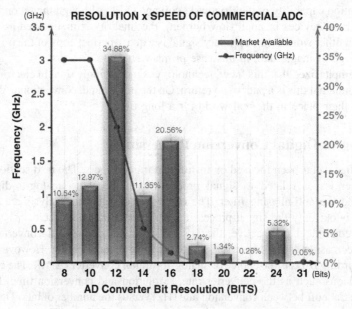

Figure 3.23 ADC resolution versus conversion rate.

The resolution of a converter is related to its number of bits B and can be calculated using the expression:

$$\Delta x(\%) = \frac{1}{2^B - 1} \times 100 \tag{3.9}$$

or, in terms of the full-scale (FS) range,

$$\Delta x = \frac{1}{2^B - 1} \times x_{FS} \tag{3.10}$$

where x represents voltage or current. However, if the dynamic range of x is x_{dr} and the conditioner circuit is appropriately designed to conform to the signal presented in the ADC, the resolution can be written:

$$\Delta x_{dr} = \frac{1}{2^B - 1} \times x_{dr}. \tag{3.11}$$

Figure 3.24 illustrates how the dynamic range can be converted to use the full resolution of the converter. The first step is subtract the input signal from the small value of the range v_a. The new interval of variation becomes the range from 0 up to $v_b - v_a$. The next step is to multiply the resultant signal in order that $k(v_b - v_a) = v_{FS}$, where v_{FS} corresponds to the maximum binary code $1111 \ldots 1_2$. A better resolution of the converter is obtained in this way.

The choice of resolution for the converter depends on the variation range of the signals. Additionally, the conversion rate will depend on the dynamics (or frequency content) of the corresponding signal. Generally, power system applications can be divided into three major areas with regards to data acquisition requirements: (1) supervision and control; (2) protection; and (3) power quality and diagnosis.

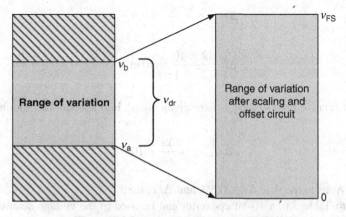

Figure 3.24   Dynamic range conversion.

Table 3.5 typical range of variation in control and supervision application.

Signal	Variation range	Minimum resolution (#bits)
Voltage (V)	$90 \leq |V| \leq 130, \quad V_N = 100$ V	10
Current (A)	$0 \leq |I| \leq 10; \quad I_N = 5$ A	12

3.6.1 Supervision and Control

A typical range for current and voltage variations in the control/supervision application is listed in Table 3.5. Despite the fact that the dynamic range of the voltage signal is $2 \times 40 = 80$ V, the common design for the voltage conditioner and converter uses the dynamic range 2×130 V. This is because a simplification was obtained by the conditioner and, in abnormal situations, the voltage can drop below 90 V. If a 10-bit converter is used, the quantization error is given by Equation (3.9), i.e.

$$\Delta v = \frac{2 \times 130}{2^{10} - 1} \approx 0.25 \text{ V}.$$

For current, the resolution is

$$\Delta i = \frac{2 \times 10}{2^{10} - 1} \approx 20 \text{ mA}.$$

For these resolutions the error introduced by the quantization processing assumes that the converter is ideal. In practice however, the converters are not ideal and an error of $\pm 1/2$ LSB (low significant bit) is commonly found in real converters. As it only has 9 bits, a 10-bit converter must be used in order to compensate for the internal errors of the converter. Resolution is therefore recalculated using 9 bits length:

$$\Delta v = \frac{2 \times 130}{2^9 - 1} \approx 0.5 \text{ V},$$

$$\Delta i = \frac{2 \times 10}{2^9 - 1} \approx 40 \text{ mA}.$$

The percent relative errors ($\Delta x_r(\%)$) are given taking into account the nominal value x_N:

$$\Delta x_r(\%) = \frac{\Delta x}{x_N} \times 100. \tag{3.12}$$

For the previous examples, $\Delta v_r c.$ 0.5% and $\Delta i_r c.$ 0.8%.

According to Table 3.6, a 10-bit converter can be used in the voltage acquisition system even when using a class A VT. For current conversion however, 10 bits cannot be used in CT

Table 3.6 Typical range of variation in control and supervision.

Signal	Variation range	Minimum resolution (#bits)
Voltage (V)	$0 \le |V| \le 150; \quad V_N = 100$ V	10 bits $\rightarrow \Delta v_r \approx 0.6\%$
		12 bits $\rightarrow \Delta v_r \approx 0.15\%$
		14 bits $\rightarrow \Delta v_r \approx 0.04\%$
Current (A)	$0 \le |I| \le 200; \quad I_N = 5A$	12 bits $\rightarrow \Delta i_r \approx 4\%$
		14 bits $\rightarrow \Delta i_r \approx 1\%$
		16 bits $\rightarrow \Delta i_r \approx 0.05\%$

classes 0.1 and 0.2 as more bits are needed for the current acquisition system. For example, if a 12-bit converter is used, the percent relative error is:

$$\Delta i_r(\%) = \frac{20}{2^{11} - 1} \times \frac{100}{5} \approx 0.2\%.$$

12-bit converters then match the current error for all CT classes. The last column of Table 3.6 summarizes the converter resolution for control and its supervision applications.

3.6.2 Protection

In protection applications, the voltage can reach up to 1.5 V_N with $V_N = 100$ V, and the current can reach up to $40 I_N$ with $I_N = 5$ A. These limits require an ADC with higher resolution than those used for control application. Typically 14- and 16-bit resolutions are used in a numerical relay. Table 3.6 summarizes the typical range variation and resolution for protection applications. The final results are obtained using equations (3.9) and (3.11). Note that, to compensate for internal errors, the number of bits used in these equations was $B - 1$. Comparing Table 3.6 with Table 3.2, it becomes clear that a 10-bit converter leads to smaller errors of the VT class E; however, 14 bits or more are needed for current.

3.6.3 Power Quality

Power quality applications are mainly concerned with voltage quality. There are several parameters that need to be calculated directly from the voltage measured, such as sag, swell, transient or flicker. Of these, the flicker measurement demands a higher precision in voltage acquisition systems and requires a high resolution in the ADC. The standard [10] specifies a resolution 0.1% over the range 10–150% of declared input voltage U_{din}. For $B = 12$, then

$$\Delta v_r(\%) = \frac{3 U_{din}}{2^{11} - 1} \times \frac{100}{U_{din}} = 0.3\%.$$

If this is true, then the converter must have 14 bits or more to reach the desired resolution.

In the case of harmonics measurements, the standards state that the error of $\pm 5\%$ is the limit for each harmonic. For example, in distribution systems where the limit of the odd harmonics is 3% of the fundamental component, the final resolution is:

$$3\% \times 5\% = \pm 0.15\%$$

which requires a converter of 11 bits or more.

Table 3.7 Accuracy and minimum resolution for the ADC (odd harmonics).

System	Limit (odd harmonics) (%)	Accuracy (%)	Resolution # bits
Distribution	3.0	±0.150	≥11 bits
Sub-transmission	1.5	±0.075	≥12 bits
Transmission	1.0	±0.050	≥12 bits

Table 3.7 summarizes the minimum resolution required for the voltage measurement of odd harmonics with a ±5% uncertainty [11]. Note that: (1) the limit for even harmonics is 25% less than for the odd harmonics; and (2) measuring the current harmonics requires a higher-resolution converter, typically >15 bits.

3.7 Mathematical Model for Noise

Voltage and current signals in a power system are not pure sinusoidal functions. There are several distortions, such as those described in Chapter 2. Some can be modeled as additive white noise, with a normal distribution of zero mean and variance equal to σ_ε^2 or, mathematically:

$$y[n] = x[n] + \varepsilon[n] \tag{3.13}$$

where $y[n]$ is the observed discrete-time signal, $x[n]$ is the desired signal and $\varepsilon[n]$ is the Gaussian white noise. Figure 3.25 illustrates the mathematical model commonly used in signal processing. There is only a small control regarding the error $\varepsilon_1(t)$. This is the function of a transducer or background error present in the system and other less significant errors. Typical values of signal-to-noise ratio SNR_A are >27 dB; 40 dB is the most widely used in practice. On the other hand, $\varepsilon_2[n]$ is the error due the quantization process and is well known in literature. For example, the signal to noise ratio due to the quantization processing SNR_{ADC} is defined:

$$SNR_{ADC} = 6.02B + 1.76 \, \text{dB} \tag{3.14}$$

A 10-bit converter therefore introduces an SNR of 61.96 dB.

The number of bits of a converter may not be defined exclusively by SNR_A. For example, if the analog SNR, $SNR_A = 40 \, \text{dB}$, a 7-bit converter leads to the SNR introduced by the ADC, $SNR_{ADC} = 44 \, \text{dB}$.

A larger resolution may be needed for the following reasons. If the converter resolution is higher, it may be possible to utilize signal processing tools in order to attenuate the noise

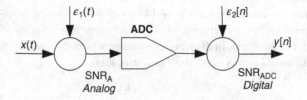

Figure 3.25 Mathematical model including noise.

process $\varepsilon_1(t)$, such as a moving average filter that can attenuate the noise from the process if $\varepsilon_1(t)$ has a zero mean. The converter must be chosen with a quantization error smaller than the measurement error.

3.8 Sampling and the Anti-Aliasing Filtering

This section addresses the questions of the anti-aliasing filter and sampling principles (for A/D converter resolution, see Section 3.7).

The sampling theorem is discussed here in a conceptual way; its mathematical analysis is developed in the next chapter. The sampling theorem states that the continuous-time signal $x(t)$ can be represented by its samples (or by discrete-time signal $x[n]$) in such a way that $x[n] = x(t)|_{t=nT_s}$, where T_s is the sampling time if the sampling frequency is $F_s > 2f_{max}$, where f_{max} is the maximum frequency in the input signal.

Example 3.1

Determine the minimum sampling frequency to analyze a voltage signal with a spectrum that contains up to the 15th harmonic.

The maximum frequency in this signal is $f_{max} = 15 \times 60$ Hz, and the system frequency is assumed to be 60 Hz. The sampling frequency or sampling rate is given by

$$F_s > 30 \times 60 \, \text{Hz}.$$

In general, if the IED needs to perform analysis up to harmonic h, the theoretical limit for the sampling rate is

$$F_s > 2 \times h \times f_1 \, \text{Hz} \tag{3.15}$$

where f_1 is the fundamental frequency in Hz.

Some textbooks establish the sampling theorem as $F_s > 2f_{max}$. The equality is in fact the mathematical limit if the signal has low energy in that frequency and the ideal low-pass filter can be built. In practice however, an ideal filter cannot be built and the sampling rate limit should be much higher than the equality. For example, using a sample three times higher than the theoretical limit relaxes the low-pass filter design, that is,

$$F_s \geq 3 \times 2 \times h \times f_1 \, \text{Hz}. \tag{3.16}$$

Since the harmonic content is not known in advance, it is necessary to guarantee that aliasing will not occur. Aliasing destroys the information in low frequency, because the high-frequency spectrum corrupts the low-frequency spectrum. Aliasing is avoided by prefiltering the analog signal through a low-pass filter. This analog low-pass filter is known as an anti-aliasing filter or guard-filter.

An illustrative example of aliasing in the time domain is presented in Figure 3.26. The input signal is composed of the fundamental and the 15th harmonic. The sampling rate used was $F_s = 16 \times 60$ Hz, which evidently does not fulfill the sampling theorem. The samples of this signal are represented by crosses and correspond to the fundamental frequency, but of smaller

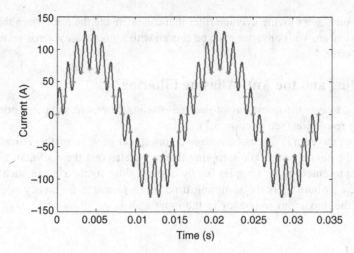

Figure 3.26 Example of aliasing in time domain.

amplitude then the original frequency. In this case it is easy to mathematically verify the aliasing error:

$$x(t) = A\sin(2\pi f_1 t) + m\sin(15 \times 2\pi f_1 t)$$
$$x[n] = x(t)|_{t=1/16f_1}$$
$$= A\sin(2\pi f_1 n/16f_1) + m\sin(15 \times 2\pi f_1 n/16f_1)$$
$$= \sin(2\pi n/16) + m\sin(-2\pi n/16 + 32\pi n/16)$$
$$= (A - m)\sin(2\pi n/16).$$

Note that a single discrete-time component appears after discretization. The frequency is the normalized fundamental frequency, now in radians $(2\pi/16)$, and the amplitude was reduced. This means that information was lost when aliasing was not avoided.

The correct design of an anti-aliasing filter is crucial to maintain the correct information. Figure 3.27 shows the parameters that must be used in the filter design. In the figure below, f_c

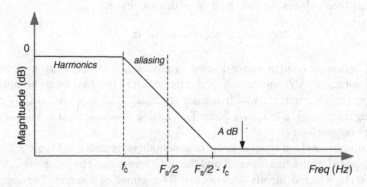

Figure 3.27 Specification of the anti-aliasing filter.

is the cutoff frequency and A is the minimum attenuation in decibels. The region between f_c and $F_s/2$ is not used to carry information, and aliasing can occur in this region. The harmonic of interest must fall in the harmonics region. Since the frequency response at the cutoff frequency has an attenuation of 3 dB, the cutoff frequency must match:

$$f_c = k \times H \times f_1 \tag{3.17}$$

where H is the higher harmonic, f_1 the fundamental frequency and k is a gain factor that guarantees a minimum distortion for harmonic H (typically $k = 1.2$). The minimum attenuation A must be higher than the SNR_{ADC}, that is,

$$A \geq 6.02B + 1.76. \tag{3.18}$$

Example 3.2

Specify the Butterworth filter that allows analysis up the 15th harmonic, assuming an ADC resolution of 14 bits.

The cutoff frequency must be

$$f_c \geq 1.2 \times 15 \times 60 = 1080$$
$$A \sim 86.04 \text{dB}.$$

The attenuation expression (in dB) for a Butterworth filter of order n in the stopband region is given by:

$$G = 20 \times n \times \log \Omega \tag{3.19}$$

where n is the filter order, Ω is the normalized frequency and G is the attenuation in decibels. For an attenuation of 86.04 dB and using a third-order Butterworth filter, Equation (3.18) yields:

$$86.04 = 60 \times \log \Omega \quad \Rightarrow \quad \Omega = 29 \text{ rad/s}.$$

The normalized frequency indicates that the cutoff frequency is equal to 1 rad/s. The last calculation shows that the attenuation reaches 86.04 for $\Omega = 29$ rad/s. This point corresponds to point $F_s - f_c$ in Figure 3.27. To denormalize the previous frequency, the results are multiplied by the true cutoff frequency which, in the present example, is $f_c \approx 1080$ Hz. We then have:

$$\Omega' = 29 \times 2\pi \times 1080$$

$$\text{or} \quad f' = 29 \times 1080 = 31.32 \text{ kHz}.$$

The sampling frequency can then be determined as

$$f_s \geq f' + f_c = 32.4 \text{ kHz}$$

which corresponds to 540 samples per cycle: a very high sampling rate for the low frequency of interest. If the order of the Butterworth filter was changed to a 5th order, than the necessary sampling rate to keep the SNR_{ADC} is:

$$f_s = 8.9 \text{ kHz, or } 148 \text{ samples per cycle.}$$

The design can be relaxed because the ADC is not ideal and cannot give ideal SNR_{ADC}. Remember that a B-bit ADC is effectively used as $B-1$ bits. When this is taken into account, the sampling rate can be reduced. For example, it is possible to use a sampling rate of 128 samples per cycle.

The sampling rate can be reduced if the magnitude characteristic of the filter is known and an internal compensation can be applied to the obtained results from the harmonics of high order. For this situation, 64 samples per cycle can be used.

3.9 Sampling Rate for Power System Application

Protection and control applications usually use signal information in low frequency. Traditionally a relay algorithm works with 16 samples per cycle, utilizing only the fundamental component and, in some applications, the second and third harmonic. New protection algorithms require higher frequencies however, and it is a common finding in technical protection literature that the algorithm may run with more than 64 samples per cycle; such devices have not yet been commercialized.

Use of the high-frequency information of the signal is common in power-quality and data-logger applications. Such an example is an impulsive phenomena that may last only a few nanoseconds (high-frequency spectrum) and demands a high-speed ADC. In general, impulsive phenomena are not captured directly by sampling the analog signal but may be noticed when an auxiliary analog filter is used to enlarge the duration of the pulse, generally a high-pass filter. Transient phenomena can last from micro-seconds to milliseconds, as listed in Table 3.8. High-frequency phenomena require a high-speed converter which can address actual speeds of 1024 samples per cycle; this cannot be found in commercial power-quality equipment. Measuring these high-frequency events requires special equipment or circuits such as a high-pass filter.

3.10 Smart-Grid Context and Conclusions

Many system events may demand equipment other than those of power frequencies that exist in the electric grid. Some of the consequent waveforms are time-varying and contain high-frequency signals that are conducted through the power lines; these need to be properly measured and, if necessary, their negative impacts compensated for. The increasing

Table 3.8 Transients phenomena.

Frequency band	Time duration
Low frequency (<50 kHz)	0.3–50 ms
Median frequency (5–500 kHz)	20 μs
High frequency (0.5–5 MHz)	5 μs

complexity of smart grids through the generation of loads injecting high-frequency components (in the range of dozens of kiloHertz), as a consequence of the massive use of power electronics for distributed generation and smart loads, requires more precise signal processing of the time-varying waveforms. Transducers are the first components in this process and their impact on measurement precision needs to be carefully considered. The acquisition system (sampling and the anti-aliasing filtering) also needs to be able to deal with the nature of the signals so that the signals can be accurately processed and analyzed.

References

1. IEC60044-2 (2003) Instrument transformers. Part 2: Inductive voltage transformers. International Electrotechnical Commission, Geneva, Switzerland.
2. IEC61869 1-9 (2007 to 2012) Instrument transformers: additional requirements. International Electrotechnical Commission, Geneva, Switzerland.
3. IEEE C57-13 (2009) Requirements for instrument transformers. IEEE standard for Performance and Test Requirements for Instrument Transformers of a Nominal System Voltage of 115kV and Above, IEEE Standards Board, Piscataway, USA.
4. Meyer, J., Stiegler, R. & Klatt, M. (2011) *Accuracy of Instrument Transformers for Voltage Harmonic Measurements.* OMICRON Instrument Transformer Measurement Forum, Brand, Austria.
5. IEC 60044-5 (2003) Instrument transformers. Part 5: Capacitive voltage transformers. International Electrotechnical Commission, Geneva, Switzerland.
6. Cigré (2001) *Protection Using Telecommunications.* Working Group 34/35.11, Final Report. Cigré, Oslo.
7. IEC 60044-1 (2003) Instrument transformers. Part 1: Current transformers. International Electrotechnical Commission, Geneva, Switzerland.
8. Kojovic, L.A. (2010) Ractical aspects of Rogowski coil applications to relaying. *IEEE PSRC Special Report.* Available: http://www.pespsrc.org.
9. Ramboz, J.D. (1996) Machinable Rogowski coil: design and calibration. *IEEE Transactions on Instrumentation and Measurement,* **45(2)**, 511–515.
10. IEC61000-4-30 (2008) Power quality measurement methods. International Electrotechnical Commission, Geneva, Switzerland.
11. Chen, S. (2003) A quantitative analysis of the data acquisition requirements for measuring power phenomena. *IEEE Transactions on Power Delivery,* **10** (4), 1575–1577.

4

Discrete Transforms

4.1 Introduction

Discrete transforms are essential in the analysis and synthesis of signal processing in power systems, particularly in the context of smart grids. This chapter presents a brief review of the most common transforms used in signal processing. The continuous and discrete time transforms are briefly reviewed and compared whenever possible. As these transforms are widely treated in several textbooks, the approach used will be directed towards power system applications. Furthermore, sampling theorem is presented as the main link between continuous and discrete time signals.

4.2 Representation of Periodic Signals using Fourier Series

In power systems an ideal voltage source is mathematically represented by the sinusoidal equation: $v(t) = A \cos(2\pi f_1 t + \theta)$ where A is the amplitude of the voltage source, f_1 is the fundamental frequency and θ is the phase. An ideal voltage source signal is a periodic signal of infinite duration. A particular function $f(t)$ is said to be periodic if there is a T value such that

$$f(t + T) = f(t). \tag{4.1}$$

The minimum value of T that verifies Equation (4.1) is referred to as the fundamental period, T. Its reciprocal is the fundamental frequency, or:

$$T = \frac{1}{f_1}.$$

In power systems, the fundamental frequency f_1 can be 50 or 60 Hz for terrestrial systems and 400 Hz for aircraft systems. When an ideal voltage source feeds a linear circuit it is well known that, in its steady state, the current and voltage at any given point of the signal circuit is sinusoidal with the same fundamental frequency. If the circuit is non-linear however, others frequencies will appear as voltage and current signals.

Power Systems Signal Processing for Smart Grids, First Edition. Paulo Fernando Ribeiro, Carlos Augusto Duque, Paulo Márcio da Silveira and Augusto Santiago Cerqueira.
© 2014 John Wiley & Sons, Ltd. Published 2014 by John Wiley & Sons, Ltd.
Companion Website: http://www.wiley.com/go/signal_processing/

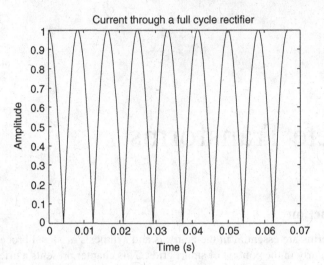

Figure 4.1 Current being drained by a single-phase rectifier (periodic and infinite signal with a purely resistive load).

As an example, consider a non-linear circuit that contains a rectifier. In the same manner as the current in a single-phase full-wave rectifier, it can mathematically be expressed:

$$i(t) = |I \cos(2\pi f_1 t + \theta)|. \tag{4.2}$$

Figure 4.1 shows the waveform of the current expressed by Equation (4.2), where $\theta = 0$ and $f_1 = 60$ Hz. The signal is periodic with a fundamental frequency of 120 Hz. Consequently, the fundamental period is $T = 1/(2f_1)$.

There are many practical reasons for expanding a given function such as Equation (4.2) as a linear combination of other functions; some of these reasons are described in Section 4.2.4. In a Fourier series expansion, a periodic function is expressed as a linear combination of harmonically related trigonometric sine and cosine functions (or as a complex exponential). Using the concept of signal expansion, each term of the summation belongs to the basis of the expansion. The basic terms are orthogonal to each other, that is, there is an inner product:

$$\langle f_i(t), f_j(t) \rangle = \delta(i - j) \tag{4.3}$$

where $\langle a, b \rangle$ represents the inner product defined in its specific space, δ is the impulse function and the functions $f_i(t)$ and $f_j(t)$ belong to the basis. Equation (4.3) reveals that $f_i(t)$ and $f_j(t)$ are orthonormal functions. That is, the inner product is 0 if the functions are different and is equal to 1 if the functions are equal.

For a Fourier series expansion, the basis is of the form:

$$\left\{ \cdots \quad \sqrt{\frac{2}{T}} \cos(2\Omega_1 t), \quad \sqrt{\frac{2}{T}} \cos(\Omega_1 t), \quad \sqrt{\frac{1}{T}}, \quad \sqrt{\frac{2}{T}} \sin(\Omega_1 t), \quad \sqrt{\frac{2}{T}} \sin(2\Omega_1 t), \quad \cdots \right\}$$

$$\tag{4.4}$$

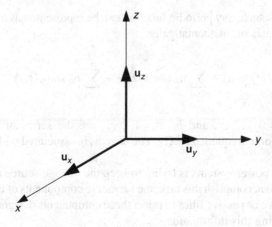

Figure 4.2 Tri-dimensional vector space.

where Ω_1 is the fundamental frequency in rad/s and T_1 its fundamental period. The constant that multiplies each trigonometric function is used to guarantee the orthonormal representation.

For a Fourier series, the inner product is defined:

$$\langle f_i(t), f_j(t) \rangle = \int_{-T/2}^{T/2} f_i(t) f_j^*(t) \mathrm{d}t \tag{4.5}$$

where * represents the conjugate of the function.

A specific function is accurately represented by a combination of the basic terms if they belong to the space of the functions spanned by its basis. The idea is similar to a vector in space, where the triple $(\mathbf{u}_x, \mathbf{u}_y, \mathbf{u}_z)$ is the basis for its tridimensional space (see Figure 4.2). Any vector in this space can be accurately represented as a linear combination of the basis, for example the vector $\mathbf{v}_i = a_{i1}\mathbf{u}_x + a_{i2}\mathbf{u}_y + a_{i3}\mathbf{u}_z$. Once the basis is defined, only the coefficient of each term of the basis is needed to represent the vector, that is, $\mathbf{v}_i = (a_{i1}, a_{i2}, a_{i3})$. The inner product in $\mathbb{R}^3$ of two vectors $\mathbf{v}_i = a_{i1}\mathbf{u}_x + a_{i2}\mathbf{u}_y + a_{i3}\mathbf{u}_z$ and $\mathbf{v}_j = a_{j1}\mathbf{u}_x + a_{j2}\mathbf{u}_y + a_{j3}\mathbf{u}_z$ is defined:

$$\langle \mathbf{v}_i(t), \mathbf{v}_j(t) \rangle = \sum_{k=1}^{3} v_{ik} \cdot v_{jk}^*. \tag{4.6}$$

Note that the basis has three terms of unitary length, represented by $\mathbf{a}_x = (1, 0, 0)$, $\mathbf{a}_y = (0, 1, 0)$ and $\mathbf{a}_z = (0, 0, 1)$. From Equation (4.6) it is easy to verify that the elements of the basis are orthonormal.

The basis in the Fourier series[1] has infinite terms and, instead of a vector, the basis is formed by a harmonically related trigonometric function where the space covered encompasses all

---

[1] The Fourier series is named after Joseph Fourier (1768–1830) who made important contributions to the study of trigonometric series, after preliminary investigations by Leonhard Euler, Jean le Rond d'Alembert and Daniel Bernoulli. Fourier introduced the series for the purpose of solving the heat equation in a metal plate, publishing his initial results in 1807 [1].

periodic functions. As such, any periodic function can be represented as a linear combination of the terms of the basis or, mathematically:

$$f(t) = a_0 + \sum_{k=1}^{\infty} a_k \cos(k\Omega_1 t) + \sum_{k=1}^{\infty} b_k \sin(k\Omega_1 t). \qquad (4.7)$$

The terms $a_k, k = 0, 1, 2, \ldots$ and $b_k, k = 1, 2, \ldots$ are the series of coefficients corresponding to the harmonic frequencies $k\Omega_1$. The term a_0 is associated with zero frequency or the DC term.

A common task in power systems is trying to keep the voltage source as a single sinusoid without harmonics. To accomplish this task, the harmonic components of the signal need to be identified and an active or passive filter to reject these components designed. A Fourier series is suitable for obtaining this information.

Another common application is related to power quality monitoring. In this case it is necessary to monitor points along a power system and record them for future analysis. The goal here is to record the signal in a way that requires minimal memory space. When the signal is expressed as a Fourier series there is no need to record each signal sample; only the series coefficients, and only those with significant energy, are required to be recorded. Of course, all this theory works well if the signal is stationary (periodic). This is not normally the case in practice; the theory can however be applied if the signal is almost stationary (quasi stationary). This means that it can be considered stationary inside its observation window. In Chapter 9 the focus is on the non-stationary approach and the analysis of such signals; in this chapter we however focus on stationary signals.

4.2.1 Computation of Series Coefficients

4.2.1.1 DC Coefficient

The DC coefficient is found simply by integrating both sides of Equation (4.7) over a fundamental period interval:

$$T a_0 = \int_{-T/2}^{T/2} f(t) \mathrm{d}t + \sum_{k=1}^{\infty} a_k \int_{-T/2}^{T/2} \cos(k\Omega_1 t) \mathrm{d}t + \sum_{k=1}^{\infty} b_k \int_{-T/2}^{T/2} \sin(k\Omega_1 t) \mathrm{d}t.$$

As the integral of an integer number of cycles of sinusoidal function is zero, the final result for the DC coefficient can be easily found as:

$$a_0 = \frac{1}{T} \int_{-T/2}^{T/2} f(t) \mathrm{d}t. \qquad (4.8)$$

4.2.1.2 Cosine Coefficients

Pre-multiplying both sides of Equation (4.7) by $\cos(j\Omega_0 t)$ and integrating yields:

$$\sum_{k=1}^{\infty} a_k \int_{-T/2}^{T/2} \cos(j\Omega_1 t)\cos(k\Omega_1 t)dt = \int_{-T/2}^{T/2} \cos(j\Omega_1 t)f(t)dt$$

$$- a_0 \int_{-T/2}^{T/2} \cos(j\Omega_1 t)dt - \sum_{k=1}^{\infty} b_k \int_{-T/2}^{T/2} \cos(j\Omega_1 t)\sin(k\Omega_1 t)dt. \tag{4.9}$$

The concept of the inner product as defined in Equation (4.3) can now be used. On the right-hand side of Equation (4.9), the inner product will be non-zero only if $k=j$. Furthermore, on the right-hand side of the equation the second term is 0 because the sinusoid function contains integer cycles at the integration interval. The last term of the equation is the inner product of two different terms of the basis, and consequently 0. Equation (4.9) therefore becomes:

$$a_j \frac{T}{2} \int_{-T/2}^{T/2} \sqrt{\frac{2}{T}}\cos(j\Omega_1 t)\sqrt{\frac{2}{T}}\cos(j\Omega_1 t)dt = \int_{-T/2}^{T/2} \cos(j\Omega_1 t)f(t)dt. \tag{4.10}$$

If the inner product in Equation (4.10) is equal to 1, the desired coefficient is:

$$a_k = \frac{2}{T} \int_{-T/2}^{T/2} \cos(k\Omega_1 t)f(t)dt. \tag{4.11}$$

In traditional studies of the Fourier series, the previous results are derived by using only the trigonometric relationship such as taking the inner product in Equation (4.9):

$$\int_{-T/2}^{T/2} \sqrt{\frac{2}{T}}\cos(k\Omega_1 t)\sqrt{\frac{2}{T}}\cos(j\Omega_1 t)dt = \frac{1}{T}\int_{-T/2}^{T/2} \cos((j+k)\Omega_1 t)\,dt + \frac{1}{T}\int_{-T/2}^{T/2} \cos((j-k)\Omega_1 t)\,dt$$

where the trigonometric identity

$$\cos(A)\cos(B) = \frac{1}{2}[\cos(A+B) + \cos(A-B)]$$

was used. If $j \neq k$ is the interval of integration, it will contain integer cycles of its sinusoid function and consequently the integral will be 0. The first integral $j = k$ will be 0 for the same reason, but the second integral will be 1, i.e.

$$\int_{-T/2}^{T/2} \sqrt{\frac{2}{T}}\cos(k\Omega_1 t)\sqrt{\frac{2}{T}}\cos(j\Omega_1 t)\mathrm{d}t = \begin{cases} 1 & \text{if } k=j \\ 0 & \text{if } k \neq j. \end{cases} \tag{4.12}$$

4.2.1.3 Sine Coefficients

Pre-multiplying both sides of Equation (4.7) by $\sin(j\Omega_1 t)$ and integrating yields:

$$\sum_{k=1}^{\infty} b_k \int_{-T/2}^{T/2} \sin(j\Omega_1 t)\sin(k\Omega_1 t)\mathrm{d}t = \int_{-T/2}^{T/2} \sin(j\Omega_1 t)f(t)\mathrm{d}t - a_0 \int_{-T/2}^{T/2} \sin(j\Omega_1 t)\mathrm{d}t$$

$$- \sum_{k=1}^{\infty} a_k \int_{-T/2}^{T/2} \sin(j\Omega_1 t)\cos(k\Omega_1 t)\mathrm{d}t.$$

By following the same procedure as in the previous section, we obtain:

$$b_k = \frac{2}{T} \int_{-T/2}^{T/2} \sin(k\Omega_1 t)f(t)\mathrm{d}t. \tag{4.13}$$

Table 4.1 provides a list of the equations needed to obtain a Fourier series of coefficients.

4.2.2 The Exponential Fourier Series

Equation (4.7) is commonly referred to in the literature as the trigonometric form of the Fourier Series. In some situations it may however be more convenient to express the same Fourier series in its exponential form: $e^{jk\Omega_1 t}$, $k = \cdots, -2, -1, 0, 1, 2, \cdots$. In such a scenario, the expansion is written:

$$y(t) = \sum_{k=-\infty}^{\infty} c_k e^{jk\Omega_1 t} \tag{4.14}$$

Table 4.1 Mathematical expression of the Fourier series coefficients.

Coefficient	Mathematical expression
a_0 (DC term)	$a_0 = \dfrac{1}{T} \displaystyle\int_{-T/2}^{T/2} f(t)\mathrm{d}t$
a_k	$a_k = \dfrac{2}{T} \displaystyle\int_{-T/2}^{T/2} \cos(k\Omega_1 t)f(t)\mathrm{d}t$
b_k	$b_k = \dfrac{2}{T} \displaystyle\int_{-T/2}^{T/2} \sin(k\Omega_1 t)f(t)\mathrm{d}t$

The basis function is formed of an infinite set of exponential, harmonically related terms:

$$S = \text{span}\left\{ \cdots \quad \sqrt{\frac{1}{T}}e^{-j2\Omega_1 t}, \quad \sqrt{\frac{1}{T}}e^{-j\Omega_1 t}, \quad \sqrt{\frac{1}{T}}, \quad \sqrt{\frac{1}{T}}e^{j1\Omega_1 t}, \quad \sqrt{\frac{1}{T}}e^{j2\Omega_1 t}, \quad \cdots \right\}. \tag{4.15}$$

The exponential function $e^{jk\Omega_1 t}$ has period T and can be easily verified:

$$e^{jk\Omega_1 t} = e^{jk\Omega_1(t+T)} = e^{jk\Omega_1 t}e^{jk\Omega_1 T}$$

$$e^{jk\Omega_1 t} = e^{jk\Omega_1 t}e^{j2\pi k} = e^{jk\Omega_1 t}$$

using the fact that $e^{j2\pi k} = \cos 2\pi k + j \sin 2\pi k = 1$.
The inner product as defined by Equation (4.5) is:

$$\left\langle \frac{1}{\sqrt{T}}e^{j\Omega_1 kt}, \frac{1}{\sqrt{T}}e^{j\Omega_1 lt} \right\rangle = \frac{1}{T}\int_{-T/2}^{T/2} e^{j\Omega_1(k-l)t}\,dt = \frac{1}{T}\frac{1}{\Omega_1(k-l)}e^{j\Omega_1(k-l)t}\Big|_{-T/2}^{T/2} \tag{4.16}$$

$$\left\langle \frac{1}{\sqrt{T}}e^{j\Omega_1 kt}, \frac{1}{\sqrt{T}}e^{j\Omega_1 lt} \right\rangle = \begin{cases} 0 & \text{for } k \neq l \\ 1 & \text{for } k = l. \end{cases}$$

The solution for $k = l$ is found by considering that

$$\frac{1}{T}\int_{-T/2}^{T/2} e^{j(k-l)\Omega_1 t}\,dt = \frac{1}{T}\int_{-T/2}^{T/2} dt = 1.$$

The task of identifying the coefficients of the complex series is extremely simplified when the basis is orthonormal. To obtain coefficient c_j, pre-multiply both sides of Equation (4.14) by $e^{-jl\Omega_1 t}$. As the integration proceeds,

$$\int_{-T/2}^{T/2} y(t)e^{+jl\Omega_1 t}\,dt = T\sum_{k=-\infty}^{\infty} c_k \frac{1}{T}\int_{-T/2}^{T/2} e^{j(k-l)\Omega_1 t}\,dt,$$

The right side of the previous equation is only non-zero when $k = j$; we then have:

$$c_k = \frac{1}{T}\int_{-T/2}^{T/2} y(t)e^{-jl\Omega_1 t}\,dt. \tag{4.17}$$

4.2.3 Relationship between the Exponential and Trigonometric Coefficients

In power systems the signal $y(t)$ is a real function. However, the parameters c_k are in general complex coefficients. If interested in finding the relationship between the complex coefficients c_k, the real coefficients a_k and b_k of the trigonometric Fourier series representation, the first step is to take the conjugate of Equation (4.14):

$$y(t) = \sum_{k=-\infty}^{\infty} c_k^* e^{-jk\Omega_1 t}.$$

By replacing k with $-k$ in the summation, we have:

$$y(t) = \sum_{k=-\infty}^{\infty} c_{-k}^* e^{jk\Omega_1 t}.$$

By comparison, it is required that $c_k = c_{-k}^*$ or, equivalently, $c_k^* = c_{-k}$. Rearranging Equation (4.14),

$$y(t) = c_0 + \sum_{k=1}^{\infty} c_k e^{jk\omega_1 t} + \sum_{k=-\infty}^{1} c_k e^{jk\omega_1 t}.$$

Changing the limits of the second summation and using the fact that $c_k^* = c_{-k}$ yields:

$$y(t) = c_0 + \sum_{k=1}^{\infty} c_k e^{jk\Omega_1 t} + c_k^* e^{-jk\Omega_1 t} = c_0 + \sum_{k=1}^{\infty} 2\Re e\{c_k e^{jk\Omega_1 t}\}. \qquad (4.18)$$

If c_k is expressed in polar form,

$$c_k = C_k e^{j\theta_k}$$

Equation (4.18) becomes

$$y(t) = c_0 + 2 \sum_{k=1}^{\infty} C_k \cos(k\omega_1 t + \theta_k).$$

Using a trigonometric identity yields:

$$y(t) = c_0 + 2 \sum_{k=1}^{\infty} C_k \cos\theta_k \cos(k\Omega_1 t) - C_k \sin\theta_k \sin(k\Omega_1 t). \qquad (4.19)$$

Finally, when comparing the coefficients of Equation (4.19) with the coefficients of the trigonometric form in Equation (4.7), we obtain:

$$\begin{aligned} a_0 &= c_0 \\ a_k &= C_k \cos\theta_k \\ b_k &= -C_k \sin\theta_k. \end{aligned} \qquad (4.20)$$

If the coefficients of the trigonometric form are known, the complex coefficients are given by:

$$c_0 = a_0$$
$$C_k = \sqrt{a_k^2 + b_k^2}$$
$$\theta_k = \tan^{-1}\left(\frac{-b_k}{a_k}\right).$$

(4.21)

Example 4.1

Plot the magnitude and phase of the Fourier coefficients of the following signal using the Euler expression:

$$y(t) = 1 + \sin \Omega_1 t + 2\cos \Omega_1 t + \cos(2\Omega_1 t + \pi)$$

Expanding each term of the above equation in terms of a complex exponential through the Euler expression yields:

$$y(t) = 1 + \frac{1}{2j}\left(e^{j\Omega_1 t} - e^{-j\Omega_1 t}\right) + \left(e^{j\Omega_1 t} + e^{-j\Omega_1 t}\right) + \frac{1}{2}\left(e^{j(2\Omega_1 t + \pi)} + e^{-j(2\Omega_1 t + \pi)}\right).$$

$$y(t) = 1 + \left(1 + \frac{1}{2j}\right)e^{j\Omega_1 t} + \left(1 - \frac{1}{2j}\right)e^{-j\Omega_1 t} + \frac{1}{2}e^{j2\Omega_1 t} + \frac{1}{2}e^{-j2\Omega_1 t}.$$

Figure 4.3a and b show the modulus and phase of the Fourier coefficients, respectively. These plots are referred to as the coefficients of the magnitude spectrum and phase spectrum. It can be observed that the magnitude coefficients are an even sequence and that phase is an odd sequence; this is because the original signal is real and can be verified by proprieties of the Fourier series.

4.2.4 Harmonics in Power Systems

The Fourier series is a mathematical tool to analytically study the frequency composition of a periodic signal. When the periodic signal in a power system is not purely sinusoidal, but

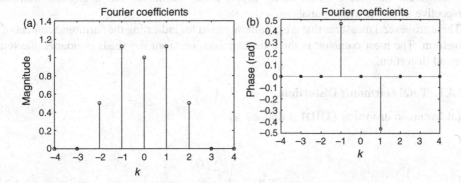

Figure 4.3 Spectrum of Fourier: (a) magnitude of Fourier coefficients; (b) phase.

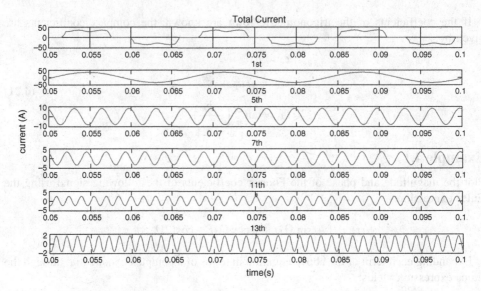

Figure 4.4 Harmonic decomposition of the current of a 12-pulse converter.

the sum of several sinusoids harmonically related, we say that there is harmonic distortion in the system. Harmonic distortion is caused by non-linear loads (typically electronic equipment) such as DC drivers, VSIs (voltage source inverters), CSIs (current source inverters) and other phenomena such as a transform saturation or arching loads. Figure 4.4 shows a typical current drained by a 12-pulse rectifier. The top plot is the distorted current and the others are the harmonic components: the fundamental (first), 5th, 7th, 11th and 13th harmonics.

An interesting fact in the figure is that there are only odd harmonics. This will always happen if both the positive and negative half-cycles of a waveform have identical shapes. The presence of even harmonics in a power system happens in only a few situations such as in half-wave rectifiers, inrush currents and arcs.

Usually the harmonics with significant energy are of a low order. Reference [2] establishes the maximum limit of 50 harmonics; higher-order harmonics are negligible from the perspective of power system analysis.

There are several measures that are commonly used for indicating the harmonic content of a waveform. The most common is the total harmonic or, from the IEEE standards, the total demand distortion.

4.2.4.1 Total Harmonic Distortion

Total harmonic distortion (THD) is defined as

$$\text{THD} = \frac{\sqrt{\sum_{n=2}^{H} G_n^2}}{G_1} \tag{4.22}$$

where G_n ($n = 1, 2, \ldots, H$) is the RMS value of the harmonic component n and H is the maximum harmonic order (typically 50). G_n is defined:

$$G_n = \frac{C_n}{\sqrt{2}}$$

where C_n is the harmonic amplitude.

The THD is defined for voltage or current and provides information about the potential heating values caused by the distortion, proportional to the fundamental component. This is because the RMS of the distorted signal is given by the square root of the sum of the individual RMS, that is,

$$\text{RMS} = \sqrt{\sum_{n=1}^{H} G_n^2} = G_1 \sqrt{1 + \text{THD}^2}. \qquad (4.23)$$

A high-voltage THD has high distortion values; its RMS will then be responsible for heating the equipment. The voltage THD is related to its nominal voltage. This is different for a current THD, which can be related to the fundamental component of the current because this current is demanded by the load. Generally the current demanded is below the nominal current projected for that system. As such, a small current may have a high THD but may not be significant enough to damage or threaten the system. For example, the current demanded by a fluorescent lamp reactor's THD is high, but the effect of a single lamp or a small set of lamps will drain a much smaller current than the nominal current. For this reason the IEEE standard 519-1992 defines the total demand distortion (TDD), described in the following section.

4.2.4.2 Total Demand Distortion

The TDD is defined in reference [3] as the total root-sum-square harmonic current distortion divided by maximum demand current load I_L, that is:

$$\text{TDD} = \frac{\sqrt{\sum_{n=2}^{H} I_n^2}}{I_L}. \qquad (4.24)$$

Note that the denominator uses the maximum demanded load current.

4.2.5 Proprieties of a Fourier Series

A representation of a Fourier series has a number of important properties that are useful for the understanding the analytical developments and concept of a Fourier series. Table 4.2 lists the main properties. Further details on how to prove these properties can be found in traditional textbooks of signals and system topics, such as references [4] and [5].

Table 4.2 Properties of the Fourier series.

Row No.	Property	Periodic signal $x(t)$ and $y(t)$	Complex Fourier series coefficients, g_k and d_k
1	Linearity	$Ax(t) + By(t)$	$Ag_k + Bd_k$
2	Time shift	$x(t - t_0)$	$g_k e^{-jk\Omega_1 t_0}$
3	Frequency shifting	$e^{jM\Omega_1 t} x(t)$	g_{k-M}
4	Conjugation	$x^*(t)$	g^*_{-k}
5	Time scaling	$x(\alpha t), \quad \alpha > 0$	g_k
6	Periodic convolution	$\displaystyle\int_T x(\tau)y(t-\tau)\mathrm{d}\tau$	$Tg_k d_k$
7	Real signal	$x(t)$ real	$g_k = g^*_{-k}$ $\lvert g_k \rvert = \lvert g_{-k} \rvert$ $\angle g_k = -\angle g_{-k}$
8	Signal symmetry	$x(t)$ real and even $x(t)$ real and odd	g_k real and even g_k purely imaginary and odd
9	Parseval's relationship for periodic signal: $\dfrac{1}{T}\displaystyle\int_T \lvert x(t) \rvert^2 \, \mathrm{d}t = \sum_{k=-\infty}^{\infty} \lvert g_k \rvert^2$		

4.3 A Fourier Transform

4.3.1 Introduction and Examples

Taking a step forward in the review on series and transforms of continuous signals, we now discuss some key points about the Fourier transform. Figure 4.5 shows the basic concept behind any transform. Roughly speaking, the objective of any transform is to change the space of representation of a signal to another space so that specific properties can be easily investigated. These can include a sparse representation (compression), filtering or separation.

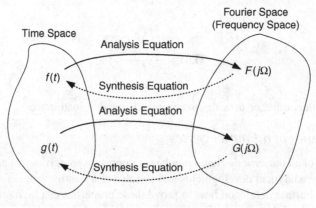

Figure 4.5 The time-space and frequency-space produced by the Fourier transform.

A Fourier transform changes the time domain space in the frequency domain space. In fact, the original space can be formed by any physical quantity such as distance, temperature and so on. Two equations govern the changes between these spaces: the *analysis* equation which takes a function in its time space and converts it into a frequency space; and the *synthesis* equation that takes a function in its frequency domain and brings it back to the time domain. Such a transformation has to be bijective, which means that there is a one-to-one correspondence between the functions from time to frequency domains and vice versa.

The analysis equation or Fourier transform is defined:

$$F(j\Omega) = \int_{-\infty}^{\infty} f(t)e^{-j\Omega t}\, dt. \tag{4.25}$$

The inverse Fourier transform or synthesis equation is defined:

$$f(t) = \frac{1}{2\pi} \int_{-\infty}^{\infty} F(j\Omega)e^{j\Omega t}\, d\Omega. \tag{4.26}$$

In general, $F(j\Omega)$ is complex and can be written:

$$F(j\Omega) = A(\Omega)e^{-j\phi(\Omega)} \tag{4.27}$$

where $A(\Omega) = |F(j\Omega)|$ is the amplitude spectrum and $\phi(\Omega) = \angle F(j\Omega)$ is the phase spectrum. The pair of functions and their Fourier transform are commonly represented as $f(t) \leftrightarrow F(j\Omega)$, where the double arrow represents the bijective characteristic of the mapping.

While the Fourier series exists for all periodic functions, the Fourier transform only exists for a set of functions that obey the convergence theorem. A sufficient, but not necessary, condition for convergence is that the function is absolutely integrable, that is:

$$\lim_{\tau \to \infty} \int_{-\tau}^{\tau} |f(t)|\, dt < \infty.$$

For example the function $f(t) = 1$ is not absolutely integrable but it has a Fourier transform equals to $2\pi\delta(\Omega)$:

$$1 \leftrightarrow 2\pi\delta(\Omega).$$

By definition, the Fourier transform of a non-absolutely integrable function cannot be obtained directly from Equation (4.25). The Fourier transform of some of the most important functions in signal analysis are non-absolutely integrable or have infinite energy. These include a constant function, a step function or a periodic function. Under these circumstances, either $F(j\Omega)$ does not exist or it does not satisfy Equation (4.26). An important function that leads to mathematical difficulties is the *impulse* function. Mathematically speaking, the impulse is not a function because it is not defined at the point of application, as shown below. However, in signal processing literature, the name 'impulse function' or the *Dirac* delta function is used frequently. The impulse function can be

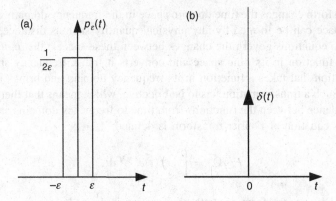

Figure 4.6 (a) Pulse function; (b) impulse function.

defined from the pulse function depicted in Figure 4.6a, when $\varepsilon \to 0$. The mathematical expression for the pulse function is:

$$p_\varepsilon(t) = \begin{cases} \dfrac{1}{2\varepsilon} & \text{for } |t| \le \varepsilon \\ 0 & \text{otherwise} \end{cases}$$

and the impulse function is defined:

$$\delta(t) = \lim_{\varepsilon \to 0} p_\varepsilon(t).$$

The Fourier transform of the pulse function is easily found:

$$\Im\{\delta(t)\} = \lim_{\varepsilon \to 0} \int_{-\infty}^{\infty} p_\varepsilon(t) e^{-j\Omega t}\, dt = \lim_{\varepsilon \to 0} \int_{-\varepsilon}^{\varepsilon} \frac{1}{2\varepsilon} e^{-j\Omega t}\, dt$$

$$= \lim_{\varepsilon \to 0} \frac{\sin \varepsilon\Omega}{\varepsilon\Omega} = \lim_{\varepsilon \to 0} \frac{\varepsilon\Omega}{\varepsilon\Omega} = 1.$$

Nonetheless, if the inverse Fourier transform of unity is attempted, we obtain

$$f(t) = \frac{1}{2\pi} \int_{-\infty}^{\infty} e^{j\Omega t}\, d\Omega = \frac{1}{2\pi} \left(\int_{-\infty}^{\infty} \cos \Omega t\, d\Omega + j \int_{-\infty}^{\infty} \sin \Omega t\, d\Omega \right).$$

The previous integrals do not converge, so we can conclude that $F(j\Omega)$ does not satisfy Equation (4.26). These difficulties can be overcome by the concept of *generalized* functions. Generalized functions are mathematically 'well-behaved' functions that allow for the solution of problems such as these. (This book will not discuss generalized functions, but the interested reader can find more details in reference [6]).

Table 4.3 Fourier transform pairs.

Row No.	$f(t)$	$F(j\Omega)$	Definition equation/tips for solution
1	$\delta(t)$	1	Definition
2	1	$2\pi\delta(\Omega)$	No
3	$\delta(t - t_0)$	$e^{-j\Omega t_0}$	Time shifting property (Table 4.4)
4	$e^{j\Omega_0 t}$	$2\pi\delta(\Omega - \Omega_0)$	Frequency shifting property (Table 4.4)
5	$\cos(\Omega_0 t)$	$\pi[\delta(\Omega - \Omega_0) + \delta(\Omega + \Omega_0)]$	Euler relationship and properties
6	$\sin(\Omega_0 t)$	$j\pi[\delta(\Omega + \Omega_0) - \delta(\Omega - \Omega_0)]$	Euler relationship and properties
7	$e^{-at}u(t)$	$\dfrac{1}{a + j\Omega}$	Definition
8	$e^{-at}\sin(\Omega_0 t)u(t)$	$\dfrac{\Omega_0}{(a + j\Omega)^2 + \Omega_0^2}$	Euler relationship and properties
9	$g_{t_0}(t) = \begin{cases} 1 & \lvert t \rvert \le t_0 \\ 0 & \lvert t \rvert > t_0 \end{cases}$	$\dfrac{2\sin\Omega t_0}{\Omega}$	Definition
10	$\dfrac{\sin\Omega_0 t}{\pi t}$	$g_{\Omega_0}(\Omega) = \begin{cases} 1 & \lvert \Omega \rvert \le \Omega_0 \\ 0 & \lvert \Omega \rvert > \Omega_0 \end{cases}$	Definition

The reader however needs to be aware that, if attempting to find a Fourier transform (or its inverse), some integrals do not converge, this does not necessarily means that the Fourier pair does not exist. A more in-depth mathematical analysis is needed to solve this. The approach here will be to present a Fourier pair without this verification. Table 4.3 lists the most common Fourier transform pair. The last column in the table indicates whether the pair obey Equations (4.25) and (4.26). If yes, the reader will be able to verify the pair through the definition equations. If not, the reader has to use the concept of *generalized* function, not presented in this book.

Example 4.2

Find the Fourier transform of the gate function:

$$g_{t_0}(t) = \begin{cases} 1 & \lvert t \rvert \le t_0 \\ 0 & \lvert t \rvert > t_0. \end{cases}$$

Figure 4.7a depicts the gate function. The Fourier transform can be obtained directly from its definition:

$$F(j\Omega) = \int_{-t_0}^{t_0} e^{-j\Omega t}\,dt = \frac{2}{2j\Omega}\left(-e^{-j\Omega t_0} + e^{j\Omega t_0}\right) = \frac{2\sin\Omega t_0}{\Omega}.$$

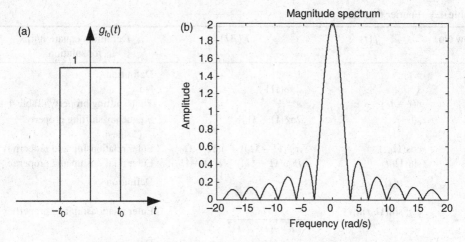

Figure 4.7 (a) Function gate and (b) magnitude spectrum.

The magnitude spectrum $A(\Omega)$ of $F(j\Omega)$ is presented in Figure 4.7b for $t_0 = 1$. The gate function is widely used in signal processing and its transform is referred to as a sinc function; note that the sinc function has 0s spread periodically.

Example 4.3

Find the Fourier transform of the function:

$$f(t) = \begin{cases} e^{-at} & \text{for } t \geq 0 \\ 0 & \text{for } t < 0 \end{cases}$$

where a is a positive and a real constant.

This function can be written as $f(t) = e^{-at}u(t)$, and the Fourier transform is

$$F(j\Omega) = \int_0^\infty e^{-(a+j\Omega)t}\, dt = \frac{1}{a + j\Omega}.$$

The pair can then be written as

$$f(t) = e^{-at}u(t) \leftrightarrow \frac{1}{a + j\Omega}.$$

The time function $f(t)$ and the magnitude spectrum of $F(j\Omega)$ are represented in Figure 4.8a and b respectively for $a = 5$. Note that the magnitude spectrum has energy over a wide range of the frequency axis. An interesting fact is that the higher the a value (the exponential vanishes quicker), the wider the magnitude spectrum (the spectrum vanishes slower). This type of signal is very common during a fault in a transmission line. It is known as the exponential

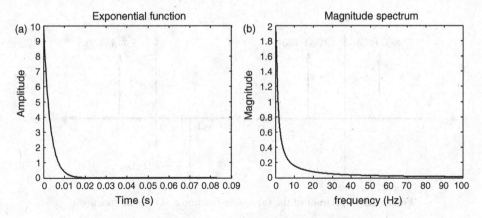

Figure 4.8 (a) Exponential signal and (b) its magnitude spectrum.

decaying DC component. As the energy of the spectrum spreads along the frequency, the estimation of the phasor (fundamental sinusoid component) is affected by the DC component and a particular method of estimation is needed. Some of the methods used for the estimation of phasors in the presence of a decaying DC are discussed in Section 7.6

4.3.2 Fourier Transform Properties

There are several important properties or theorems of the Fourier transform which are useful for signal processing applications. These properties can be used to determine the Fourier transform of a signal without the need to apply the analysis or synthesis equations. The tips in Table 4.3 show that these properties can be used as a shortcut to obtain the Fourier pair. This section will present the main theorems and demonstrate their application in examples.

Table 4.4 presents the main theorems of a Fourier transform. In this table a, b, t_0 and Ω_0 are arbitrary real constants.

Table 4.4 Fourier Transform properties.

Row No.	Properties	$f(t) \leftrightarrow F(j\Omega)\, g(t) \leftrightarrow G(j\Omega)$
1	Linearity	$af(t) + bg(t) \leftrightarrow aF(j\Omega) + bG(j\Omega)$
2	Duality	$F(jt) \leftrightarrow 2\pi f(-\Omega)$
3	Time shifting	$f(t - t_0) \leftrightarrow F(j\Omega)e^{-j\Omega t_0}$
4	Time scaling	$f(at) \leftrightarrow \frac{1}{|a|}F\left(\frac{j\Omega}{a}\right)$
5	Frequency shifting	$e^{j\Omega_0 t}f(t) \leftrightarrow F(j\Omega - j\Omega_0)$
6	Time convolution	$f(t) * g(t) \leftrightarrow F(j\Omega)G(j\Omega)$
		where $f(t) * g(t) = \int_{-\infty}^{\infty} f(\tau)g(t - \tau)\mathrm{d}\tau$
7	Frequency convolution	$f(t)g(t) \leftrightarrow F(j\Omega) * G(j\Omega)$
		where $F(j\Omega) * g(j\Omega) = \frac{1}{2\pi}\int_{-\infty}^{\infty} F(jv)g(j\Omega - jv)\mathrm{d}v$
8	Parseval's formula	$\int_{-\infty}^{\infty} |f(t)|^2\, \mathrm{d}t = \frac{1}{2\pi}\int_{-\infty}^{\infty} |F(j\Omega)|^2\, \mathrm{d}\Omega$

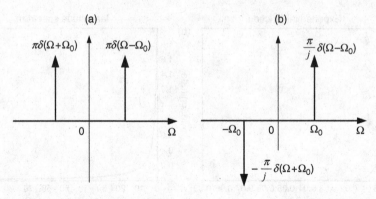

Figure 4.9 Spectrum of the (a) cosine function and (b) sine function.

Example 4.4

Find the Fourier transform of the following function:

$$f(t) = A \cos \Omega_0 t$$

Solution:
Using the Euler formula,

$$f(t) = \cos \Omega_0 t = \frac{e^{j\Omega_0 t} + e^{-j\Omega_0 t}}{2}$$

and the linear property of Table 4.4,

$$F(j\Omega) = \frac{1}{2}\Im\{e^{j\Omega_0 t}\} + \frac{1}{2}\Im\{e^{-j\Omega_0 t}\}.$$

Using the Fourier transform pair in Table 4.4 (pair 4) we obtain:

$$F(j\Omega) = \pi\delta(\Omega - \Omega_0) + \pi\delta(\Omega + \Omega_0).$$

Figure 4.9 depicts the plot of the Fourier transform of the cosine and sine function respectively. Note that the sinusoid functions are represented as impulse functions. This fact leads us to an important conclusion: any periodic function can be written as a sum of its sinusoids (Fourier series), and the Fourier transform of any periodic function can be represented as a set of its impulse functions.

4.4 The Sampling Theorem

Power system signals are continuous time signals. However, discrete-time signal processing algorithms are used more often for processing power system signals. For example, the substitution of analog protection relays by numerical relays which contain a digital signal processor (DSP) running a digital signal processing algorithm. The modern harmonic analyzers are all digital equipment, running DSP algorithms such as the fast Fourier transform (FFT). We therefore need to convert the continuous-time signals to a discrete-time signal, and there needs to be certainty that the DSP algorithm will produce the same results as its analog processing. The link between the analog and digital world is given by the sampling theorem, derived in the following. The sampling theorem states that a band-limited signal $x(t)$ for which

$$X(j\Omega) = 0 \quad \text{for} \quad |\Omega| \geq \frac{\Omega_s}{2} \tag{4.28}$$

where $\Omega_s = 2\pi F_s = 2\pi/T_s$ is the sampling frequency in rad/s, F_s is the sampling frequency in Hz and T_s is the sampling period, can be uniquely determined from its samples $x(nT_s)$.

To demonstrate the sampling theorem, we refer to the analog signal as $x_a(t)$ and the samples of the sampling signal $x[n]$. We then have:

$$x[n] = x_a(nT_s). \tag{4.29}$$

The Fourier transform of $x_a(t)$ as defined by Equation (4.25) is

$$X_a(j\Omega) = \int_{-\infty}^{\infty} x_a(t)e^{-j\Omega t}\, dt. \tag{4.30}$$

Mathematically, the sampling signal can be represented by the product of the analog signal and a periodic impulse train $p(t)$, where

$$p(t) = \sum_{n=-\infty}^{\infty} \delta(t - nT_s) \tag{4.31}$$

$$x_p(t) = x_a(t)p(t) = \sum_{n=-\infty}^{\infty} x_a(nT_s)\delta(t - nT_s). \tag{4.32}$$

Figure 4.10 depicts the mathematical representation of the uniform sampling processing. In Figure 4.10a an ideal impulse modulator is represented. The analog signal (Figure 4.10b) is multiplied by the train of impulse (Figure 4.10c), resulting in the sampled signal (Figure 4.10d).

Applying the Fourier transform definition in Equation (4.32), taking into account that $x_a(nT_s)$ is a constant in t, yields

$$X_p(j\Omega) = \sum_{n=-\infty}^{\infty} x_a(nT_s)\Im\{\delta(t - nT_s)\}. \tag{4.33}$$

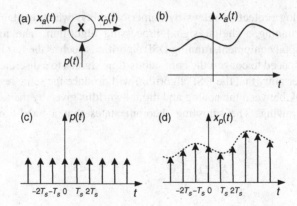

Figure 4.10 Generation of a sampled signal: (a) ideal sampling model; (b) analog signal; (c) impulse train; (d) sampled signal.

According to Table 4.3, line 4,

$$X_p(j\Omega) = \sum_{n=-\infty}^{\infty} x_a(nT_s)e^{-j\Omega nT_s}. \tag{4.34}$$

Using Equation (4.29),

$$X_p(j\Omega) = \sum_{n=-\infty}^{\infty} x[n]e^{-j\Omega nT_s}. \tag{4.35}$$

Equation (4.35) is the Fourier transform of the sampled signal, which is very close to the discrete-time Fourier transform (DTFT). However, before defining the DTFT we need to understand what happens with the spectrum of the sampled signal $X_p(j\Omega)$ compared to the spectrum of the original signal $X_a(j\Omega)$. To arrive at the relationship between the two spectrum representations, the fact that the impulse train is a periodic function of fundamental period T_s needs to be remembered; consequently, it can be rewritten as an exponential Fourier series:

$$p(t) = \frac{1}{T_s} \sum_{k=-\infty}^{\infty} e^{j\Omega_s kt}. \tag{4.36}$$

Equation (4.32) can then be rewritten:

$$x_p(t) = x_a(t)p(t) = \frac{1}{T_s} \sum_{k=-\infty}^{\infty} e^{j\Omega_s kt} x_a(t). \tag{4.37}$$

Using the frequency shifting property in Table 4.4, we arrive at the following Fourier transform relationship:

$$X_p(j\Omega) = \frac{1}{T_s} \sum_{k=-\infty}^{\infty} X_a(j\Omega - k\Omega_s). \tag{4.38}$$

$X_p(j\Omega)$ is therefore a periodic function of Ω_s, consisting of a sum of shifted and scaled replicas of $X_a(j\Omega)$. It is referred to as the *baseband spectrum*, shifted by integers that are multiples of Ω_s and scaled by $1/T_s$. The frequency range $-\Omega_s/2 \leq \Omega \leq \Omega_s/2$ is referred to as the *baseband* or Nyquist band.

Equation (4.38) describes the relation between the spectrum of the original signal and the sampled signal, and is the basis for defining the sampling theorem. Figure 4.11 illustrates the effect of the time-domain sampling in its frequency domain. Figure 4.11a depicts the spectrum of the original signal, which is assumed to be a band-limited signal with no energy above frequency Ω_m. Figure 4.11b depicts the spectrum of the sampling signal, assuming $\Omega_s > 2\Omega_m$. The correspondent spectrum is periodic in frequency and repeats at each multiple of Ω_s; only two replicas of the baseband spectrum are represented. Note that there is no mix of individual replicas and it is possible to obtain the original signal if the sampled signal is passed through a low-pass filter with rejection frequencies equal to $\Omega_s - 2\Omega_m$. Figure 4.11c depicts the case

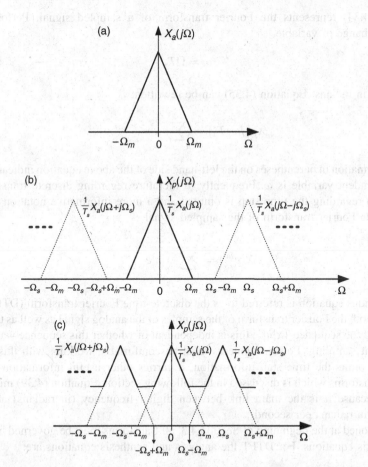

Figure 4.11 Illustration of the effect of time-sampling in the frequency domain: (a) spectrum of original signal; (b) spectrum of the sampled signal when $\Omega_s > 2\Omega_m$; and (c) spectrum of the sampled signal when $\Omega_s < 2\Omega_m$.

when $\Omega_s < 2\Omega_m$. As observed in this case, there is an overlap between the individual replicas and part of the original information is lost; consequently, the original signal cannot be recovered by filtering. This effect of spectrum overlap is known in literature as *aliasing*.

The above result is more commonly known as the sampling theorem or Nyquist sampling theorem, which can be summarized as follows. Given a band-limited signal $x(t)$,

$$X(j\Omega) = 0 \quad \text{for} \quad |\Omega| \geq \Omega_m$$

can be uniquely determined from its samples values $x(nT_s)$ if $\Omega_s > 2\Omega_m$. This relationship is also known as the Nyquist theory. The highest frequency in the signal is known as the Nyquist frequency, since it determines the minimum sampling frequency that must be used to sample the signal and recover it from the illustration.

4.5   The Discrete-Time Fourier Transform

Equation (4.34) represents the Fourier transform of a sampled signal. Performing the following change of variable,

$$\omega = \Omega T_s \tag{4.39}$$

where ω is in radians, Equation (4.35) can be rewritten:

$$X_p(j\omega/T_s) = \sum_{n=-\infty}^{\infty} x[n]e^{-j\omega n}.$$

The information in parentheses on the left-hand side of the above equation indicates that the new independent variable is ω. Frequently in literature regarding discrete transforms, the information regarding the time step is omitted and a more informative notation is used to represent the Fourier transform of the sampled signal:

$$X(e^{j\omega}) = \sum_{n=-\infty}^{\infty} x[n]e^{-j\omega n}.$$

The previous equation is referred to as the discrete-time Fourier transform (DTFT) which represents both the Fourier transform of the samples of the analog signal as well as the Fourier transform of the sequence $\{x[n]\}$; this is independent of whether this sequence was obtained from current sampling. Power engineers must be comfortable working with this notation; although it omits the time step information, it carries other useful information about the discrete transforms which is discussed in the following section. Equation (4.39) must be kept in mind because it is the main link between digital frequency (in radians) and analog frequency (in radians per second).

As mentioned at the beginning of Section 4.3, all transforms must be governed by analysis and synthesis equations. For DTFT the analysis and synthesis equations are:

$$X(e^{jw}) = \sum_{n=-\infty}^{\infty} x[n]e^{-j\omega n}, \quad \text{analysis equation or DTFT} \tag{4.40}$$

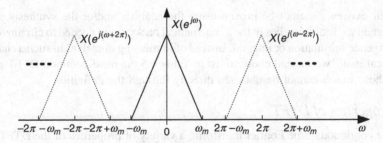

Figure 4.12 DTFT spectrum.

$$x[n] = \frac{1}{2\pi} \int_{-\pi}^{\pi} X(e^{j\omega})e^{j\omega n}d\omega, \quad \text{synthesis equation or IDTFT.} \tag{4.41}$$

The synthesis equation is also known as the inverse discrete-time Fourier transform or IDTFT. In a DTFT equation, the independent variable is ω and the notation $e^{j\omega}$ carries useful information such as the periodicity of the DTFT, which is $2\pi k$ where k is an integer. We have

$$X(e^{j(\omega+2\pi k)}) = \sum_{n=-\infty}^{\infty} x[n]e^{-j(\omega+2\pi k)n} = \sum_{n=-\infty}^{\infty} x[n]e^{-j\omega n}e^{-2\pi kn} = \sum_{n=-\infty}^{\infty} x[n]e^{-j\omega n}$$

where we have used the fact that

$$e^{-2\pi kn} = \cos(2\pi kn) - j\sin(2\pi kn) = 1$$

for integers k and n.

If both axes of Figure 4.11b are multiplied by T_s, then $\Omega_s/f_s = 2\pi$ and the factor $1/T_s$ is omitted. Figure 4.12 represents the Fourier spectrum of $X(e^{j\omega})$; note the periodicity of the spectrum.

Other information carried by the notation $e^{j\omega}$ is that the DTFT is a complex function of ω and can be represented in a rectangular (real and imaginary) or polar form. The polar form is commonly used:

$$X(e^{j\omega}) = |X(e^{j\omega})|e^{j\theta(\omega)} \tag{4.42}$$

where $|X(e^{j\omega})|$ is called the magnitude function or magnitude spectrum and $\theta(\omega) = \arg\{X(e^{j\omega})\}$ is the phase function or phase spectrum. Furthermore, the spectrum produced by sequence $\{x[n]\}$ is a continuous function in ω.

4.5.1 DTFT Pairs

The existence of the DTFT depends on the convergence of Equation (4.40), and the summation must be finite. If the DTFT pair exists, then the IDTFT must satisfy Equation (4.41). As mentioned in Section 4.3, some pairs do not appear to obey these equations, that is, their existence cannot be proven using the analysis and/or the synthesis equations directly. Again, the limitation lies in the mathematical background needed to circumvent some non-convergence summation or integral. Instead of immersing ourselves in such a challenging mathematical issue, we instead chose to list in Table 4.5 the most common DTFT pairs and highlight those which cannot be obtained directly through the definition.

4.5.2 Properties of DTFT

As with the continuous-time Fourier transform, a variety of properties of the DTFT provide further insight into the transform and are often useful during the evaluation of transforms and their inverse. Table 4.6 lists the main properties of the DTFT.

4.6 The Discrete Fourier Transform (DFT)

In practice, the frequency analysis of discrete-time signals is most conveniently performed on a digital signal processor rather than through analog processing. By definition, the DTFT is performed over an infinite sequence length, and frequency is a continuous variable. These constraints make the processing of such sequences not feasible in a digital processor. To overcome these limitations, the discrete Fourier transform (DFT) is used. The analysis and synthesis equations are defined:

$$X[k] = \sum_{n=0}^{N-1} x[n]e^{-j2\pi kn/N} \qquad \text{analysis equation or DFT} \qquad (4.43)$$

$$x[n] = \frac{1}{N}\sum_{k=0}^{N-1} X[k]e^{j2\pi kn/N} \qquad \text{synthesis equation or IDFT.} \qquad (4.44)$$

where the synthesis equation is also known as the inverse discrete Fourier transform (IDFT). Equation (4.43) is the direct DFT and the summation is performed over N samples of $x[n]$. $X[k]$ is the discrete Fourier transform of $x[n]$. The DFT has the same length as the IDFT; that is, in Equation (4.43) $n = 0, 1, \ldots, N-1$ and in Equation (4.44) $k = 0, 1, \ldots, N-1$.

The first connection between the DFT and DTFT can be made considering a sequence of length N:

$$X(e^{j\omega}) = \sum_{n=0}^{N-1} x[n]e^{-j\omega n}$$

$$X[k] = X(e^{j\omega})|_{\omega=2\pi k/N}. \qquad (4.45)$$

The DFT is obtained by the uniform sampling of the DTFT at N equally spaced frequencies $\omega_k = 2\pi k/N$ where $0 \le k \le N-1$ on the ω axis from $0 \le \omega < 2\pi$. The ω axis can be represented as shown in Figure 4.13 for a sequence of length $N = 32$ samples. In this figure, only the k values on the upper semicircle are marked but its variation is $0 \le k \le 31$. The value of $2\pi/N$ is known as the frequency resolution or frequency bin, and represents the smaller value for ω_k. This representation is very useful, because it highlights the facts that: (1) the periodicity of

Table 4.5 Commonly used discrete time Fourier transform pairs.

Row No.	Sequence	DTFT	Definition equation/tips for solution
1	$\delta[n]$	1	No (IDTF does not converge for $n = 0$)
2	1	$\displaystyle\sum_{k=-\infty}^{\infty} 2\pi\delta(\omega+2\pi k)$	No
3	$u[n]$	$\dfrac{1}{1-e^{-j\omega}} + \displaystyle\sum_{k=-\infty}^{\infty} \pi\delta(\omega+2\pi k)$	No
4	$e^{j\omega_0 t}$	$\displaystyle\sum_{k=-\infty}^{\infty} 2\pi\delta(\omega-\omega_0+2\pi k)$	Frequency shifting properties (Table 4.6)
5	$\cos(\omega_0 n)$	$\pi\displaystyle\sum_{k=-\infty}^{\infty} \{\delta(\omega-\omega_0+2\pi k) + \delta(\omega+\omega_0+2\pi k)\}$	Euler relationship and properties
6	$\sin(\omega_0 t)$	$\dfrac{\pi}{j}\displaystyle\sum_{k=-\infty}^{\infty} \{\delta(\omega-\omega_0+2\pi k) + \delta(\omega+\omega_0+2\pi k)\}$	Euler relationship and properties
7	$\alpha^n u[n], |\alpha|<1$	$\dfrac{1}{1-\alpha e^{-j\omega}}$	Definition
8	$(n+1)\alpha^n u[n], \alpha<1$	$\dfrac{1}{(1-\alpha e^{-j\omega})^2}$	Property Table 4.6
9	$x[n] = \begin{cases} 1 & |n| \le N_0 \\ 0 & |n| > N_0 \end{cases}$	$\dfrac{\sin\left(\omega\left(N_0+\frac{1}{2}\right)\right)}{\sin(\omega/2)}$	Definition
10	$h[n] = \dfrac{\sin\omega_c n}{\pi n}$	$H(e^{j\omega}) = \begin{cases} 1 & 0 \le |\omega| \le \omega_c \\ 0 & \omega_c \le |\omega| \le \pi \end{cases}$	Definition

Table 4.6 Discrete-time Fourier transform properties.

Row No.	Properties	$x[n] \leftrightarrow X(e^{j\omega})$ $y[n] \leftrightarrow Y(e^{j\omega})$
1	Linearity	$ax[n] + by[n] \leftrightarrow aX(e^{j\omega}) + bY(e^{j\omega})$
2	Time shifting	$x[n - n_0] \leftrightarrow e^{-j\omega n_0}X(e^{j\omega})$
3	Frequency shifting	$e^{j\omega_0 n}x[n] \leftrightarrow X(e^{j(\omega-\omega_0)})$
4	Convolution	$x[n] * y[n] \leftrightarrow X(e^{j\omega})Y(e^{j\omega})$ where $x[n] * y[n] = \sum\limits_{k=-\infty}^{\infty} x[k]y[n-k]$
5	Time reversal	$x[-n] \leftrightarrow X(e^{-j\omega})$
6	Differentiation in frequency	$nx[n] \leftrightarrow j\dfrac{\mathrm{d}X(e^{-j\omega})}{\mathrm{d}\omega}$
7	Symmetry	$x[n]\text{real} \rightarrow \begin{cases} X(e^{j\omega}) = X^*(e^{-j\omega}) \\ \|X(e^{j\omega})\| = \|X(e^{-j\omega})\| \\ \sphericalangle X(e^{j\omega}) = -\sphericalangle X(e^{-j\omega}) \end{cases}$ $x[n]$ real and even $\rightarrow X(e^{j\omega})$ real and even $x[n]$ real and odd $\rightarrow X(e^{j\omega})$ purely imaginary and odd
8	Multiplication	$x[n]y[n] \leftrightarrow \dfrac{1}{2\pi}\int\limits_{2\pi} X(e^{j\theta})Y(e^{j(\omega-\theta)})\mathrm{d}\theta$
9	Parseval's formula	$\sum\limits_{n=-\infty}^{\infty} \|x[n]\|^2 = \dfrac{1}{2\pi}\int\limits_{2\pi} \|X(e^{j\omega})\|^2\,\mathrm{d}\omega$

$X[k]$ is N; and (2) the value of $F_s/2$ corresponds to π. The second piece of information is important when the frequency in Hertz corresponding to each value of $X[k]$ is required:

$$\pi \rightarrow F_s/2$$
$$\omega_k \rightarrow f.$$

From the previous relationship,

$$f = \frac{\pi k}{N}F_s. \tag{4.46}$$

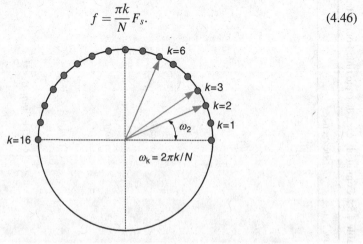

Figure 4.13 Correspondence between the frequency DTFT and DFT.

A common representation of the DFT pair uses the simplification of the exponential term:

$$W_N = e^{-j2\pi/N}. \tag{4.47}$$

The DFT and IDFT can then be represented:

$$X[k] = \sum_{n=0}^{N-1} x[n] W_N^{kn} \qquad \text{analysis equation or DFT} \tag{4.48}$$

$$x[n] = \frac{1}{N} \sum_{k=0}^{N-1} X[k] W_N^{-kn} \qquad \text{synthesis equation or IDFT.} \tag{4.49}$$

The DFT is a mathematical tool to obtain the Fourier spectrum by using computational processing, different to the DTFT where the computation is performed manually. However, as in the following examples, some DFTs are solved manually for a better understanding of some important points. The first step is to find the DFT of N samples of a sinusoid sequence with the digital frequency of $\omega_r = 2\pi r/N$.

Example 4.5

Find the N-points DFT of the length sequence: $x[n] = \cos(2\pi rn/N)\ 0 \le n \le N - 1$ and $0 \le r \le N - 1$, r is an integer number.

Using the Euler identity, we can write

$$x[n] = \frac{1}{2}\left(e^{j2\pi rn/N} - e^{-j2\pi rn/N}\right).$$

In finding the DFT of the single N-points of the exponential $x_1[n] = e^{j2\pi rn/N}$,

$$X_1[k] = \sum_{n=0}^{N-1} e^{j2\pi(r-k)n/N} = \frac{1 - e^{j2\pi(r-k)}}{1 - e^{j2\pi(r-k)/N}} \tag{4.50}$$

when using the formula of the sum of the first N terms of a geometric series. The above expression is 0 for $r - k \ne lN$, l an integer, because $e^{j2\pi(r-k)} = 1$ and the denominator is non-zero. However, when $r - k = lN$, we have an indetermination of the type 0/0. Applying L'Hôpital's rule to eliminate the indetermination yields:

$$X_1[k] = \begin{cases} N, & \text{for } r - k = lN, \quad l \text{ an integer} \\ 0, & \text{otherwise.} \end{cases} \tag{4.51}$$

By using this result in the initial equation, taking into account that the DFT is a linear transform, we have

$$X[k] = \begin{cases} N/2, & \text{for } k = r \\ N/2, & \text{for } k = N - r \\ 0, & \text{otherwise.} \end{cases} \tag{4.52}$$

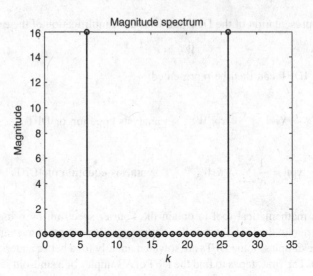

Figure 4.14 Magnitude spectrum of 32 samples of cosine sequence ($N = 32$; $r = 6$).

Figure 4.14 depicts the DFT of the previous signal for $N = 32$ and $r = 6$. Note that there are two pulses at indices $r = 6$ and $N - r = 26$. This is consistent with Table 4.5 line 5 where the DTFT of a sinusoid signal of infinite length is a train of impulses, with the first two centered at frequencies ω_0 and $-\omega_0$. If ω_0 corresponds to a bin, that is, $\omega_0 = 2\pi r/N$, then the DFT will correctly represent the DTFT of the sequence $x[n] = \cos\omega_r n$; otherwise, the DFT will not be able to accurately represent the DTFT.

The magnitude plot presented in Figure 4.14 is symmetric regarding the central sample, which is true if the sequence is real. Additionally, the phase is anti-symmetric regarding the central sample. The property that states this condition is:

$$X[k] = X^* \left[\langle -k \rangle_N \right] \tag{4.53}$$

where the symbol $\langle \rangle_N$ means 'modulo N'. The modulo N operator functions in the sequence index in order for the results to be kept within the interval 0 to $N-1$. If $r = \langle n \rangle_N$, then

$$r = n + lN, \tag{4.54}$$

where l is an integer such $0 \leq r \leq N - 1$. If the index is negative as in Equation (4.53,) $l = 1$ yields:

$$r = k + N \quad \text{for } k < 0.$$

We then have the following symmetry in the previous example:

$$X[1] = X^*[31]$$
$$X[2] = X^*[30]$$
$$\vdots$$
$$X[15] = X^*[17]$$
$$X[16] = X^*[16]$$

The central term $k = 16$ must be real in order for Equation (4.53) to be satisfied. Figure 4.13 was generated using the MATLAB® code:

```
N=32;
r=6;
tet=2*pi/N;
n=0:N-1;
x=cos(tet*r*n);
figure(1)
X=fft(x);
stem(n,abs(X),'k','LineWidth',3)
title({'Magnitude spectrum'},'FontSize',16,'FontName','Verdana')
xlabel({'k'},'FontSize',14,'FontName','Verdana')
ylabel({'Magnitude'},'FontSize',14,'FontName','Verdana')
```

using the FFT algorithm to compute the DFT. The FFT is an elegant algorithm for the computation of the DFT that drastically reduces the related computational effort. While the DFT requires the N^2 complex, multiplications and $N(N-1)$ complex additions, the FFT requires $N \log_2 N$ complex multiplications and $(N/2)\log_2 N$ complex additions.

Example 4.6

Compute the N-point DFT of the sequence:

$$x[n] = \begin{cases} 1, & 0 \leq n \leq N-1 \\ 0, & \text{otherwise} \end{cases}$$

$$X[k] = \sum_{n=0}^{N-1} e^{-j2\pi nk/N} = \frac{1 - e^{-j2\pi k}}{1 - e^{-j2\pi k/N}}, \quad 0 \leq k \leq N-1.$$

In the expression above there is an indetermination for $k = 0$. By application of the L'Hôpital rule,

$$X[k] = \begin{cases} N, & k = 0 \\ 0, & \text{otherwise.} \end{cases}$$

Example 4.7

Computing the inverse DFT. In several applications we need to find the sequence from the DFT. For this purpose we can use the IDFT Equation (4.44) or, more easily, apply the inverse

FFT algorithm (IFFT). The command in MATLAB® is ifft(X). A common error can however occur at the time of building the vector $X[k]$ because it must obey the symmetry property described by Equation (4.53).

4.6.1 Sampling the Fourier Transform

Equation (4.45) shows the DFT as a sampling frequency of the DTFT. The question that needs to be answered is whether the sequence obtained using the inverse DFT is the same as that obtained by using the inverse of DTFT. If the sequences are the same, then we can obtain any point of the DTFT using only samples of the DFT. As such, consider a sequence $x[n]$ and its DTFT $X(e^{j\omega})$. The sampling of the DTFT at N equally spaced points $\omega_k = 2\pi k/N$, $k = 0, 1, \cdots N - 1$ and the IDFT of these samples will give the inverse sequence $y[n]$, $n = 0, 1, \cdots N - 1$. This sequence can be equal or not to $x[n]$. Note that there was no imposition of any restriction to the length of $x[n]$.

$$Y[k] = X(e^{j2\pi k/N}) = \sum_{l=-\infty}^{\infty} x[l]e^{-j2\pi kl/N} \tag{4.55}$$

The N-points inverse DFT of $Y[k]$ from Equation (4.44) is

$$y[n] = \frac{1}{N}\sum_{k=0}^{N-1} Y[k]e^{j2\pi nk/N}, \quad n = 0, 1, \cdots N - 1. \tag{4.56}$$

Substituting Equation (4.55) into Equation (4.56),

$$\begin{aligned} y[n] &= \frac{1}{N}\sum_{k=0}^{N-1}\sum_{l=-\infty}^{\infty} x[l]e^{-j2\pi kl/N}e^{j2\pi nk/N} \\ &= \sum_{l=-\infty}^{\infty} x[l]\left[\frac{1}{N}\sum_{k=0}^{N-1}e^{j2\pi k(n-l)/N}\right]. \end{aligned} \tag{4.57}$$

The last summation is the identity presented in Equation (4.51). Making use of this identity yields

$$y[n] = \sum_{r=-\infty}^{\infty} x[n+rN], \quad n = 0, 1, \cdots N - 1. \tag{4.58}$$

Equation (4.58) is the desired relation between $y[n]$ and $x[n]$. It indicates that the sequence $y[n]$ is obtained from $x[n]$ by adding an infinite number of shifted replicas of $x[n]$ to get the results only from the interval $n = 0, 1, \ldots, N - 1$. The replicas are shifted by an integer multiple of N.

4.6.2 Discrete Fourier Transform Theorems

There are a number of important theorems for the DFT that are useful in digital signal processing; the most important are listed in Table 4.7.

Table 4.7 Discrete Fourier transform properties.

Row No.	Properties	$x[n] \leftrightarrow X[k]$ $y[n] \leftrightarrow Y[k]$
1	Linearity	$ax[n] + by[n] \leftrightarrow aX[k] + bY[k]$
2	Circular time shifting	$x[\langle n - n_0 \rangle_N] \leftrightarrow W_N^{-kn_0} X[k]$
3	Circular frequency shifting	$W_N^{-k_0} x[n] \leftrightarrow X[\langle k - k_0 \rangle_N]$
4	N-point circular convolution	$x[n] \odot y[n] \leftrightarrow X[k]Y[k]$
		where $x[n] \odot y[n] = \sum_{k=0}^{N-1} x[k]y[\langle n - k \rangle_N]$
5	Modulation	$x[n]y[n] \leftrightarrow \frac{1}{N} \sum_{m=0}^{N-1} X[m]Y[\langle k - m \rangle_N]$
6	Parseval's formula	$\sum_{n=0}^{N-1} |x[n]|^2 = \frac{1}{N} \sum_{k=0}^{N-1} |X[k]|^2$

4.7 Recursive DFT

In real-time applications where the DFT components need to be computed at each new sample, the recursive DFT represents an efficient way to compute these components. To derive the basic expression to the kth component, consider Figure 4.15 which depicts two consecutive data windows. Data window $n-1$ is defined as $\mathbf{x}_{n-1} = \{x[n - N], x[n - N + 1] \ldots x[n - 2], x[n - 1]\}$ and data window n as $\mathbf{x}_n = \{x[n - N + 1], x[n - N + 2] \ldots x[n - 1], x[n]\}$.

The DFT for the kth harmonic for the window n is given by:

$$X_k[n] = \sum_{m=0}^{N-1} x[n - m] W_N^{k(N-m-1)}, \quad k = 0, 1, \ldots, N - 1. \tag{4.59}$$

Changing the variables yields:

$$X_k[n] = \sum_{l=n-N+1}^{n} x[l] W_N^{k(N+l-n-1)}. \tag{4.60}$$

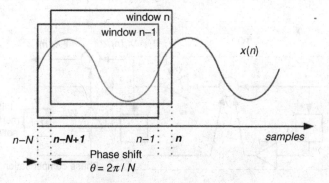

Figure 4.15 Two consecutive windows.

The DFT for window $n-1$ is easily obtained as:

$$X_k[n-1] = \sum_{l=n-N}^{n-1} x[l] W_N^{k(N+l-n)}. \tag{4.61}$$

Multiplying both sides by W_N^{-k} results in:

$$W_N^{-k} X_k[n-1] = \sum_{l=n-N}^{n-1} x[l] W_N^{k(N+l-n-1)}. \tag{4.62}$$

Re-arranging the second term of Equation (4.62) yields:

$$W_N^{-k} X_k[n-1] = \sum_{l=n-N+1}^{n} x[l] W_N^{k(N+l-n-1)} + W_N^{-k}(x[n-N] - x[n]) \tag{4.63}$$

by making use of the identity of $W_N^{kN} = 1$. As such, the summation in Equation (4.63) can be easily identified as $X_k[n]$ and the final expression for a recursive DFT is:

$$X_k[n] = W_N^{-k}(X_k[n-1] + x[n] - x[n-N]). \tag{4.64}$$

The block diagram for Equation (4.64) is illustrated in the shaded part of Figure 4.16. Note that the input signal is a real signal and the output of the filter is a complex signal, because of the complex multiplier W_N^k. Another important point to mention is the phase of $X_k[n]$. As observed from Figure 4.15, the phase is shifted forwards by θ radians every new sampling, which means that the phase refers to the window n instead of the initial reference. To move the reference back to the initial point, we need to subtract θ radians from the phase of each new DFT sample. This can be accomplished using the final section of the block diagram of Figure 4.16. In this part of the figure, the initial value for the memory is assumed equal to 1.

Figure 4.17 shows the phase response of $X_k[n]$ and $\hat{X}_k[n]$ for a signal $x[n] = \cos[\theta n - 1.5]$.

Using the DFT method, the frequencies of the components calculated by Equation (4.64) are functions the window size N. The frequency of the component of order $k=1$, $X_1[n]$, is

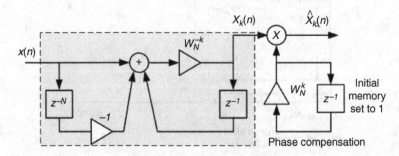

Figure 4.16 Block diagram for recursive DFT.

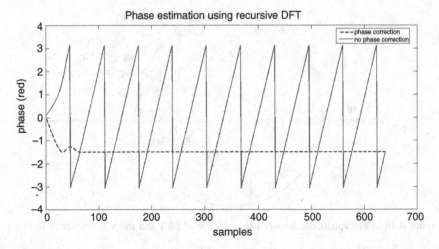

Figure 4.17 Phase estimation using DFT: solid line is the phase of $X_k(n)$ and dashed line is the phase of $X^k(n)$.

given by:

$$f_1 = \frac{1}{NT_s} = \frac{F_s}{N} \tag{4.65}$$

and the other components have frequencies:

$$f_k = kf_1, \quad \text{where} \quad k = 2, 3, \ldots, N. \tag{4.66}$$

As an example, let $x[n]$ be a real periodic signal with fundamental frequency f_0 sampled with L samples per cycle, that is, using a sampling frequency of $F_s = Lf_0$, where L is an integer. This process is called 'synchronous sampling' because the sampling frequency is an integer multiple of the fundamental frequency of the signal. The sampling period can then be written as $T_s = T_0/L$, where T_0 is the fundamental period of $x[n]$. From Equation (4.65), we have:

$$f_1 = \frac{1}{N(T_0/L)} \tag{4.67}$$

if the window size is equal to the number of samples per cycle (i.e. $N = L$); then $f_1 = f_0$.

Example 4.8

Consider that $x[n]$ was sampled at a rate of 16 samples per cycle. Determine the window length L in order that all the terms can be estimated using DFT.

$$x[n] = A_1 \cos(\omega_0 n) + A_{1.5} \cos(1.5\omega_0 n) + A_3 \cos(3\omega_0 n)$$
$$\omega_0 = 2\pi f_0 T_s. \tag{4.68}$$

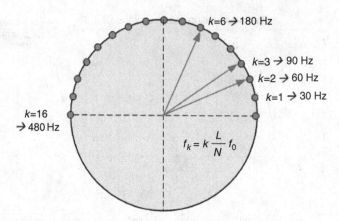

Figure 4.18 Correspondence between the index k of DFT and the real frequency in Hertz.

The signal in this example is composed of a fundamental component, the third harmonic and an interharmonic. If f_0 is 60 Hz, then the frequency of the interharmonic is 90 Hz. Choosing $N = 2L$ leads to Equation (4.67), and hence $f_1 = 30$ Hz. The value of k for the components in the previous signal corresponds to $k = 2$, 3 and 6, respectively. Figure 4.18 depicts an illustrative representation between the DFT index k, the real frequency in Hz and the angle in radians. The angle of π radians corresponds to half of the sampling frequency ($F_s/2$ Hz), and corresponds to $k = N/2$. In this situation the DFT gives the correct estimation for each component of the signal. However, instead of taking $N = 2L$ let $N = L$; $f_1 = 60$ Hz and the interharmonic term cannot then be estimated and, moreover, the fundamental and the third harmonic estimations will not be accurate because the interhamonic will interfere with the accuracy of the final results. This effect is known as *picket-fences* [7], and is caused by interharmonics in the signal (see Chapter 8).

The frequency resolution of a DFT is related to its window's length. If synchronous sampling of a signal is assumed, the small frequency component can be obtained from DFT:

$$f_1 = \frac{f_0}{N_c} \qquad (4.69)$$

where N_c ($= N/L$) is the number of integer cycles of the fundamental frequency. In the IEC 61100-3-60 standard the number of cycles used for a 60 Hz power system is 12 (10 for 50 Hz system). The frequency resolution is 5 Hz, which means that the DFT is computed every 5 Hz up to the maximum frequency given by $F_s/2$. The price to be paid for this small frequency resolution and the reduction of the picket-fences effect is the increase of the convergence time for the estimation of each component. In the expression given by Equation (4.64) the convergence takes N_c cycles. This time is prohibitive for typical control and protection applications. Chapter 7 will focus on accurate and fast estimators.

4.8 Filtering Interpretation of DFT

The DFT can be interpreted as a process of modulation and filtering, allowing an easy interpretation of some common phenomena that occur in DFT as well as introducing a new method for the computation of the DFT that can overcome some of its limitations. The starting point for this interpretation is the DFT definition (Equation (4.59)), which can be simplified as:

$$X_k[n] = W_N^{-k} \sum_{m=0}^{N-1} x[n-m] e^{j2\pi km/N}.$$ (4.70)

The constant W_N^{-k} is responsible for the modification of the phase of the term generated by the summation in Equation (4.70). For the simplification of the analysis this constant is left out, that is:

$$\breve{X}_k[n] = \sum_{m=0}^{N-1} x[n-m] e^{j2\pi km/N}.$$ (4.71)

Equation (4.71) can be interpreted as the convolution of the sequence $x[n]$ and the function

$$h[n] = \upsilon[n] e^{j2\pi kn/N}$$ (4.72)

where $\upsilon[n]$ is a rectangular window of length N. Equation (4.72) represents the impulse response of a complex filter. Figure 4.19 shows the DFT filtering interpretation of Equation (4.71). Note that despite $x[n]$ being a real sequence, the output of the filter is a complex signal. The modulus of $\breve{X}_k[n]$ is the modulus of the kth DFT term taking the $\mathbf{x}_n$ window, and its phase is the phase of the DFT unless a phase correction needs to be applied according to Equation (4.70).

Figure 4.20 depicts the magnitude response for $h_k[n], k = 0, 1, 2$. These plots were generated considering $L = N = 16$, that is, the sampling rate of $F_s = 16 \times 60$ Hz. In this case $k = 0$, 1, 2 corresponds to DC, the fundamental and the second harmonic component, respectively. The first filter is readily identified as the moving average filter (MAF) (Section 5.3) the other filters are complex filters since their magnitude frequency response is not a even function. It can be observed that the fundamental filter rejects all harmonic components, including the DC component. The same is observed for the other filters, for example the $h_k[n]$ rejects all components except for the component of frequency f_k.

A fundamental doubt may occur at this moment. If for example $x[n] = A \cos(w_1 n)$, how can the absolute value of $\breve{X}_1[n]$ be constant and equal to A? Isn't the idea of filter processing a specific range of frequencies that passes throughout the filter and blocks others? If so,

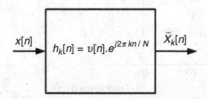

Figure 4.19 DFT filtering interpretation.

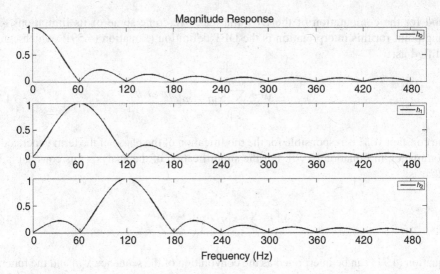

Figure 4.20 Magnitude response for DFT complex filter for DC, fundamental and second harmonic filter.

shouldn't the output of $h_1[n]$ filter be a sinusoidal signal? The answer is no, because the filter is complex. Figure 4.21 illustrates what happens when $x[n]$ passes through the filter. Using the Euler identity,

$$x[n] = \frac{A}{2}e^{j\omega_1 n} + \frac{A}{2}e^{-j\omega_1 n} \tag{4.73}$$

or, in the frequency domain,

$$X\left(e^{jw}\right) = \frac{A}{2}\delta(\omega - \omega_1) + \frac{A}{2}\delta(\omega + \omega_1). \tag{4.74}$$

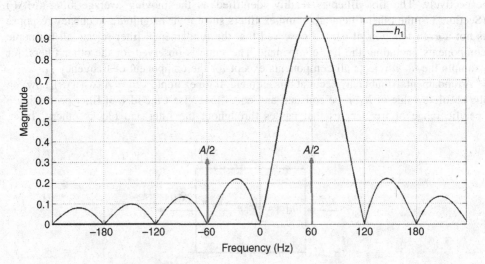

Figure 4.21 The filtering operation of a complex filter.

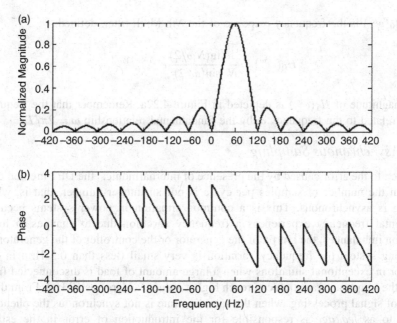

Figure 4.22 Frequency response of the filter: (a) magnitude response and (b) phase response.

The second impulse of Equation (4.74) is the left impulse in Figure 4.21. Note that it falls where the filter has zero gain and then it is filtered. The output of the filter is only the first term of Equation (4.74) or, in the time domain,

$$y[n] = \frac{A}{2}e^{j\omega_1 n}. \qquad (4.75)$$

The filter output is a rotating phasor, but its module is the desired amplitude of the signal divided by 2 and its phase is the phase of the input signal.

The spread effect due the presence of an interharmonic can be understood by observing Figures 4.21 and 4.22. Consider for example a signal composed of the fundamental and second harmonics plus an interharmonic with frequency of 117 Hz. By observing the frequency response of the filters in Figure 4.22, it can be noted that the gains of this filter are not 0 in the frequency of 117 Hz, so the estimation of each component is affected by the interharmonics (mainly the second harmonic) because of the lower attenuation for this component.

4.8.1 Frequency Response of DFT Filter

Note that filters $h_k[n]$ are obtained by multiplying the basic window $v[n]$ by a complex exponential function. This corresponds to the modulation theorem, so the frequency response of each kth filter is obtained as

$$H_k(e^{j\omega}) = H_0(e^{j(\omega - \omega_k)}) \qquad (4.76)$$

where $H_0(e^{j\omega})$ is the frequency response of the Nth MAF filter, defined:

$$H_0(e^{j\omega}) = \frac{1}{N}\frac{\sin(N\omega/2)}{\sin(\omega/2)}e^{-j(N-1)\omega/2}. \tag{4.77}$$

The magnitude of $H_1(e^{j\omega})$ is depicted in Figure 4.22a. Remember that the frequency in Hertz is related to the frequency ω by the fundamental relationship $\omega = 2\pi f T_s$.

4.8.2 Asynchronous Sampling

In addition to the error caused by the presence of interharmonics, the DFT method can also fail when the number of samples per cycle is not an integer number, that is, when the sampling is asynchronous. This is a common situation in power systems because the fundamental frequency experiences a frequency variation due to changes in load and generation imbalances, the inertia of the generator or the controller of the generator. When in a strong system the frequency variation is very small (less than 0.5 Hz); in a weak system or in exceptional situations when a large amount of load is disconnected from the system, the frequency variation can reach higher values (as large as 10 Hz). From the point of view of signal processing, when the sampling rate is not synchronous the phenomenon referred to as *leakage* is responsible for the introduction of error in the estimation components.

An easy interpretation of errors caused by asynchronous sampling can be obtained from Figure 4.21. Assume the input signal is a single cosine given by $x(t) = A\cos(2\pi f t)$ where $f = f_1 + \Delta f$ is the off-nominal frequency, then the discrete signal using an integer multiple of f_1 as sampling rate is given by:

$$x[n] = \frac{A}{2}e^{j(\omega_1 + \Delta\omega)n} + \frac{A}{2}e^{-j(\omega_1 + \Delta\omega)n}. \tag{4.78}$$

After passing this signal by $H_1(e^{j\omega})$, the steady-state output is given by

$$\breve{X}[n] = H_1(e^{j(\omega_1 + \Delta\omega)})\frac{A}{2}e^{j(\omega_1 + \Delta\omega)n} + H_1(e^{j(-\omega_1 - \Delta\omega)})\frac{A}{2}e^{-j(\omega_1 + \Delta\omega)n} = X_{f_+}[n] + X_{f_-}[n]. \tag{4.79}$$

Equation (4.79) is the addition of two phasors, and is obtained by the multiplication of the real phasor by the frequency response of the filter in $f_+ = f_1 + \Delta f$ and $f_- = -(f_1 + \Delta f)$ frequencies. While $X_{f_+}[n]$ rotates counterclockwise, $X_{f_-}[n]$ rotates clockwise. This sum is illustrated in Figure 4.22, where it is obvious that the amplitude of resultant phasor $\breve{X}[n]$ will oscillate with frequency $2f$. In addition, the mean value of amplitude is $|X_{f_+}|$ while the amplitude of oscillation is $|X_{f_-}|$.

Figure 4.23a shows the magnitude estimation using DFT for off-nominal frequencies. The frequency deviation ranges from -5 Hz to 5 Hz. The sampling rate is $F_s = 32 \times 60$ Hz. This figure shows that the error in magnitude is very small and can be neglected for practical applications. The problem however resides in the phase estimation. In this case, if the reference frequency is 60 Hz but the local frequency presents a deviation, the phase difference between the reference signal (a cosine signal of 60 Hz and zero phase, $\cos(\omega_1 n)$) and a real

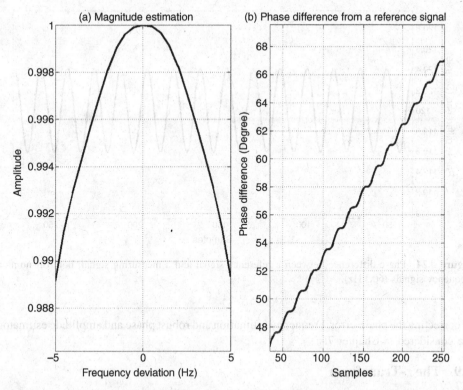

Figure 4.23 Error due to off-nominal frequency using DFT: (a) magnitude estimation for frequency deviation from −5 Hz to 5 Hz; (b) phase difference between a reference signal (at 60 Hz) and an off-nominal signal (61 Hz).

signal increases with time. Figure 4.23b presents the phase difference for a real signal given by $\cos(2\pi\omega n + 45°)$ and a frequency deviation of 0.5 Hz. The x axis is shown from sample 32 in order to discard the transient response produced by the filtering processing. Note that the phase difference increases continually. This example uses a reference signal that assumes the exact system frequency, when the signal is acquired at the same site however, the signals will undergo the same frequency variation and the phase difference will be as depicted in Figure 4.24, where a oscillation behavior is observed. In this example the phase difference is measured between the signals (for a frequency deviation of 0.5 Hz):

$$x_r[n] = \cos(\omega n)$$
$$x_m[n] = \cos(\omega n + 120°). \tag{4.80}$$

The error depicted in Figure 4.24 is insignificant. The error related to off-nominal frequency is significant when the frequency deviation is bigger, which only occurs in weak power systems, or in phasor measurement unit (PMU) applications when distant sites can undergo different frequency deviations. In both situations the frequency estimation is an essential part

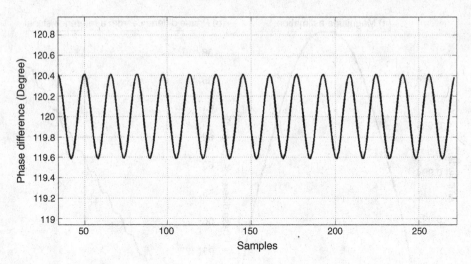

Figure 4.24 Phase difference between a reference signal and a measuring signal, both off-nominal frequency signals (60.5 Hz).

of correcting the phase error. Frequency estimation and robust phase and amplitude estimators are considered in Chapter 7.

4.9   The z-Transform

The Laplace transform is generally used to simplify the analysis of continuous differential equations in the time domain. The application of the Laplace transform to a ordinary differential equation describing the input–output relation changes the integral-differential relation to an algebraic relation in the variable s. Correspondingly, discrete-time systems deal with difference equations relating input–output and the use of the z-transform converts the changes in the difference relationship to an algebraic relationship in the z-domain. The s-transform and the z-transform are parallel techniques with several similarities. Figure 4.25 presents the plane s and the plane z; while the s-plane is arranged in a rectangular

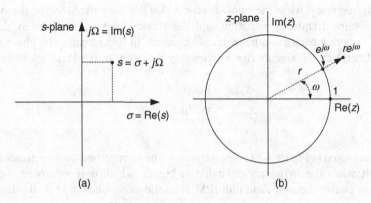

Figure 4.25   The s- and z-plane.

coordinate system, the z-plane uses a polar format. While the $j\Omega$ axis (the imaginary axis) carries information about the Fourier transform, the unity circle carries information about the DTFT.

We can obtain the Fourier transform from the Laplace transform by the substitution:

$$X(j\Omega) = X(s)|_{s=j\Omega}. \tag{4.81}$$

Similarly, we can derive the DTFT from the z-transform through the substitution:

$$X(e^{j\omega}) = X(z)|_{z=e^{j\omega}}. \tag{4.82}$$

Both derivations are only possible if the region of convergence (ROC) of the s-transform contains the $j\Omega$ axis and the ROC of the z-transform contains the unity circle $e^{j\omega}$. This will be true for most cases, but there are some exceptions that must be considered.

The analysis and synthesis equations of the Laplace transform are well known, but are repeated here for convenience:

$$X(s) = \int_{-\infty}^{\infty} x(t)e^{-st}\,dt, \qquad \text{analysis equation (Laplace transform)} \tag{4.83}$$

$$x(t) = \frac{1}{2\pi j}\int_{\sigma-j\infty}^{\sigma+\infty} X(s)e^{st}\,dt \qquad \text{synthesis equation (inverse transform).} \tag{4.84}$$

The s-transform pair is denoted $x(t) \leftrightarrow X(s)$.

This synthesis equation states that $x(t)$ can be recovered from its Laplace transform evaluated for a set of values of $s = \sigma + j\omega$ in the ROC, with σ fixed and ω varying from $-\infty$ to ∞.

The z-transform analysis and synthesis equations are defined:

$$X(z) = \sum_{n=-\infty}^{\infty} x[n]z^{-n} \qquad \text{analysis equation (z-transform)} \tag{4.85}$$

$$x[n] = \frac{1}{2\pi j}\oint_{C} G(z)z^{n-1}\,dz \qquad \text{synthesis equation (inverse z-transform).} \tag{4.86}$$

The z-transform pair is denoted $x[n] \leftrightarrow X(z)$.

The synthesis equation states the sequence $x[n]$ can be recovered from its z-transform evaluated along a contour $z = re^{j\omega}$ in the ROC with the radius fixed and ω varying over a 2π interval.

The evaluation of the inverse transform (Laplace or z-transform) requires the use of contour integration in the complex plane. Thankfully, the synthesis equation is rarely needed in the evaluation of the inverse transform; the same technique as the partial fractions used to compute the inverse of an s-transform can be used to compute the inverse z-transform. To use the partial fractions expansion, the requirement in the z-transform is a rational function of z.

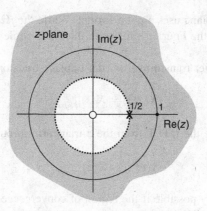

Figure 4.26   Example 4.9: ROC.

Example 4.9

Find the z-transform of $x[n] = \alpha^n u[n]$.

The z-transform is obtained using the analysis equation:

$$X(z) = \sum_{n=0}^{\infty} \left(\alpha z^{-1}\right)^n.$$

The above equation represents the sum of the terms of a geometric series with infinite terms, which only converges if $\left|\alpha z^{-1}\right| < 1$. Then,

$$X(z) = \frac{1}{1 - \alpha z^{-1}}, \quad \text{for} \quad \left|\alpha z^{-1}\right| < 1. \tag{4.87}$$

The condition for the convergence defines the ROC:

$$|z| > |\alpha|$$

Figure 4.26 illustrates the ROC for $\alpha = 1/2$; the dark region is the ROC. Note that the value $\alpha = 1/2$ is the root of the denominator of Equation (4.87); the roots of the denominator of a rational function of z are known as function *poles* and are marked in the figure with the symbol $\times$. The zeroes of the numerator are the zeroes of the function and are marked in the figure by 0s. Note that the zero of Equation (4.87) is located at the origin of the z-plane.

4.9.1   Rational z-Transforms

A single-input, single-output (SISO) discrete-time system is described in the time domain by a difference equation. The general format of the SISO discrete-time system is given by the recursive equation:

$$y[n] = \sum_{i=0}^{M} b_i x[n-i] - \sum_{i=1}^{N} a_i y[n-i] \tag{4.88}$$

where $y[n]$ is the output, $x[n]$ is the input and b_i $(i = 0, 1, \ldots, M)$ and a_i $(i = 1, 2, \ldots, N)$ are the system coefficients. This equation represents the algorithm that will be programmable in the digital signal processor or synthesized in hardware (for example in a field programmable gate array or FPGA), and it describes the relationship between the input and output. Just as continuous systems are controlled by differential equations, discrete time systems are controlled or described by the difference equations. However, several important aspects of the system cannot be directly observed from looking at the difference equation, such as the stability of the system and the frequency response characteristics. These aspects are better understood and observable using the z-domain. To change Equation (4.88) for the z-domain, we must first derive the important property of the z-transform of a time shifting sequence. Let $X[z]$ be the z-transform of $x[n]$; we want to find the z-transform of $x[n - n_0]$.

$$X_1(z) = \sum_{n=-\infty}^{\infty} x[n - n_0]z^{-n}.$$

Letting $n - n_0 = m$ we obtain:

$$X_1(z) = \sum_{m=-\infty}^{\infty} x[m]z^{-(m+n_0)} = z^{-n_0} \sum_{m=-\infty}^{\infty} x[m]z^{-m} = z^{-n_0}X(z).$$

This property can be represented:

$$x[n - n_0] \leftrightarrow z^{-n_0}X(z). \tag{4.89}$$

Appling the z-transform in Equation (4.88) and making use of the property above yields:

$$Y[z] = \sum_{i=0}^{M} b_i z^{-i} X[z] - \sum_{i=1}^{N} a_i z^{-i} Y[z].$$

The transfer function $H(z) = Y(z)/X(z)$ can be rewritten:

$$H[z] = \frac{\sum_{i=0}^{M} b_i z^{-i}}{\sum_{i=0}^{N} a_i z^{-i}}, \tag{4.90}$$

where $a_0 = 1$. In power system applications, the transfer function is 'proper' or 'strictly proper', which means $N \geq M$. Furthermore, the poles are all single, that is, there are no multiple poles in Equation (4.90). Equation (4.90) can be factored in different forms, the expansion in partial fractions being the most practical form. The expansion in partial fractions, considering simple poles and proper functions, can be written [8]:

$$H[z] = \sum_{l=1}^{N} \frac{\rho_l}{1 - \lambda_l z^{-l}}, \tag{4.91}$$

where $z = \lambda_l$ are the poles of $H(z)$, $1 \leq l \leq N$, λ_l are distinct and the constants ρ_l are called residues, defined

$$\rho_l = \left(1 - \lambda_l z^{-1}\right) H(z)\big|_{z=\lambda_l}. \tag{4.92}$$

Example 4.10

Find the impulse response of the system with transfer function:

$$H(z) = \frac{1 - 0.8z^{-1} - 0.15z^{-2}}{1 - 1.3z^{-1} + 0.4z^{-2}}.$$

The impulse response is obtained by taking the inverse z-transform of

$$Y(z) = H(z)X(z)$$

Table 4.8 shows the z-transform of the unit impulse sequence as $\delta[n] \leftrightarrow 1$. To find the impulse response, the transfer function must be written in the form of Equation (4.91). The MATLAB® command '*residuez*' returns the residues and poles of the function; an example of the required code is as follows:

```
b = [1 -0.8 -0.15];     %numerator coefficients
a = [1 -1.3 0.4];       %denominator coefficients
[r,p] = residuez(b,a);  % r is the vector contained the residue
                        % p is the vector contained the poles
```

which would generate the result: $r = [-0.625 \ 2.0]$ and $p = [0.8 \ 0.5]$.

Table 4.8 Some commonly used z-transform pairs.

Row No.	Sequence	z-transform	ROC
1	$\delta[n]$	1	All values of z
2	$u[n]$	$\dfrac{1}{1-z^{-1}}$	$|z| > 1$
3	$\alpha^n u[n]$	$\dfrac{1}{1-\alpha z^{-1}}$	$|z| > |\alpha|$
4	$r^n \cos(\omega_0 n)u[n]$	$\dfrac{1 - (r\cos(\omega_0))z^{-1}}{1 - (2r\cos(\omega_0))z^{-1} + r^2 z^{-2}}$	$|z| > |r|$
5	$r^n \cos(\omega_0 n)u[n]$	$\dfrac{(r\sin(\omega_0))z^{-1}}{1 - (2r\cos(\omega_0))z^{-1} + r^2 z^{-2}}$	$|z| > |r|$

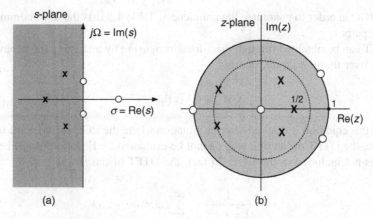

Figure 4.27 Stability region in the (a) s-plane and (b) z-plane.

The partial-fraction expansion is:

$$Y(z) = \frac{-0.625}{1 - 0.8z^{-1}} + \frac{2.0}{1 - 0.5z^{-1}}.$$

To compute the inverse transform, the results of Example 4.10 and the fact that the z-transform is linear are used, to obtain:

$$\alpha^n u[n] \leftrightarrow \frac{1}{1 - \alpha z^{-1}}, \quad \text{the ROC is } |z| > |\alpha|,$$

$$y[n] = [-0.625(0.8)^n + 2(0.5)^n]u[n].$$

4.9.2 Stability of Rational Transfer Function

The previous example shows that the poles are responsible for the exponential terms in the time domain. If the modulus of the pole is <1, that is, the pole is inside the unit circle, the impulse response vanishes as the time (n) increases. This type of stability is known as BIBO (bounded input produces bounded output) criterions. In the z-domain, a linear and time-invariant system is BIBO-stable if and only if the poles of the transfer function are inside the unit circle.

Figure 4.27 depicts the parallel idea between the stability of the s-plane and the z-plane. While the poles must be on the left side in the s-plane, in the z-plane the poles must be inside the unit circle. There is no restriction to the position of the zeroes once they do not affect the system stability; they can be within any location of the plane.

4.9.3 Some Common z-Transform Pairs

Remember that the basic idea behind any transform is that the mapping from the time domain to the transform domain must be bijective. Both the Laplace transform and z-transform have to define the ROC in order to guarantee the uniqueness. Table 4.8 lists the most commonly used z-transform pairs.

The DTFT can be obtained through the z-transform using by analyzing the behavior of the z-transform over the unit circle, that is

$$X(e^{j\omega}) = X(z)|_{z=e^{j\omega}}. \tag{4.93}$$

However this equation is true only for the situation where the ROC includes the unit circle. For example, the DTFT of the unit step cannot be evaluated by Equation (4.93) because the unit circle is not included in the ROC. In fact, the DTFT of the $u[n]$ is:

$$u[n] \overset{DTFT}{\longleftrightarrow} \frac{1}{1 - e^{-j\omega}} + \sum_{k=-\infty}^{\infty} 2\pi\delta(\omega + 2\pi k).$$

Note that if applying Equation (4.93) the second term of the DTFT will not appear by the suggested transformation. On the other hand, the z-transform of $\alpha^n u[n]$ for $|\alpha| < 1$ contains the unit circle and the DTFT is obtained as:

$$X(e^{j\omega}) = X(z)|_{z=e^{j\omega}} = \frac{1}{1 - \alpha z^{-1}}\bigg|_{z=e^{j\omega}} = \frac{1}{1 - \alpha e^{-j\omega}},$$

i.e. the value in line 7 of Table 4.5.

Table 4.9 Useful properties of the z-transform.

Row No.	Property	Sequence $x[n], y[n]$	z-Transform $X[z], Y[z]$	ROC $\mathfrak{R}_x, \mathfrak{R}_y$
1	Conjugation	$x^*[n]$	$X^*[z^*]$	$\mathfrak{R}_x$
2	Time-reversal	$x[-n]$	$X[1/z]$	$1/\mathfrak{R}_x$
3	Linearity	$\alpha x[n] + \beta y[n]$	$\alpha X[z] + \beta Y[z]$	$\mathfrak{R}_x \cap \mathfrak{R}_y$
4	Time-shifting	$x[n - n_0]$	$z^{-n_0} X[z]$	$\mathfrak{R}_x$ except possibly the point $z = 0$ or $z = \infty$
5	Multiplication by exponential sequence	$\alpha^n x[n]$	$X[z/\alpha]$	$|\alpha|\mathfrak{R}_x$
6	Convolution	$x[n] * y[n]$	$X[z]Y[z]$	$\mathfrak{R}_x \cap \mathfrak{R}_y$
7	Modulation	$x[n]y[n]$	$\dfrac{1}{2\pi j} \oint_C X(v)Y(z/v)v^{-1}\, dv$	$\mathfrak{R}_x\mathfrak{R}_y$
8	Parseval's relation	$\displaystyle\sum_{n=-\infty}^{\infty} x[n]y^*[n] = \dfrac{1}{2\pi j} \oint_C X(v)Y(z/v)v^{-1}\, dv$		

4.9.4 z-Transform Properties

Some useful properties of the z-transform are as listed in Table 4.9. Demonstrations, which can be found in the majority of digital signal textbooks, are beyond the scope of this book.

4.10 Conclusions

Discrete transforms are essential in the understanding, analysis and synthesis of power systems and, in particular, the complex networks of the future. This chapter provided a review of the most common transforms used in signal processing and the theoretical foundation of the different transforms. The proprieties of Fourier series, Fourier transforms, sampling theorem, DTFT, DFT and recursive DFT were discussed and illustrated. The approach used here was relevant for the applications of electrical power systems, in particular to harmonic distortions.

References

1. Masson, P.J., Silveira, P.M., Duque, C. and Ribeiro, P.F. (2008) Fourier series: Visualizing Joseph Fourier's imaginative discovery via FEA and time-frequency decomposition. *ICHQP 13th International Conference on Harmonics and Quality of Power*, 1–5.
2. IEC (2002) Electromagnetic Compability (EMC) Part 4–7. International Electrotechnical Commission.
3. IEEE Standard 519 (1992) IEEE recommended practices and requirements for harmonic control in electrical power systems. IEEE Industry Society/Power Energy Society.
4. Oppenheim, A.V., Willsky, A.S. and Nawab, H. (1997) *Signals and Systems*, Prentice Hall.
5. Lathi, B.P. (2004) *Linear Systems and Signals*, Oxford University Press, Oxford.
6. Antoniou, A. (1993) *Digital Filters, Analysis, Design and Applications*, McGraw Hill.
7. Baghzouz, Y.J., Burch, R.F., Capasso, A. *et al.* (1998) Time varying harmonics. Part I: characterizing measured data. *IEEE Transactions on Power Delivery*, **13** (3), 938–944.
8. Mitra, S.K. (2005) *Digital Signal Processing: A Computer-Based Approach*, 4th edition, McGraw-Hill.

5

Basic Power Systems Signal Processing

5.1 Introduction

This chapter focuses on basic power systems signal processing techniques and highlights the main operations, applications and methods as related to digital filters, moving average, trapezoidal integration and special digital systems for use in common power systems applications such as the estimation of the differentiator, the RMS, time-domain harmonic distortions and notch filters. Although the smart-grid context will introduce many time-varying variables in the behavior of the electric power network, the utilization of the classical linear and time-invariant systems will continue to be the main tool to analyze and design signal processing algorithms for the future grid. These basic concepts are essential for the understanding of more advanced concepts, such as those related to time-varying behavior.

5.2 Linear and Time-Invariant Systems

Linear and time-invariant (LTI) discrete-time systems constitute the main class of systems used in power systems applications. An LTI system can be described by a differential equation with constant coefficients, as discussed in Chapter 4. If

$$y[n] = \sum_{i=0}^{M} b_i x[n-i] - \sum_{i=1}^{N} a_i y[n-i], \tag{5.1}$$

then in the z-domain the transfer function $H(z) = Y(z)/X(z)$ is given by:

$$H[z] = \frac{\displaystyle\sum_{i=0}^{M} b_i z^{-i}}{\displaystyle\sum_{i=0}^{N} a_i z^{-i}}. \tag{5.2}$$

Power Systems Signal Processing for Smart Grids, First Edition. Paulo Fernando Ribeiro, Carlos Augusto Duque, Paulo Márcio da Silveira and Augusto Santiago Cerqueira.
© 2014 John Wiley & Sons, Ltd. Published 2014 by John Wiley & Sons, Ltd.
Companion Website: http://www.wiley.com/go/signal_processing/

Equation (5.1) describes the behavior of the LTI system in the discrete-time domain through the difference equation. It is the direct form, used to implement the system in digital signal processors or field-programmable gate array (FPGA). On the other hand, the z-domain representation is useful in the design of LTI systems such as the digital filter. Furthermore, it can be used to obtain additional information regarding behavior of the system described by its differential equation, such as stability and simplification.

The second summation in Equation (5.1) represents the recursive element where the past samples of the system output $y[n-1], y[n-2], \cdots, y[n-N]$ are used to obtain the output sample $y[n]$. In digital filter language, recursive filters are called IIR (infinite impulse response) filters. This is due to the fact that its impulse response, in the mathematical sense, is infinite. An LTI system is considered non-recursive only if the first summation of Equation (5.1) is present. Alternatively, in digital filter language the system is known as FIR (finite impulse response). The impulse response of a FIR filter can be easily determined by substituting $x[n] = \delta[n]$, and computing the output $y[n]$ of (5.1):

$$h[n] = y[n] = \sum_{i=0}^{M} b_i \delta[n-i]. \tag{5.3}$$

Under such circumstances the impulse response will be its filter coefficients, that is, $h[n] = \{b_0, b_1, \cdots, b_M\}$ composed of a finite number of terms. Figure 5.1 shows the impulse response of a FIR and IIR filter, computed using the MATLAB® function impulse_response.m.

The FIR filter has 17 coefficients (order 16th) and was designed using the MATLAB® function fir1($M-1$, ω_c) where ($M-1$) is the filter order and ω_c is the normalized cutoff frequency ranging from 0 to 1 ($\omega_c = 0.2$ in the present project). The unit value for ω_c corresponds to $F_s/2$ Hz. The impulse response is plotted in Figure 5.1a, where the value of each sample is equal to the correspondent filter coefficient. The impulse response for the IIR filter is presented in Figure 5.1b. The IIR filter was designed using the MATLAB® command $[b,a] = \text{cheby1}(N,R, \omega_p)$, where N is the filter order, R is the peak-to-peak ripple in the pass-band (in decibels) and ω_p is the pass-band-normalized frequency. For this example the

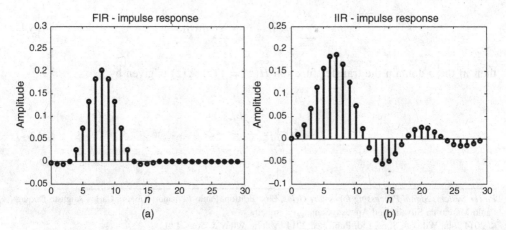

Figure 5.1 Impulse response of (a) 16th-order FIR filter and (b) 4th-order Chebyshev filter.

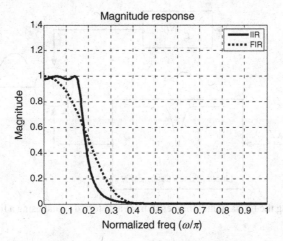

Figure 5.2 Magnitude response for a Chebyshev filter (continuous line) and FIR filter (dotted line).

command was $[b,a] = \text{cheby1}(4,0.2,0.5)$. Note that the impulse response of the FIR filter has exactly 16 non-zero terms, which is the length of the FIR filter. On the other hand, the impulse response for the IIR has more than 30 non-zero terms. The name 'infinite impulse response' was given since, although the energy of the impulse response of a stable IIR filter tends to zero, mathematically it has infinite non-zero terms as n increases.

Figure 5.2 represents the magnitude response of both filters. The FIR has 17 coefficients and the IIR filter has 8 coefficients. Consequently the FIR filter requires a higher computational effort: 17 multiplications (in fact a linear phase FIR filter can be implemented with about half the products, in this case 9 products). The IIR filter requires 8 multiplications and consequently is computationally more efficient. Likewise, the IIR magnitude response in Figure 5.2 is of better quality than the FIR filter. In fact, if the FIR filter should match the IIR filter in its magnitude response for the cutoff frequency, stop-band attenuation, ripple in pass-band and so on, the order needs to be higher. For example, the FIR order specification for the previous example must be higher than 50 in order to match the IIR response characteristics.

From the previous paragraph it can be seen that the FIR filter is computationally more demanding than the IIR filter. However, FIR filters have some attractive advantages: (a) they are always stable and (b) they can be designed to have a linear phase. The first advantage is due to the fact that all poles of FIR filters are located at position zero in the z-plane. This implies that the ROC includes all of the z-plane, except the point zero. It is worth mentioning that both FIR and IIR filters have as many poles as zeros; some of them can be located at zero or infinite. However, the infinite pole can only occur for the IIR filter, for example the transfer function given by

$$H(z) = \left(1 + 2z^{-1}\right)/\left(1 - 0.25z^{-2}\right)$$

has two poles (0.5 and -0.5), one zero at $z = -2$ and the other at infinity.

The linear phase is the most important characteristic of a FIR filter. This chapter will prove that the phase response is linear in frequency, if the impulse response of a FIR filter is symmetric or anti-symmetric. For practical applications in power systems, a linear

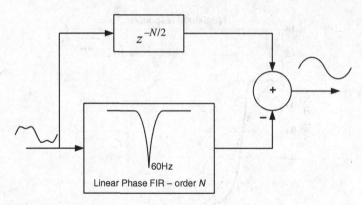

Figure 5.3 Extracting the fundamental frequency from a distorted signal using a linear phase notch FIR filter.

phase means that the output is delayed for an integer number of samples (filter order must be even) related to the input. One such application is depicted in Figure 5.3, where a FIR filter is designed as a notch filter and centered at 60 Hz. The ideal notch filter is able to eliminate a single frequency (60 Hz, for example), leaving the other components unchanged. An ideal filter cannot be of course implemented, so an approximation is made. A FIR notch filter can be designed in such a fashion to have a linear phase and a sharp stop band, as illustrated in the figure. If the final objective is not to eliminate the fundamental component but instead to find it, the structure in Figure 5.3 can produce this result also. Note that the path delay of $N/2$ samples in the upper section of the configuration was added to compensate for the filter delay.

5.2.1 Frequency Response of LTI System

When the transfer function of a stable LTI system is described by a z-transform, its frequency response can be easily obtained by using the relation already established in Chapter 4 (Equation (4.82)):

$$H(e^{j\omega}) = H(z)|_{z=e^{j\omega}}. \tag{5.4}$$

If the function is stable, then all pole locations of a causal LTI system are located within the unit circle. As such, the ROC includes the unit circle and Equation (5.4) is valid. A useful geometric interpretation of this frequency response can be made when considering the $H(z)$ factorization:

$$H(z) = z^{N-M} \frac{b_0 \prod_{k=1}^{M} (z - z_k)}{a_0 \prod_{k=1}^{N} (z - p_k)} \tag{5.5}$$

or, in the frequency domain,

$$H(e^{j\omega}) = e^{j(N-M)\omega} \frac{b_0 \prod_{k=1}^{M} \left(e^{j\omega} - \rho_k e^{j\phi_k}\right)}{a_0 \prod_{k=1}^{N} \left(e^{j\omega} - r_k e^{j\theta_k}\right)}. \tag{5.6}$$

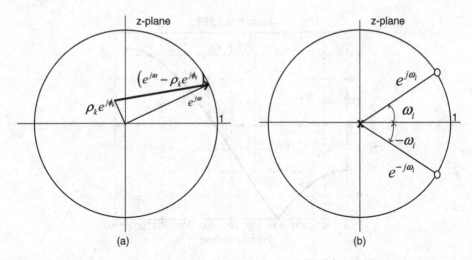

Figure 5.4 (a) Geometric interpretation of the difference between two vectors; (b) example with two zeros in the unit circle.

Note that the zeros and poles are given as a polar representation, and the magnitude response is obtained as follows:

$$|H(e^{j\omega})| = |e^{j(N-M)\omega}| \frac{|b_0|}{|a_0|} \frac{\prod_{k=1}^{M}|(e^{j\omega} - \rho_k e^{j\phi_k})|}{\prod_{k=1}^{N}|(e^{j\omega} - r_k e^{j\theta_k})|}. \tag{5.7}$$

The basic term of Equation (5.6) $(e^{j\omega} - \rho_k e^{j\phi_k})$ is the difference between two vectors, as illustrated in Figure 5.4. As the unit vector $e^{j\omega}$ runs over the unit circle, the magnitude is obtained through the product of the modulus of the difference vectors $(e^{j\omega} - \rho_k e^{j\phi_k})$ as referring to the zeros and divided by the product of the modulus of the difference vectors in reference to the poles $(e^{j\omega} - r_k e^{j\theta_k})$. The phase can be found through the addition of the numerator phases and subtracting the denominator phases of each term.

For a better understanding of the zero, pole location and frequency response relationships, consider a parametric FIR filter whose transfer function is given by:

$$H(z) = \frac{1 - 2\cos(\omega_i)z^{-1} + z^{-2}}{|1 - 2\cos(\omega_i)z_1^{-1} + z_1^{-2}|} \tag{5.8}$$

where $z_1 = e^{j\omega_1}$. As described in reference [1] and [2], this structure rejects the frequency ω_i and has a unit gain at the frequency ω_1. It can be used to eliminate the harmonic ω_i, while it preserves the unit gain at the fundamental frequency ω_1. Note that the denominator is a constant, as it is evaluated at a specific point $z_1 = e^{j\omega_1}$. The zeros of the numerator are given by

$$z_k = \cos\omega_i \pm j\sin\omega_i = e^{\pm j\omega_i}.$$

Figure 5.4b shows the allocation of zeros for this filter. This structure has a zero magnitude at frequency ω_i, because the difference between the vectors $(e^{j\omega} - r_k e^{j\theta_k})$ is zero. The

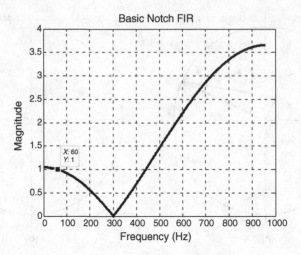

Figure 5.5   Basic notch FIR filter magnitude response.

magnitude response plotted in Figure 5.5 assumes the following: fundamental frequency of $f_1 = 60$ Hz, the 5th harmonic is to be rejected and its sampling frequency $F_s = 32 \times 60$ Hz. In terms of digital representation:

$$\omega_1 = 2\pi f_1/F_s \ \ = 0.1963 \text{ rad,}$$
$$\omega_5 = 10\pi f_1/F_s = 0.9817 \text{ rad.}$$

From Figure 5.5, the unit gain frequency is 60 Hz and the zero gain frequency is 300 Hz. This filter can be considered a digital notch filter of the first order. Again, a magnitude higher than unity at low and high frequency bands is not a good shape for a notch filter. A notch filter should have a unit gain outside the rejected frequency. However, in several cascades of this basic structure, each canceled harmonic produces an interesting parametric filter. Such a filter is useful to reject harmonics when the fundamental frequency varies as discussed in reference [1].

5.2.2 Linear Phase FIR Filter

One of the main characteristics of a FIR filter is that it can be designed to have a linear phase. A linear phase in the frequency domain corresponds to a fixed delay in the time domain, as illustrated in Figure 5.2. The impulse response of a filter with a linear phase must obey symmetric or anti-symmetric characteristics. There are four type of linear phase FIR filters depending on the filter order (even or odd), and its symmetry conditions. The most general linear phase FIR filter is a Type 1 filter, a filter that has a impulse response that is symmetric and has an odd length $(N+1)$ (the filter order N is even). The impulse response $h[n]$ of a Type 1 filter satisfies the condition:

$$h[n] = h[N - n]. \tag{5.9}$$

Figure 5.6a exemplifies a Type 1 impulse response for $N = 8$. Note the symmetry that occurs for sample $N/2$. To demonstrate that a Type 1 FIR filter has a linear phase, we express

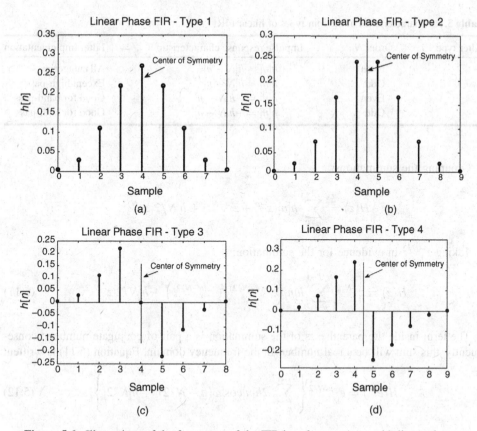

Figure 5.6 Illustrations of the four types of the FIR impulse responses with linear phase.

the z-transform of the Type 1 FIR filter in the form:

$$H(z) = \sum_{n=0}^{N} h[n]z^{-n} = \sum_{n=0}^{N/2-1} h[n]z^{-n} + h[N/2]z^{-N/2} + \sum_{n=N/2+1}^{N} h[n]z^{-n}. \qquad (5.10)$$

Using Equation (5.9) in the last summation yields:

$$H(z) = \sum_{n=0}^{N/2-1} h[n]z^{-n} + h[N/2]z^{-N/2} + \sum_{n=N/2+1}^{N} h[N-n]z^{-n}.$$

Changing the index in the second summation, such as $m = N - n$:

$$H(z) = \sum_{n=0}^{N/2-1} h[n]z^{-n} + h[N/2]z^{-N/2} + \sum_{n=0}^{N/2-1} h[n]z^{-N}z^{n}.$$

Table 5.1 Characteristics of the four types of linear FIR.

Filter type	Order N	Impulse response characteristic	Filter implementation
1	Even	$h[n] = h[N - n]$	All categories
2	Odd	$h[n] = h[N - n]$	Except high-pass
3	Even	$h[n] = -h[N - n]$	Good for band-pass
4	Odd	$h[n] = -h[N - n]$	Good for high-pass

Grouping the summations:

$$H(z) = \sum_{n=0}^{N/2-1} h[n](z^{-n} + z^{(n-N)}) + h[N/2]z^{-N/2}.$$

Taking $z^{-N/2}$ in evidence for the summation:

$$H(z) = z^{-N/2} \sum_{n=0}^{N/2-1} h[n]\left(z^{-(n-N/2)} + z^{(n-N/2)}\right) + h[N/2]z^{-N/2}. \tag{5.11}$$

The term inside the parentheses of the summation is a pair of conjugate numbers; consequently, this sum will be a real number. In the frequency domain, Equation (5.11) is written:

$$H(e^{j\omega}) = e^{-j\omega N/2}\left\{\sum_{n=0}^{N/2-1} 2h[n]\cos[\omega(n - N/2)] + h[N/2]\right\}. \tag{5.12}$$

The term inside the parentheses is a real value, since the impulse response $h[n]$ is assumed to be real. The phase of Equation (5.12) is given by:

$$\sphericalangle H(e^{j\omega}) = -\omega N/2. \tag{5.13}$$

Note that the phase is a linear function of its frequency. Also, from Equation (5.11), the term $z^{-N/2}$ represents a delay in time of $N/2$ samples. For N even, the delay is an integer and can be directly implemented in a discrete-time algorithm.

Other categories of impulse responses for a linear phase FIR are illustrated in Figure 5.6b–d. The Type 1 function is the most general function because it can implement any type of filter. However, there are some restrictions to the use of the function classes 2, 3 and 4. For example, the Type 3 class filter has zeros placed at $z = \pm1$ [2] and consequently cannot be used to implement low-pass or high-pass filters; it may be a good choice for band-pass filter purposes, however. Table 5.1 lists the main characteristics of each of linear phase FIR filter.

5.3 Basic Digital System and Power System Applications

5.3.1 Moving Average Systems: Application

In several power system applications, mainly in regards to the estimation of electrical parameters (e.g. frequency, magnitude and phase), it is usual that the assumed-constant estimation presents variations caused by additive noise or by other undesired components that can produce error in the estimation process. In most cases, these variations (produced by high-frequency components or white noise of zero mean) can be filtered out by a low-pass filter. For this, the most frequently used is the moving average filter (MAF) due to its simplicity and low computational effort. The MAF, a FIR filter, is defined:

$$y[n] = \frac{1}{M} \sum_{k=0}^{M-1} x[n-k].\tag{5.14}$$

Its impulse response given by $h[n] = 1/M, (n = 0, 1, \ldots, M-1)$. Consequently, it has a linear phase (Type 1 filter if M is odd). This equation can be rewritten:

$$y[n] = \frac{1}{M} \left(\sum_{k=1}^{M} x[n-k] + x[n] - x[n-M] \right).$$

From Equation (5.14),

$$y[n-1] = \frac{1}{M} \sum_{k=0}^{M-1} x[n-1-k] = \frac{1}{M} \sum_{l=1}^{M} x[n-l],$$

and finally

$$y[n] = y[n-1] + \frac{1}{M} (x[n] - x[n-M]).\tag{5.15}$$

Equation (5.15) seems to be a recursive filter. Isn't the MAF a non-recursive filter by definition, however? That apparent paradox can be explained by writing Equation (5.15) in the z-domain:

$$H(z) = \frac{1}{M} \frac{1-z^{-M}}{1-z^{-1}}.\tag{5.16}$$

The zeros of the transfer function are located at positions $z = e^{j2\pi k/M}, k = 0, 1, \ldots, M-1$. The pole in $z = 1$ is canceled by the zero in $z = 1$. In fact the difference is that Equation (5.14) has an order of $M-1$ and consequently must only have $M-1$ zeros (all over the unit circle), where the $M-1$ poles are located at the origin of the z-plane. The magnitude response is obtained from Equation (5.16), making $z = e^{j\omega}$. From the graphical interpretation it is possible to anticipate the zero gain of the transfer function at points $z = e^{j2\pi k/M}$, $k = 1, \cdots, M-1$. At $z = 1$ there is an indetermination in Equation (5.16) that, after applying the L'Hopital' rule, leads to $H(1) = 1$. Then the magnitude response of the transfer function (5.16) will have zeros at frequencies multiples of $2\pi/M$. If $M = 32$ and $Fs = 32 \times 60$. The zeros of the filters correspond to their harmonic components:

$$f_k = \frac{1}{M} k \times F_s = \frac{1}{M} k \times 60M = 60k \text{ Hz, for } k = 1, \ldots, M-1.\tag{5.17}$$

As well as the application of a MAF as a low-pass filter to smooth the estimator output, a common application may be for harmonic estimation or decomposition. In fact, the MAF can be seen as a rectangular window used widely in the estimation and decomposition process. The following two sections introduce two common applications of this filter in power systems.

5.3.2 RMS Estimation

The digital estimation of the root mean square (RMS) value is frequently used in power system applications, from protection to monitoring. The mathematical definition of the RMS for a continuous-time periodic signal is:

$$V_{\mathrm{RMS}} = \sqrt{\frac{1}{T_1} \int_{T_1} x(t)^2 \, \mathrm{d}t}. \tag{5.18}$$

The integral over a single period can be approximated using the rectangular integration formula or the trapezoidal formula. If the sampling frequency is coherent, that is, the number of samples over a single period is an integer, then a rectangular approximation for the integral will lead to an accurate result:

$$V_{\mathrm{RMS}}[n] = \sqrt{\frac{1}{M} \sum_{i=0}^{M-1} x[n-i]^2}. \tag{5.19}$$

The RMS is computed in Equation (5.19) using a sliding window of M samples, taking the samples $\{ x[n-M+1], \ldots, x[n-1], x[n] \}$. The expression within the square root can be recognized as the MAF; Equation (5.19) can consequently be rewritten in its pseudo-recursive formula:

$$V_{\mathrm{RMS}}^2[n] = V_{\mathrm{RMS}}^2[n-1] + \frac{1}{M}\left(x^2[n] - x^2[n-M]\right). \tag{5.20}$$

Figure 5.7 presents the computational block diagram of Equation (5.20). This RMS structure has a transient of M samples, which means that the estimated value will only be correct after M samples. This transient is due to the impulse response of the moving average whose length is equal to the number of samples in the mean, that is, the length of the impulse

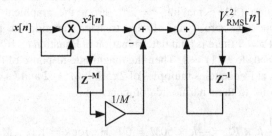

Figure 5.7 Filter to compute RMS value of a periodic signal.

response. M is normally chosen to be equal to the number of samples in a single period of the fundamental frequency. However, if the RMS must be evaluated for longer window, a simple averaging of $V_{RMS}^2[n]$ of the total number of samples inside the selected window is recommended. After this, the root mean square can be used. For example, the standard IEC 61000-4-30 states that the RMS must be taken over 12 cycles of 60 Hz fundamental frequency (10 cycles for 50 Hz systems).

The previous algorithm for the computation of the RMS value is based on the approximation of the integral by a summation. A second interpretation to obtain the RMS value is based on the filtering approach. Without loss of generality, let the input signal be of the form:

$$x[n] = A \cos(\omega_1 n) + B \cos(3\omega_1 n).$$

This signal is the discrete-time signal composed of the fundamental component and the third harmonic. Taking the square of $x[n]$ and applying trigonometric properties yields:

$$x[n]^2 = \frac{1}{2} \left[A^2 + B^2 + A^2 \cos(2\omega_1 n) + B^2 \cos(6\omega_1 n) \right] + \tag{5.21}$$
$$+ AB[(\cos(2\omega_1 n) + \cos(4\omega_1 n))].$$

From theoretical analysis, the RMS value of a periodic signal is equal to:

$$X_{RMS} = \sqrt{\frac{\sum_{i=0}^{H} A_i^2}{2}} \tag{5.22}$$

where A_i $(i = 1, 2, \ldots, H)$ is the amplitude of the harmonic, H the highest harmonic and A_0 the DC component. From Equation (5.21), the verification of the RMS can be obtained by filtering $x^2[n]$ through a low-pass filter with stop-band frequency lower than ω_1. If the sampling frequency is synchronous, the MAF is perfect for this form of processing as it is able to reject all harmonic components, including the fundamental. The output of the filter, $x_f^2[n]$, will be the constant term:

$$x_f^2[n] = \frac{1}{2} \left(A^2 + B^2 \right)$$

as required. The filter interpretation allows for the design of different RMS estimators (see Figure 5.8). For example, if the fundamental frequency is in the time-varying domain, the MAF cannot completely reject the harmonics. This is due to the fact that the harmonic frequency will be shifted with regards to the original zero filter location. Another variant of the RMS estimator is one that has only a half-cycle transient. In this case, instead of obtaining samples over one cycle, samples need to be collected over a half cycle. For example, if the synchronous sampling rate is 64 samples per cycle and the MAF order is 32, then the transient will last half a cycle; however, the zeros location of the MAF will be located every $2f_1$ instead of f_1 $(f_1 = 60 \text{ or } 50 \text{ Hz})$. This estimator can be used if the input signal contains only odd harmonics. However, the presence of any even harmonic will generate a high error in the estimation.

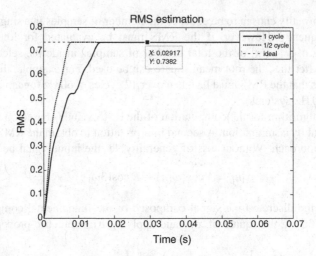

Figure 5.8 RMS estimator: one cycle (continuous line), half-cycle (dotted line) and ideal value.

5.3.3 Trapezoidal Integration and Bilinear Transform

The accuracy of the rectangular integration method for integral approximation will depend on the shape of the curve, and will sometimes be needed for a small time step. A well-known and superior approximation technique is the trapezoidal rule. The trapezoidal rule is frequently used in power systems not only for integral approximations but also for transforming a transfer function from the Laplace domain to the z-domain. This transformation is referred to as the *bilinear* transform. Let the trapezoidal approximation of the integral be:

$$y[n] = \int_{-\infty}^{nT_s} x(t)\mathrm{d}t = \int_{-\infty}^{(n-1)T_s} x(t)\mathrm{d}t + \int_{(n-1)T_s}^{nT_s} x(t)\mathrm{d}t. \tag{5.23}$$

The first integral on the right side of Equation (5.23) is then recognized as $y[n-1]$ and the approximation of the second integral by trapezoidal rule leads to

$$y[n] = y[n-1] + \frac{T_s}{2}\left(x[n] + x[n-1]\right). \tag{5.24}$$

Note that the sampling period (the integration step) from the index of $x[n]$ was omitted. As such, in the z-domain Equation (5.24) becomes

$$H(z) = \frac{Y(z)}{X(z)} = \frac{T_s}{2}\frac{1+z^{-1}}{1-z^{-1}}. \tag{5.25}$$

Recalling the Laplace operator,

$$Y(s) = \frac{1}{s}X(s) \leftrightarrow y(t) = \int_{-\infty}^{t} x(t)\mathrm{d}t,$$

we arrive at the well-known bilinear transform:

$$s = \frac{2}{T_s}\frac{1 - z^{-1}}{1 + z^{-1}}.\qquad(5.26)$$

The transfer function in the z-domain can be obtained from $H(s)$ through a substitution of variable [3]:

$$H(z) = H(s)|_{s=\frac{2}{T_s}\frac{z-1}{z+1}}.\qquad(5.27)$$

The main characteristics of a bilinear transform include:

- the ability to map the left s-plane in the inside unit circle of the z-plane; this means that the stability of a transfer function is preserved by the bilinear transformation;
- the $j\Omega$ axis in the s-plane (imaginary axis) is mapped over the unit circle ($e^{j\omega}$) in the z plane;
- the order of the transfer function is preserved.

The mapping is not linear, that is, the relationship between analog frequency Ω and digital frequency ω is not linear. The relationship can be obtained directly from Equation (5.26) changing $s = j\Omega$ and $z = e^{j\omega}$ and results in

$$j\Omega = \frac{2}{T_s}\frac{1 - e^{-j\omega}}{1 + e^{-j\omega}} = j\frac{2}{T_s}\tan(\omega/2)$$

or

$$\Omega = \frac{2}{T_s}\tan(\omega/2).\qquad(5.28)$$

Equation (5.28) is plotted in Figure 5.9 for $T_s = 2$ and $-\pi < \omega < \pi$. Note that the mapping is highly non-linear, since the complete negative imaginary axis in the s-plane is mapped in the lower half of the unit circle over the range $-\pi < \omega < 0$. The same happens with the positive

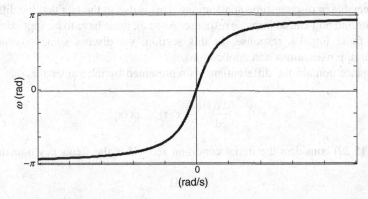

Figure 5.9 Mapping of the angular analog frequency (rad/s) versus angular digital frequency (rad).

imaginary axis in the s-plane that is mapped out in the upper half of the unit circle $0 < \omega < \pi$. This introduces a distortion of the frequency axis, referred to as warping. If the range of the frequency of interest is very low, then the mapping can however be considered linear; note that for a low range $\tan(\omega/2) \approx \omega/2$ yields:

$$\Omega \approx \frac{2}{T_s}\omega/2 \Rightarrow \omega \approx \Omega T_s.$$

For some classes of analog systems it is possible to prewarp the original specification in such a way that after applying the bilinear transform the distortion leads the discrete-time system specification towards the desired operation. The simplest example is the design of the digital notch filter from its analog counterpart. If the notch filter must reject the frequency Ω_0, the exact digital frequency should be $\omega_0 = \Omega_0 T_s$. However, as the bilinear transformation is not a linear frequency mapping, the analog notch frequency needs to be prewarped using Equation (5.28). The analog notch filter should then have its frequency equal to

$$\Omega_0' = \frac{2}{T_s}\tan(\Omega_0 T_s/2).$$

After applying the bilinear transform, the digital notch filter will represent its frequency at the correct position.

The prewarping approach is generally used in IIR digital filter design using an analog filter as prototype. MATLAB$^{\circledR}$ automatically uses the prewarping approach during the design process. Unfortunately, in power system applications there are transfer functions for which this procedure cannot be used. In those cases the solution is to reduce the sampling time or the time step. This is carried out in electrical magnetic transient simulators such as the alternative transient program (ATP) and real-time simulation computer-aided design (RSCAD), when the trapezoidal integration method to generate the discrete-time system is instead used.

5.3.4 Differentiators Filters: Application

Differentiation is a mathematical operation widely used in the continuous-time world and frequently implemented in discrete-time systems. As will be shown, the ideal differentiator is not implementable in discrete-time applications. This is due to the fact that the differentiator has an infinite and non-causal impulse response. As such, these have to be approximated for a causal and finite impulse response. In this section, we discuss some commonly used differentiator approximations and applications.

In the Laplace domain the differentiator is represented by the s operator, that is,

$$y(t) = \frac{dx(t)}{dt} \leftrightarrow Y(s) = sX(s). \tag{5.29}$$

Equation (5.29) considers the initial condition zero. If in the frequency domain

$$s \leftrightarrow j\Omega \tag{5.30}$$

then the frequency spectrum of an ideal differentiator is a purely imaginary function, and its magnitude response is linear. The ideal discrete-time differentiator is then defined by the frequency response:

$$H_{\text{DIF}}(e^{j\omega}) = j\omega, \quad |\omega| < \pi. \tag{5.31}$$

The impulse response of the ideal differentiator can be found by using the inverse discrete-time Fourier transform (IDTFT),

$$h_{\text{DIF}}[n] = \frac{j}{2\pi} \int_{-\pi}^{\pi} \omega e^{j\omega n} \, d\omega.$$

Solving this integral leads to

$$h_{\text{DIF}}[n] = \begin{cases} 0, & n = 0 \\ \dfrac{\cos \pi n}{n}, & |n| > 0. \end{cases} \tag{5.32}$$

The above impulse response is non-causal and infinite, so it is not realizable in practice. However, it can be approximated by truncating the impulse response sequence to a finite length and shifting the truncated coefficients to convert the system to a causal one. For example, taking 31 (filter order $N = 30$) terms of Equation (5.32) for $-N/2 \le n \le N/2$ and shifting the sequence $N/2$ samples right, we obtain the impulse response shown in Figure 5.10. As can be observed, the impulse response is a linear phase FIR filter of Type 3 (see Table 5.1).

It is useful to demonstrate what happens with the phase characteristic for a Type 3 FIR filter. Following the same steps as in Section 5.2.1,

$$H(z) = z^{-N/2} \sum_{n=0}^{N/2-1} h[n] \left(z^{-(n-N/2)} - z^{(n-N/2)} \right). \tag{5.33}$$

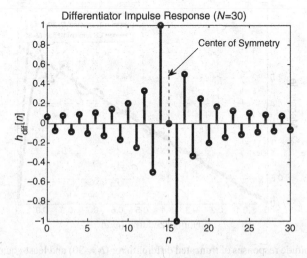

Figure 5.10 Differentiator impulse response for $N = 30$.

Now, the terms inside the summation are purely imaginary numbers. In the frequency domain,

$$H(e^{j\omega}) = e^{-j\omega N/2} \sum_{n=0}^{N/2-1} -j2h[n]\sin(\omega(n-N/2)),$$

which can be rewritten:

$$H(e^{j\omega}) = e^{j\pi/2}e^{-j\omega N/2} \sum_{n=0}^{N/2-1} 2h[n]\sin(\omega(N/2-n)). \tag{5.34}$$

Note that the phase of Equation (5.34) is a linear phase with the added constant $\pi/2$. In this way, after compensating for the linear phase delay of $N/2$, the output of the system will be multiplied by the complex number j and the summation will approximate the linear amplitude expression in Equation (5.31).

Figure 5.11 shows the magnitude response of the filter with an impulse response given by Figure 5.10. The correspondent curve is labeled 'Truncated $N = 30$'. Note that the magnitude response oscillates around the ideal differentiator magnitude response. The filter order can be increased to reduce the oscillation in the low-frequency region. In the high-frequency region however, the amplitude of the maximum peak cannot be reduced due to the *Gibbs* phenomenon. This phenomenon can be reduced if instead of directly truncating the ideal impulse response it is multiplied by using a non-rectangular window function. Another approximation method used is the least-square method. The MATLAB® function *firls* designs a linear phase FIR filter using the least-square error minimization. It is a generic function that designs several categories of linear phase FIR filters, including a differentiator. For a differentiator

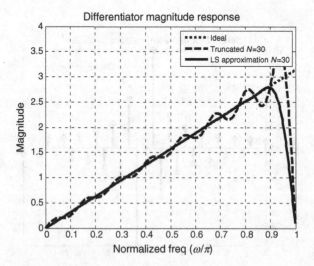

Figure 5.11 Magnitude responses of truncated differentiator ($N = 30$) and least-squares approximation method.

design, the basic formula:

$$h\_dif = \text{firls}(N, F, A, \text{'differentiator'})$$

can be used. This function will return a length $N + 1$ impulse response with linear phase (real, antisymmetric coefficients) differentiator FIR filter, which has the best approximation to the desired frequency response described by F and A in the least-squares sense. F is a vector of frequency band with pair edges of ascending order, and normalizes the frequency 0 to 1 where 1 corresponds to the Nyquist frequency or half of the sampling frequency. A is a real vector, of the same size as F, and specifies the desired amplitude of the frequency response. For example, the specification

$$h\_dif = \text{firls}(30, [0\ 0.9], [0\ 0.9 * pi], \text{'differentiator'}) \tag{5.35}$$

can be used to design a differentiator filter. In Figure 5.11 the continuous curve is the magnitude response obtained from the above command (labeled 'LS approximation $N = 30$'). As observed, the magnitude response is very close to the ideal response in almost all frequency ranges. There is of course a small ripple, also in the band pass, but its magnitude is very small.

As mentioned, the differentiator designed above has a linear phase. When testing its performance in the time domain, the form of the input signal can be assumed to be $x(t) = \cos(\Omega_1 t) + 0.3\cos(5\Omega_1 t)$, where $\Omega_1 = 120\pi$ rad/s. Its derivative function is

$$y(t) = -\Omega_1 \sin(\Omega_1 t) - 1.5\Omega_1 \sin(5\Omega_1 t). \tag{5.36}$$

When the discrete-time signal $x[n]$ is obtained by sampling the continuous signal at the sampling frequency of $F_s = 32 \times 60$ Hz and $x[n]$ is passed through the differentiator filter obtained by Equation (5.35), the result is plotted together. By the delayed discrete-time version of Equation (5.36), Figure 5.12 results. This figure shows that after the filter transient, the estimated value is close to the ideal value given by Equation (5.36).

5.3.5 Simple Differentiator

In the opening paragraphs of Section 5.3.4 the design of a differentiator for wide band was described; the least-square method in particular is able to generate good differentiators for wide-band issues. A wide-band differentiator is one whose frequency response is linear in almost the whole frequency range. In several applications however, a narrow band differentiator is sufficient. The simplest differentiator is one that approximates a derivative by a first-order differential equation:

$$y(t) = \frac{dx(t)}{dt} \leftrightarrow y[n] = \frac{x[n] - x[n-1]}{T_s}. \tag{5.37}$$

The equivalent digital filter has a transfer function

$$H(z) = \frac{1}{T_s}(1 - z^{-1}). \tag{5.38}$$

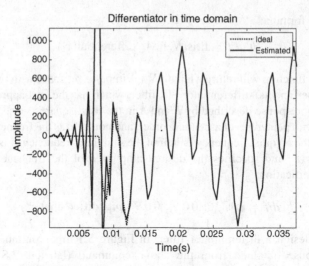

Figure 5.12 Differentiator performance in time domain.

The impulse response has, as required, an antisymmetric characteristic. However, as the filter order is odd, the linear phase corresponds to a non-integer delay in time. In fact, the time delay is half the sample and, as such, it cannot be compensated for directly. To overcome this issue of non-integer delay, the *central-difference differentiator* can be used. Its transfer function is [4]:

$$H(z) = \frac{1}{2T_s}\left(1 - z^{-2}\right). \tag{5.39}$$

A general expression for differentiators is [4]:

$$h[n] = \frac{-3n}{M(M+1)(2M+1)}, \qquad -M \leq n \leq M. \tag{5.40}$$

The causal filter of order $N = 2M$ is obtained after multiplying the non-causal filter by z^{-M}. As can be verified, the impulse response of the filter is antisymmetric.

In the same reference [6] Hamming describes a narrowband differentiator called the '*low-noise Lanczos*' filter. The name 'low-noise' is because it has low gain at high frequencies and is consequently less sensitive to noise. Furthermore, it is a narrow-band differentiator as its mathematical response is the ideal response (only) for low frequencies. This impulse response of low-noise Lanczos is:

$$h_{sL}[n] = \begin{bmatrix} -1 & 8 & \underset{\uparrow}{0} & -8 & 1 \end{bmatrix}/12 \tag{5.41}$$

where the arrow indicates the coefficient corresponding to $h_{sL}[0]$.

Figure 5.13 introduces the magnitude response of simple differentiators. The magnitude response of the ideal differentiator is plotted together with its first order differentiator Fd, central differential differentiator Cd and Lanczos differentiator Ld. As observed from the

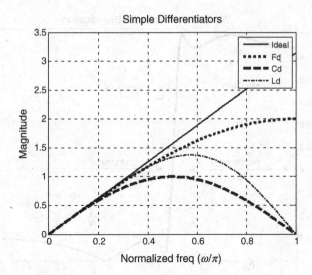

Figure 5.13 Simple differentiators magnitude response. Continuous line: ideal differentiator; dotted line: first-order differentiator; dashed line: central difference differentiator; and dashed-dotted line: Lanczos differentiator.

figure, the first-order differentiator is the one that best approximates the ideal magnitude response. However, for practical applications this may not be a good characteristic as the differentiator tends to amplify high-frequency noises and, under real-world conditions, the signal contains much high-frequency noise. With this in mind, the Lanczos differentiator is the best among them because it approximates the ideal differentiator up to a given frequency; it goes to zero thereafter, attenuating the high-frequency noises.

5.4 Parametric Filters in Power System Applications

Five steps are involved in the design of a digital filter:

1. specification of the filter requirements
2. calculation of the filter coefficients (approximation)
3. choice of the structure to implement the filter (realization)
4. analysis of the effects of finite word length on filter performance
5. implementation of this filter in software (digital signal processor or DSP) or hardware (field-programmable gate array or FPGA).

Of course, power engineering may not use all five steps; it is mostly steps 4 and 5 that involve specific knowledge. Increasingly however, power engineering has to work with digital filters designed for solving a great number of problems which may include protection, monitoring, control and power systems analysis. This section will deal with parametric filters. Before this however, it is worth looking into the issue of filter specification parameters. The approximation topic (step 2) is left for Section 5.5 where we cover the basics of MATLAB® functions to approximate the filters. The topic regarding realization (step 3) is not considered in this book.

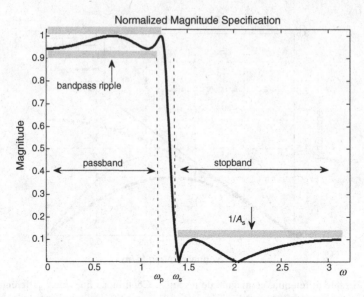

Figure 5.14 Magnitude specification of a realizable filter.

5.4.1 Filter Specification

Figure 5.14 shows the magnitude specification of a low-pass filter. In the figure three regions can be identified: (a) pass; (b) stop; and (c) transition bands. The transition band is the area linking the pass-band and stop-band regions. The ideal low-pass filter would have an infinitely sharp cutoff from pass-band to stop-band and unit gain in the pass-band with zero gain in the stop-band. An ideal filter is physically impossible; the ideal conditions therefore need to be relaxed in order to obtain a realizable filter. Figure 5.14 illustrates the type of 'relaxing' that needs to be achieved by the filter analysis. The magnitude of the pass-band does not usually need to be a constant or unitary; a certain ripple is tolerable. Furthermore, it is not necessary for the gain at the stop-band to be zero, but it cannot be higher than $1/A_s$ where A_s is the attenuation in the stop-band as shown in Figure 5.14. The transition region between pass-band and stop-band is defined by the two frequencies ω_p and ω_s ($\omega_s > \omega_p$). The frequency ω_p defines the end of the pass-band and ω_s the beginning of the stop-band. The width of the transition band is therefore $\omega_s - \omega_p$.

The digital filter specifications are often given in terms of the loss function $G(e^{j\omega}) = 1/H(e^{j\omega})$, generally expressed in dB, that is, $G(\omega) = -20log_{10}|H(e^{j\omega})|$ in dB. The peak pass-band ripple R_p and the minimum stop-band attenuation R_s, is given in dB. For example, the elliptic digital filter designed in MATLAB® uses the command where N is the order of the low-pass digital elliptic filter with R_p decibels of peak-to-peak ripple (in the pass-band) and a minimum stop-band attenuation of R_s decibels. The function returns are the coefficients of numerator b and the denominator a of the transfer function. The pass-band-edge frequency ω_p must be $0.0 < \omega_p < 1.0$, with 1.0 corresponding to half the sample rate (π rad or $F_s/2$).

There are other filter specifications in MATLAB®. For example, in the Butterworth filter design the specifications are simpler: $[b, a] = \text{butter}(N, \omega_n)$. The frequency ω_n is the cutoff frequency that corresponds to the frequency where the magnitude is $-3\,\text{dB}$ in the transfer function.

In Section 5.6 the design of digital filter using MATLAB® will be further discussed.

5.4.2 First-Order Low-Pass Filter

The single parametric low-pass transfer function is given by:

$$H_{\text{LP}}(z) = \frac{k(1 + z^{-1})}{1 - \alpha z^{-1}} \tag{5.42}$$

where the parameter α is used to determine the cutoff frequency. Note the filter has a zero in $z = -1$. Consequently, the magnitude response is zero in the frequency $\omega = \pi$. The maximum value occurs at $z = 1$, so

$$H_{\text{LP}}(e^{j0}) = \frac{2k}{1 - \alpha}.$$

In order to have unit gain at DC frequency $k = (1 - \alpha)/2$ is chosen, and the final expression is:

$$H_{\text{LP}}(z) = \frac{1 - \alpha}{2} \frac{1 + z^{-1}}{1 - \alpha z^{-1}}. \tag{5.43}$$

The frequency response is given by:

$$H_{\text{LP}}(e^{j\omega}) = \frac{1 - \alpha}{2} \frac{1 + e^{-j\omega}}{1 - \alpha e^{-j\omega}}.$$

To obtain the α parameter in order to have the cutoff frequency in ω_c we consider $|H_{\text{LP}}(e^{j\omega_c})|^2 = 1/2$. After algebraic manipulation we have:

$$\alpha = \frac{1 - \sin(\omega_c)}{\cos(\omega_c)}. \tag{5.44}$$

Note that the α parameter is the pole of the transfer function, so for filter stability $|\alpha| < 1$. The analysis of Equation (5.44) reveals that $|\alpha| < 1$ for $0 < \omega < \pi$.

5.4.3 First-Order High-Pass Filter

A simple parametric IIR high-pass filter has a transfer function given by:

$$H_{\text{HP}}(z) = \frac{1 + \alpha}{2} \frac{1 - z^{-1}}{1 - \alpha z^{-1}} \tag{5.45}$$

where $|\alpha| < 1$ for stability. The relation between α and the 3 dB cutoff frequency ω_c is also given by Equation (5.44). Figure 5.15 presents the magnitude response for the low-pass and

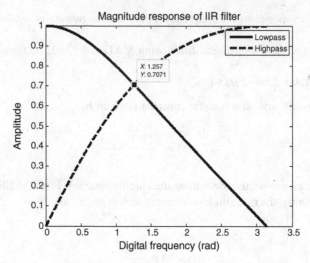

Figure 5.15 Magnitude response of parametric low-pass and high-pass IIR filter.

high-pass parametric IIR filters with a cutoff frequency of $0.4\,\pi$. The cutoff frequency is highlighted in the figure.

5.4.4 Bandstop IIR Digital Filter (The Notch Filter)

A bandstop filter is used to reject a frequency band located inside the extreme frequencies (DC and π rad). The transfer function is given by:

$$H_{\mathrm{BS}}(z) = \frac{1+\alpha}{2} \frac{1 - 2\beta z^{-1} + z^{-2}}{1 - \beta(1+\alpha)z^{-1} + \alpha z^{-2}} \tag{5.46}$$

which is a second-order parametric bandstop that has been used in several power system applications. The β parameter can be set up in order to control the location of the zeros of the filter over the unit circle, in order to reject a specific frequency. In fact, for $|\beta| < 1$ the zeroes of the filter can be identified as:

$$z_k = \beta \pm j\sqrt{1 - \beta^2}, \quad k = 1, 2.$$

It is easy to demonstrate that $|z_k| = 1$ and

$$\beta = \cos \omega_0 \tag{5.47}$$

where ω_0 is the frequency that will be rejected, also referred to as the notch frequency. A filter that completely rejects a single frequency is referred to as a *notch filter*. The parameter α is referred to as the notch factor, which controls the 3 dB bandwidth. It is defined as the difference between the cutoff frequency $\omega_{c2} - \omega_{c1}$, where $\omega_{c2} > \omega_{c1}$ and ω_{c1} and ω_{c2} are the frequencies for $|H_{\mathrm{BS}}(e^{j\omega_i})|^2 = 1/2$, $(i = 1, 2)$. The 3 dB bandwidth is

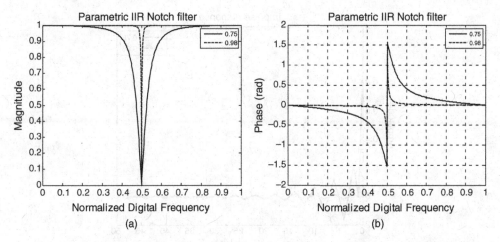

Figure 5.16 Notch filter frequency response: (a) magnitude response; (b) phase response.

given by [2]:

$$B_{3dB} = \omega_2 - \omega_1 = \cos^{-1}\left(\frac{2\alpha}{1 + \alpha^2}\right). \tag{5.48}$$

Additionally, $|\alpha| < 1$ for filter stability. As can be observed from Equation (5.48), the closer α is to the unit value, the sharper (more selective) is the notch filter. On the negative side, the filter transient increases as α approaches the unit value. In a practical project, this tradeoff needs to be controlled.

Figure 5.16a displays the magnitude responses for $\omega_0 = 0.5\pi$, $\alpha = 0.75$ and $\alpha = 0.98$. The phase response can be seen in Figure 5.16b. Note that the delay caused by the filter is close to zero as the frequency moves away from the notch frequency. This behavior is better for the notch filter with notch factors equal to 0.98.

From Figure 5.16 it is clear that the notch filter with a notch factor equal to 0.98 presents a better frequency response. The problem is that the poles of the filter approach the unit circle as the notch factor approaches 1, which can lead to filter instability due to problems with the quantization. When working with finite arithmetic in processors or FPGA, the quantized pole can in fact overpass the unit circle and become unstable. As such, care is needed when implementing a notch filter with too high a notch factor. The other problem is regarding the filter transient response that can be measured through its impulse response duration. In general, the sharper the transition band, the longer the transient response. However, unaccounted-for impulse responses can occur on both notch filters, as depicted in Figure 5.16. These responses are plotted in Figure 5.17. Despite the fact that the notch filter with $\alpha = 0.98$ keeps oscillating for a longer period than the notch filter with $\alpha = 0.75$, the latter has a stronger transient response. As such, the former filter can be a better choice for several practical applications.

The transient response of a filter is an important subject in several power system applications. In protection application for example, the transient filter response must be as short as possible in order for the protection system to point out the fault condition as quickly as possible. Unfortunately, most textbooks on the subject of digital signal processing do not

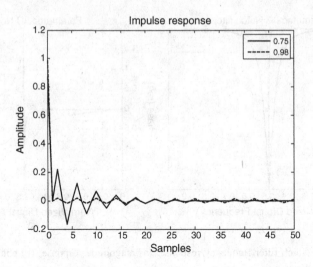

Figure 5.17 Impulse response of notch filter for notch factor equal to 0.75 and 0.98.

devote enough attention to these issues; in a general scenario the concern is related to the steady-state response of the filters and the phase delay caused by them, as they are associated with telecommunications issues.

The poles-zeros plot for both notch filters is presented in Figure 5.18. As observed, the pole is closer to the unit circle for a notch factor of 0.98, which reveals that the impulse response will vanish slowly as compared to the first notch filter. In order to understand the impulse response transient, the inverse transform of Equation (5.46) needs to be found. This can be obtained using the residuez MATLAB® command. The residue for both cases is obtained from the commands:

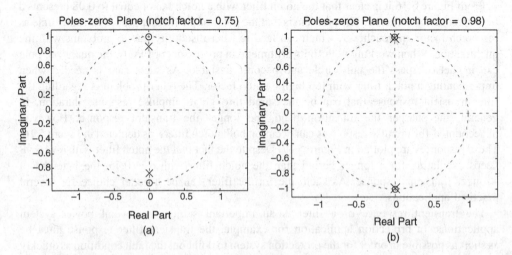

Figure 5.18 Poles-zeros plane: notch filter with notch factor of (a) 0.75; and (b) 0.98.

Table 5.2 Parameters of the partial expansion of the notch filter example.

Case i; (notch factor)	Residues (R_i)		Poles (P_i)		K_i
$i=1$; (0.75)	-0.1458	-0.1458	$0.8660j$	$-0.8660j$	1.1667
$i=2$; (0.98)	-0.0101	-0.0101	$0.9899j$	$-0.9899j$	1.0102

```
[ R1,P1,K1] =residuez(bnt1,ant1); %notch factor 0.75
[ R2,P2,K2] =residuez(bnt2,ant2); %notch factor 0.98
```

The numerator and denominator coefficients for the notch filter are given by bnt1, ant1 for notch factor 0.75 and bnt2, ant2 for notch factor 0.96. The outputs of the residue function are: the residues of the partial expansion function R_i; the poles of the partial expansion function P_i; and the coefficients of the impulse response K_i (where $i = 1, 2$). These coefficients are zero only when the transfer function is 'strictly proper'. A strictly proper function is one where the degree of the numerator is less than the degree of the denominator.

For the general function with simple poles and proper functions (numerator and denominator degree can be equal), the partial expansion is given by:

$$H(z) = K_i + \frac{R_{i_0}}{1 - P_{i_0}z^{-1}} + \cdots + \frac{R_{i_N}}{1 - P_{i_N}z^{-1}}$$

and the impulse response is given by

$$h[n] = K_i\delta[n] + R_{i_0}P_{i_0}^n u[n] + \cdots + R_{i_N}P_{i_N}^n u[n].$$

Table 5.2 presents the values for each parameter (K, R and P) obtained from MATLAB® for the notch filter. From row 5 of Table 4.9 of Chapter 4, after mathematical manipulation:

$$h_i[n] = K_i\delta[n] - \rho_i r_i^n \sin\left(\frac{\pi}{2}n\right)u[n], \quad i = 1 \text{ or } 2 \tag{5.49}$$

where $\rho_i = |R_i|$ and $r_i = |P_i|$. Comparing Equation (5.49) to Figure 5.17 it can be observed that for $n = 0$ the first value is equal to K_i. Note that r_1^n tends to zero quicker because of the magnitude of the pole. However, the residue ρ_1 is much higher (more than ten times) than residue ρ_2; for this reason the transient response of the first notch filter is stronger at the beginning of the transient response than the second notch transient response.

5.4.5 Total Harmonic Distortion in Time Domain (THD)

An interesting application of a notch filter is the determination of the total harmonic distortion (THD) in the time domain. As defined by Equation (4.22), the THD is:

$$\text{THD} = \frac{\sqrt{\sum_{n=2}^{H} G_n^2}}{G_1}, \tag{5.50}$$

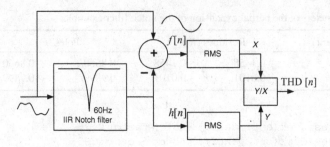

Figure 5.19 Computation of the THD in time domain.

where G_n $(n = 1, 2, \ldots, H)$ is the RMS value of the harmonic component n and H is the maximum harmonic. Figure 5.19 is the block diagram used to compute the THD in the time domain. The IIR notch filter is used to extract the fundamental component. As the phase delay is close to zero for the harmonic components, phase compensation is not necessary. This is different in the case of the FIR notch filter (Figure 5.3). The sequence $f[n]$ is composed only of the fundamental component, while the sequence $h[n]$ represents the signal with the fundamental component extracted. If the signal contains only harmonics, this sequence will be an all-harmonic combination without the fundamental. The RMS of both signals is computed using the sequence shown in Figure 5.7. Dividing the RMS of $h[n]$ (labelled in the figure by Y) by the RMS of $h[n]$ (labelled X), results in the THD of the input signal as a function of n. If the signal is stationary, this value will be constant after the filter transients.

As an example, consider the signal described by:

$$x[n] = 28 \cos(\omega_0 n) + 3 \cos(3\omega_0 n) + 3.5 \cos(5\omega_0 n) + 2 \cos(7\omega_0 n). \qquad (5.51)$$

The theoretical THD for this signal is THD $= 17.9462$. In Figure 5.20 the exact and the estimated values are presented (sampling rate used was 64 per cycle and the notch parameter

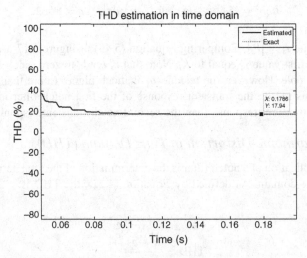

Figure 5.20 THD time domain estimation through IIR notch filter.

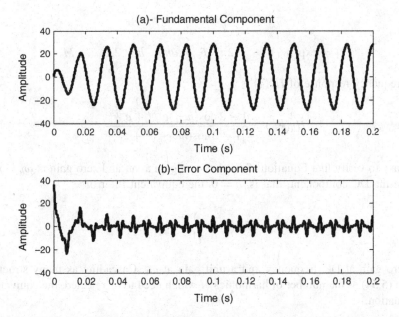

Figure 5.21 Notch decomposition: (a) fundamental component and (b) error component.

used was 0.98). After the filter transient (about 2 cycles), the THD estimator reaches an accurate value. The final error is less than 0.15%.

5.4.6 Signal Decomposition using a Notch Filter

Another interesting application for notch filters in power systems is for signal decomposition. When using a notch filter it is possible to divide the signal into two components: (a) fundamental component and (b) error signal. The error signal is the signal obtained after the fundamental component extraction. This is an attractive approach for power quality event detection and classification. Consider again Figure 5.19; the signal $h[n]$ can be used as the error signal for a general class of events or disturbances. The error signal is denoted $er[n]$. Using the same signal as that described by Equation (5.51), the fundamental signal and the error signal are depicted by Figure 5.21.

The decomposed signals are commonly applied for detection or classification systems based on neural network structures, in which the filter delay can be easily dealt with. This type of decomposition is very easy to implement and requires only a low computational effort. An interesting application of this decomposition in a protection system is described in reference [5].

5.5 Parametric Notch FIR Filters

A second-order parametric FIR section that rejects the harmonic h and has unit gain in the fundamental frequency f_1 is of the form [1]:

$$H_h(z) = \frac{1 - 2\cos(h\omega_1)z^{-1} + z^{-2}}{\left|1 - 2\cos(h\omega_1)z_1^{-1} + z_1^{-2}\right|} \tag{5.52}$$

where

$$z_1 = \exp(jw_1); \, w_1 = 2\pi f_1 T_s; \quad and \quad h = 2, 3, \ldots, M. \tag{5.53}$$

The frequency response of $H_h(z)$ is

$$H_h(e^{j\omega}) = \frac{1 - 2\cos(h\omega_1)e^{-j\omega} + e^{-j2\omega}}{\left|1 - 2\cos(h\omega_1)z_1^{-1} + z_1^{-2}\right|}. \tag{5.54}$$

It is easy to verify that Equation (5.54) has unit gain at ω_1 and zero gain at $\omega_h = h\omega_1$. To eliminate the DC component, that is, $h = 0$, the equivalent filter of

$$H_0(z) = \frac{1 - z^{-2}}{\left|1 - z_1^{-2}\right|} \tag{5.55}$$

has a zero gain at DC frequency and a unit gain at ω_1. Cascading as many structures of Equation (5.54) as the number of harmonics M which need to be rejected, the equivalent FIR filter equation,

$$H(z) = H_0(z) \prod_{h=2}^{M} H_h(z) \tag{5.56}$$

is a parametric filter; once ω_1 is known, or $\cos(\omega_1)$, the parameters are easily modified.

As an example, the magnitude response of the equivalent filter for $F_s = 1920\,\mathrm{Hz}$, $f_1 = 59\,\mathrm{Hz}$ and $M = 16$ is depicted by Figure 5.22, which can be obtained from MATLAB® program *parametric_fir_filters.m*. Note in Figure 5.22 that the unit gain is 59 Hz and the zero gains are 118 Hz, 177 Hz, etc. Observe that the zero gain falls out in multiples of 60 Hz.

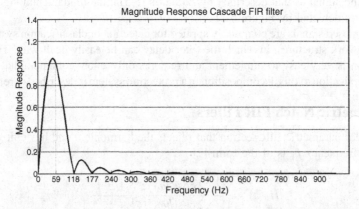

Figure 5.22 The magnitude response for cascade of second-order FIR filters.

5.6 Filter Design using MATLAB® (FIR and IIR)

The signal processing toobox of MATLAB® includes a variety of functions for the design of both IIR and FIR digital filters, some of which are mentioned in this chapter. As well as the m-file functions, MATLAB® offers dialog boxes for the design of filters and their analysis. The filter design and analysis tool (FDAtool) is called from the command line using: fdatool. FDAtool is a powerful user interface for fast design and analysis of filters by setting filter specifications, importing filters from the MATLAB® workspace or by adding, moving or deleting poles and zeros. Analysis tools include the magnitude and phase responses and pole-zero plots.

FDAtool allows users to: design advanced filters which cannot be developed using the signal processing toolbox software; quantize double-precision filters designed in this graphic user interface (GUI) using the design mode; quantize double-precision filters imported into this GUI using the import mode; analyze quantized filters; scale second-order section filters; and transform both FIR and IIR filters from one response to another.

Figure 5.23 depicts the FDAtool GUI which allows the effortless design of many different types of digital filters. The filter and design category can be specified as well as the pass-band and stop-band frequencies. Further, attenuation in the pass-band and stop-band can be specified to give the desired filter responses by changing the values on the bottom right corner of the window. FDAtool designs the lowest-order filter which achieves these attenuations. Readers are invited to explore this very easy and useful tool.

5.7 Sine and Cosine FIR Filters

Sine and cosine FIR filters, also known as discrete Fourier transform (DFT) filters, are widely used in power system applications because they have unit gains at the fundamental component

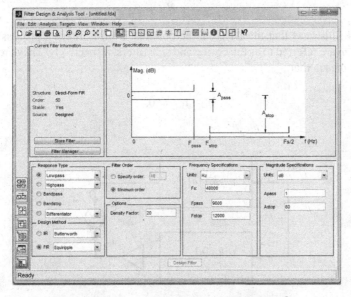

Figure 5.23 FDAtool design from MATLAB®.

and zero gain at DC and harmonic components. These characteristics are needed in applications where the fundamental phasor is required to be estimated, such as in protection applications. The impulse response of these filters is developed in the following.

Equation (4.59) rewritten below is the basic expression of a DFT filter:

$$X_k[n] = W_N^{-k} \sum_{m=0}^{N-1} x[n-m] \, W_N^{-km}, (k = 0, 1, \dots, N-1). \tag{5.57}$$

Equation (5.57) is the convolution expression where the impulse response of the FIR filter is complex and given by:

$$\mathbf{h}_k = \begin{bmatrix} W_N^{-k} & W_N^{-2k} & W_N^{-3k} & \cdots & W_N^{-Nk} \end{bmatrix}. \tag{5.58}$$

When Equation (5.57) is separated into real and imaginary parts, the two filter coefficients found have impulse responses:

$$\mathbf{h}_{ck} = \frac{2}{N} \begin{bmatrix} \cos(\theta k) & \cos(2\theta k) & \cos(3\theta k) & \cdots & \cos(N\theta k) \end{bmatrix} \tag{5.59}$$

and

$$\mathbf{h}_{sk} = \frac{2}{N} \begin{bmatrix} \sin(\theta k) & \sin(2\theta k) & \sin(3\theta k) & \cdots & \sin(N\theta k) \end{bmatrix}, \tag{5.60}$$

where

$$\theta = \frac{2\pi}{N}. \tag{5.61}$$

Figure 5.24 illustrates that the DFT can be obtained from a filtering process. The filters $h_{ck}[n]$ and $h_{sk}[n]$ are referred to as the cosine and sine filters, respectively. They are orthogonal filters, which means that:

$$\mathbf{h}_{ck} \cdot \mathbf{h}_{sk}' = 0. \tag{5.62}$$

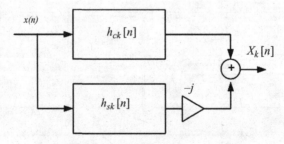

Figure 5.24 Filtering interpretation of DFT.

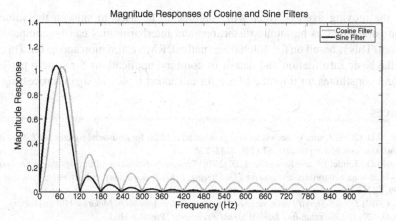

Figure 5.25 Magnitude response of cosine and sine filters.

Although the cosine and sine filters are used together to obtain the DFT of the kth term, they can also be used separately in several power system applications due to their properties of (1) unit magnitude at frequency, i.e.

$$f^1 = kF_s/N; \tag{5.63}$$

and (2) zero magnitude at frequencies with multiples of f^1, i.e.

$$f^0 = rF_s/N, (r = 0, 1, 2, \ldots, N, r \neq k). \tag{5.64}$$

If F_s is conveniently chosen, the filter can eliminate the harmonic components. For example, if $F_s = Nf_1$, then $f^0 = rf_1, (r \neq 1)$. Figure 5.25 shows the magnitude response for $N = 32$ and $f_1 = 60$ Hz. Note that the sine filter has better rejection characteristics than the cosine filter, despite the fact that both are able to eliminate the harmonic components.

As observed from Equations (5.63) and (5.64), changing the unit and zero points of gain is possible through the adjustment of the sampling frequency or the size of the window. This adjustment is important because the system frequency generally undergoes some variation. Estimation algorithms that track the frequency system to change the filters characteristics are introduced in Chapter 7. The option of changing the sampling frequency is generally not used for practical reasons, and the adjustment of the length of the window is an option for several estimation algorithms. Note that the modification of N changes the impulse response of the filters defined by Equations (5.59) and (5.60). For real-time applications it is better to save the coefficients in the memory according to the range of variation of N. Another way is to update the coefficients by using the parametric FIR as discussed in Section 5.5.

5.8 Smart-Grid Context and Conclusions

Although smart-grid applications mean the introduction of many time-varying variables in the system behavior, the basic method for analysis and design will continue to be based on linear and time-invariant concepts. Several methods presented in this chapter can be applied to the future smart grid in an attempt to adapt the conventional method to the new context. For example, the moving average filter will continue to be used to smooth the value of the estimation parameters as harmonic distortions and interharmonics can be computed in the time domain. This is based on the notch filter method, RMS estimation and so on. This chapter provides the basic information and details of common applications for system performance analysis and constitutes an important base for advanced topics in signal processing.

References

1. Kusljevic, M.D. (2008) A simple method for design of adaptive filters for sinusoidal signals. *IEEE Transactions on Instrumentation and Measurements*, **57** (10), 2242–2249.
2. Kusljevic, M.D., Tomic, J.J. and Javanovic, L.D. (2010) Frequency estimation of three-phase power system using weighted-least-square algorithm and adptive FIR filtering. *IEEE Transactions on Instrumentation and Measurements*, **59**, 322–329.
3. Mitra, S. (2010) *Digital Signal Processing: A Computer Based Approach*, McGraw-Hill Publishing.
4. Lyons, R.G. (2011) *Understanding Digital Signal Processing*, Prentice Hall.
5. Carvalho, J.R., Coury, D.V., Duque, C.A. and Jorge, D.C. (2011) Development of detection and classification stages for a new distance protection approach stages for a new distance protection approach. *Power and Energy Society General Meeting. IEEE*, 1–7.
6. Hamming, R. (1998) *Digital Filters*, Dover.

6

Multirate Systems and Sampling Alterations

6.1 Introduction

Different power systems applications may require different sampling rates. An example may be when a data file is recorded from a real power system which is set to acquire a signal at a given sampling rate F_s and, for the evaluation of a new algorithm, the sampling rate needs to be changed to F_s'. In this case the sampling rate can be altered using an off-line algorithm. There are applications that demand on-line processing however; such is the case for some phasor and harmonics estimators. Under such circumstances, the algorithm resamples the signal in real time to obtain new samples as if obtained using a synchronous sampling method. Another application is when a digital system is processed using different sampling rates at different parts of the system, for example to reduce computational cost or make the deployment simpler. Such an example is the modern protective relay that uses a high sampling rate for data recording tasks and smaller sampling rates for relaying algorithms. Both the modern protective relay and the discrete wavelet transform (DWT) use digital devices to change the sampling rate during stages of synthesis and analysis.

The above-mentioned are just a few examples in which the sampling rate may require to be changed. In the following section the basics of such signal processing principles are described. Firstly, the basic digital signal processing blocks used to increase and decrease the sampling rate are introduced, then the decimator and interpolator devices are discussed. Subsequently, examples will illustrate how to change this sampling rate to a fractional value and finally to an arbitrary value. Real-time sampling rate alterations are also considered.

6.2 Basic Blocks for Sampling Rate Alteration

In digital signal processing the main devices used for sampling rate alteration are the down-sampler and the up-sampler, depicted by Figure 6.1a and b, respectively. The down-sampler device reduces the sampling rate by a factor of M, achieved by selecting every Mth sample from the original signal and discarding the remainder. The equation that describes this

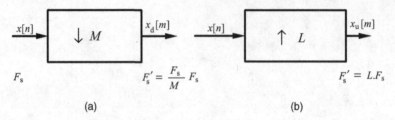

(a)   (b)

Figure 6.1 Change sampling frequency device: (a) down-sampler and (b) up-sampler.

operation in time domain is given by (6.1).

$$x_d[m] = x[Mm]. \tag{6.1}$$

The up-sampler device is responsible for the increase of the sampling rate by a factor of L. The output sequence can be written in the time domain as a relation of the input sequence as:

$$x_u[m] = \begin{cases} x[m/L], & m = 0, \pm L, \pm 2L, \ldots \\ 0, & \text{otherwise.} \end{cases} \tag{6.2}$$

6.2.1 Frequency Domain Interpretation

The frequency domain interpretation is very helpful for an understanding of the mechanism of sampling frequency alteration. The first step is to write Equation (6.2) in the z domain, obtained by direct application of the z-transform definition:

$$X_u(z) = \sum_{m=-\infty}^{\infty} x_u[m] z^{-m}. \tag{6.3}$$

The non-zero terms in the summation are those for which m is an integer multiple of L. As such, Equation (6.3) can be rewritten:

$$X_u(z) = \sum_{n=-\infty}^{\infty} x[n] z^{-nL} = X(z^L). \tag{6.4}$$

The reader can easily derive the above equation by writing down some terms of the summation.

The steps to arrive at the relationship between the down-sampled sequence and the original is a little more awkward and is left as an exercise for the reader. The reader may refer to chapter 13 of reference [1] to check the steps of the demonstration. The final expression is given by:

$$X_d(z) = \frac{1}{M} \sum_{k=0}^{M-1} X\left(W_M^k z^{1/M}\right) \tag{6.5}$$

where $W_M = e^{-j2\pi/M}$.

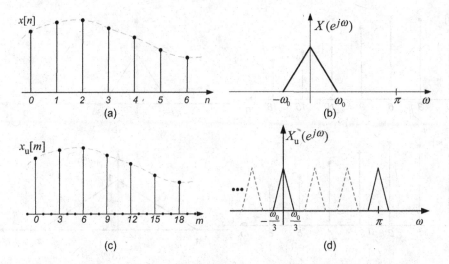

Figure 6.2 The up-sampling operation: (a) original sequence in discrete-time domain; (b) frequency spectrum of the original sequence; (c) up-sampling sequence in the time domain; and (d) up-sampling sequence in the frequency domain.

6.2.2 Up-Sampling in Frequency Domain

A frequency response is obtained from the z response, by taking $z = e^{j\omega}$. Then Equation (6.4) in the frequency domain is given by:

$$X_u(e^{j\omega}) = X(e^{j\omega L}).$$
(6.6)

Figure 6.2 illustrates the up-sampling operation in time and frequency domains for a particular up-sampling factor of $L = 3$. Figure 6.2a depicts a discrete-time sequence $x[n]$, and Figure 6.2b is the corresponding frequency response. Figure 6.2c is the up-sampled sequence in the discrete-time domain. The up-sampler device adds $L - 1$ zeros between each pair of the original samples, increasing the sampling rate by L. Figure 6.2d shows what happens in the frequency domain. The original spectrum is compressed by a factor L and $L - 1$ images of the original spectrum appear between 0 and π frequency. (Remember that the frequency content of a digital signal is evaluated within the range $-\pi$ to π.) The high-frequency component that appears in the new spectrum is coherent with the zero samples added to the original signal.

6.2.3 Down-Sampling in Frequency Domain

If considering the spectrum of a down-sampled signal, Equation (6.5) in the frequency domain is:

$$X_d(e^{j\omega}) = \frac{1}{M} \sum_{k=0}^{M-1} X\left(e^{j\frac{(\omega - 2\pi k)}{M}}\right).$$
(6.7)

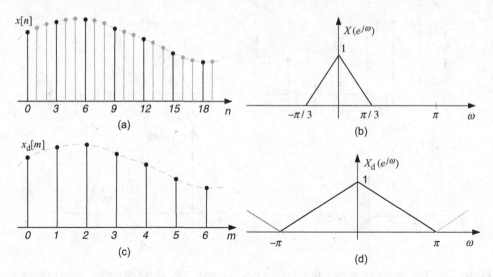

Figure 6.3 The down-sampling operation: (a) original sequence in discrete-time domain; (b) frequency response of the original sequence; (c) down-sampling sequence in the time domain; and (d) down-sampling sequence in the frequency domain.

For $M = 3$:

$$X_d\left(e^{j\omega}\right) = \frac{1}{3}\left[X\left(e^{j\frac{\omega}{3}}\right) + X\left(e^{j\frac{(\omega-2\pi)}{3}}\right) + X\left(e^{j\frac{(\omega-4\pi)}{3}}\right)\right]. \tag{6.8}$$

Equation (6.8) shows that the frequency response of the decimated signal is the mean of three uniformly shifted and stretched versions of the original spectrum. The illustration in Figure 6.3 depicts a particular case in which the original signal is band limited in the rage of $-\pi/3$ to $\pi/3$. In Figure 6.3a and b the original signal is in the discrete-time and frequency domains, respectively. In Figure 6.3c and d the decimated signal is in the discrete-time and frequency domain.

The reader is challenged to carefully analyze Figure 6.3d. The spectrum of the first term in the summation of Equation (6.8) is exactly as presented in the figure. The spectrum of the second term is equal to the first, shifted by 2π. The frequency response of the discrete signal is however periodic in 2π, where the shifted version coincides with the spectrum of the first term. The third term in Equation (6.8) is a 4π-shifted version of the first term and consequently coincides with the spectrum of the first term. The scaling term in the equation makes the frequency response sum 1 at 0 frequency. In this case there was no distortion because the spectrum of the original signal was band-limited to $\pm\pi/3$. Aliasing due to factor of M down-sampling is absent if and only if the signal is band-limited to $\pm\pi/M$.

6.3 The Interpolator

In Figure 6.2d it can be noted that the spectrum of the up-sampled signal presents undesired images. These need to be removed in order to obtain an interpolated signal. This is carried out through a digital low-pass filter with a pass-band edge frequency equal to ω_0/L and stop-band

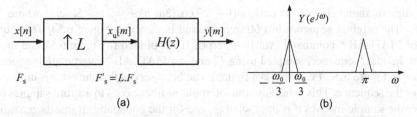

Figure 6.4 The interpolator: (a) interpolator structure; and (b) spectrum of the interpolated signal.

edge frequency equal to ω_s. Figure 6.4a shows the up-sampler followed by a low-pass filter. This configuration is known as the *interpolator structure*. In Figure 6.4b the spectrum of the interpolated signal is shown. Note that the images were removed. As the digital π frequency corresponds to $F'_s/2 = LF_s/2$, the digital frequency ω_0/L corresponds to $\omega_0 F_s/2\pi$,

$$\pi \leftrightarrow LF_s/2$$
$$\omega_0/L \leftrightarrow f_0 = \frac{\omega_0 F_s}{2\pi}. \tag{6.9}$$

The relation (6.9) and Figure 6.4b show that the interpolated signal is equivalent to the sample in the original continuous-time signal and has a sampling rate of LF_s.

Whenever a digital signal needs to be interpolated because it is sampled at a different sampling rate than that required by the application, the MATLAB® function *interp.m* can be used. For off-line applications the *interp* function should be used, and its basic form is: $y = $ interp(x, L). The *interp* function resamples the data at a higher rate using internal low-pass interpolation and the following algorithm, as described in the MATLAB® help pages:

1. The input vector is expanded to the correct length by inserting $L - 1$ zeros between the original data values.
2. A special symmetric FIR filter is designed that allows the original data to pass through unchanged and interpolates between so that the mean-square errors between the interpolated points and their ideal values are minimized.
3. The filter is applied to the input vector in order to produce the interpolated output vector.

The coefficients of the symmetric filter B can be obtained by using the *interp* command in the form $[y,B] = $ interp(x,L), which returns the coefficients of the interpolation filter B.

Example 6.1

Consider the continuous-time signal given by

$$x(t) = \cos(\Omega_0 t) + 0.3 \cos(3\Omega_0 t) + 0.2 \cos(5\Omega_0 t),$$

where $\Omega_0 = 120\pi$ rad/s. This signal was originally sampled using the sampling rate of $F_s = 900$ Hz. The signal is to be resampled at frequency of $F'_s = 5F_s$.

The digital signal is $x[n] = \cos(\omega_0 n) + 0.3 \cos(3\omega_0 n) + 0.2 \cos(5\omega_0 n)$, where $\omega_0 = 120\pi/F_s$. The original sequence has 60 samples and is shown in Figure 6.5a. The sequence using the MATLAB® command $[xint] = \text{interp}(x,10)$ is plotted in Figure 6.5b and the error between the ideal sequence, sampled using F'_s and the MATLAB® interpolated sequence, is presented in Figure 6.5c. From this last figure it can be observed that the error is higher at the border of the sequence. This is because the interpolator filter needs to acquire samples before and after the sample indexes it is interpolating, and for the boundaries it inserts zeros for the samples outside the limits.

6.3.1 The Input–Output Relation for the Interpolator

The input–output relationship of the interpolator structure of Figure 6.4 in the time domain is developed in the following. Equation (6.2) can be rewritten

$$x_u[Lm] = x[m], \quad (m = 0, \pm 1, \pm 2, \ldots). \tag{6.10}$$

The output of the interpolator is given by the convolution summation:

$$y[m] = \sum_{k=-\infty}^{\infty} h[m-k]x_u[k] \tag{6.11}$$

and if the non-zero terms in Equation (6.11) are multiples of L, then

$$y[m] = \sum_{k=-\infty}^{\infty} h[m-Lk]x_u[Lk]. \tag{6.12}$$

By using Equation (6.10) the input–output relationship for the interpolator is given by:

$$y[m] = \sum_{k=-\infty}^{\infty} h[m-Lk]x[k]. \tag{6.13}$$

The input–output relationship in the z domain is given by:

$$Y(z) = H(z)X(z^L). \tag{6.14}$$

6.3.2 Multirate System as a Time-Varying System and Nobles Identities

It must be emphasized that an important characteristic of multirate systems is that they are time-varying systems. The characteristic responsible for this is the presence in multirate systems of the up- and/or down-sampler devices. The reader is invited to demonstrate the time-varying behavior of the down-sampler and up-sampler. At this point, it is of utmost importance to remind the reader that the *cascade of time-varying system is not interchangeable*. This property is valid only for linear-time-invariant systems. This means that

$$H(z)X(z^L) \neq X(z^L)H(z).$$

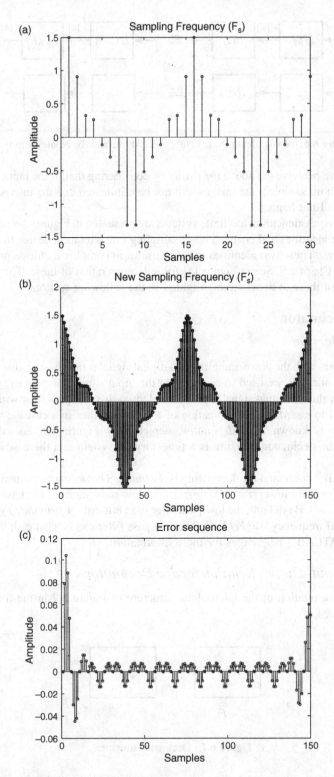

Figure 6.5 Plots of the interpolator: (a) input sequence at low sampling rate; (b) the interpolated sequence $L = 5$; and (c) the error sequence.

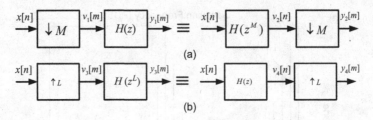

(a)

(b)

Figure 6.6 Noble identities: (a) Noble identity 1; and (b) Noble identity 2.

The reader can perceive the above inequality by considering that, if the input signal is pre-filtered and then up-sampled, the images will not be eliminated and the interpolation of the zero samples will not happen.

Two cascade equivalencies in multirate systems are presented in Figure 6.6a and b. They are known as noble identities and enable a basic sampling rate alteration device to be applied in multirate systems. These two identities are very useful in complex multirate processing and will be used in Chapter 9. See reference [1] for a demonstration of these. The time-varying characteristics of these systems are highlighted in the following sections.

6.4 The Decimator

6.4.1 Introduction

The main problem with the down-sampling of a digital signal is the aliasing, discussed before. To prevent this aliasing, we need to be sure that the input signal has no energy higher than ω_0/M. To verify this, the input signal is pre-filtered through a low-pass filter with a stop-band frequency equal to ω_0/M. The combination of a low-pass filter in a cascade with a down-sampler device is known as a *decimator*, depicted by Figure 6.7. As commented in Section 6.3.2, the decimator structure is a time-varying system and the cascade cannot be replaced.

The MATLAB® command for decimating is *decimate*. The *decimate* command resamples the input data signal at a lower rate ($1/M$ times), after low-pass filtering. The basic command is $y = \text{decimate}(x, M)$. By default, the low-pass filter is an 8th-order *Chebyshev Type I low-pass filter* with cutoff frequency $0.8(F_s/2)/L$. The low-pass filter can be changed; the reader can explore the MATLAB® help pages for more information.

6.4.2 The Input–Output Relation for the Decimator

The input–output relation of the interpolator structure of Figure 6.7 in the time domain is presented below. Let

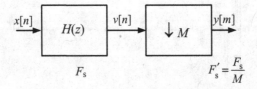

Figure 6.7 Decimator structure.

$$v[n] = \sum_{k=-\infty}^{\infty} h[n-k]x[k]. \tag{6.15}$$

If the output of decimator is given by

$$y[m] = v[Mm] \tag{6.16}$$

then

$$y[m] = \sum_{k=-\infty}^{\infty} h[Mm-k]x[k]. \tag{6.17}$$

The input–output relationship in the z domain is given by:

$$Y(z) = \frac{1}{M} \sum_{k=0}^{M-1} H\left(z^{1/M}W_M^k\right)X\left(z^{1/M}W_M^k\right). \tag{6.18}$$

6.5 Fractional Sampling Rate Alteration

The interpolator and decimator are used to change the sampling rate by an integer factor; L and M must be integers in these structures. However, in practical applications we may need to change the sampling rate by a non-integer factor. If the sampling rate alteration is a fractional number, then the adjustment can be obtained by cascading an interpolator with a decimator structure as depicted in Figure 6.8.

In Figure 6.8 $H_i(z)$ and $H_d(z)$ are the interpolator and decimator filters, respectively. They correspond to the time-invariant part of the system. Since they operate at the same sampling rate, they can be associated with one single equivalent filter. Although the design of the interpolator and decimator filters are outwith the scope of this book, it is worthwhile noting that, instead of designing the filters separately and then connecting them to obtain the equivalent filter, the best approach is to design the equivalent cascade directly.

6.5.1 Resampling Using MATLAB®

The MATLAB® command used for alteration/adjustment of the fractional sampling rate is the *resample* function. Its general form is $y = \text{resample}(x, L, M)$. The resample command resamples the sequence on vector x L/M times. The resampling function uses an internal filter; as such, the user does not need to worry about the design of the equivalent filter. The equivalent filter is a FIR filter and the phase delay is internally compensated. The output sequence y is L/M times the length of x. L and M must be integers. In the resampling process, as in interpolator or decimator processing, the samples before and after the limit of the

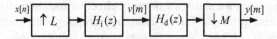

Figure 6.8 Fractional interpolator.

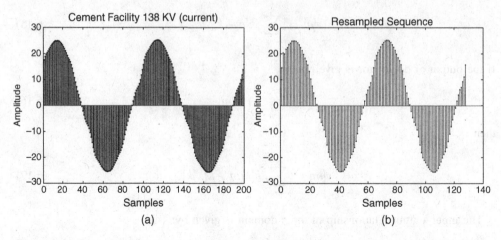

Figure 6.9 Example of fractional resampling: (a) original sequence; and (b) resampled sequence.

sequence are assumed to be zeros. The process can therefore present large errors at the boundaries of the sequences.

Example 6.2   Resampling a signal using fractional sampling rate alterations

The digital signal shown in Figure 6.9a represents the current of an industrial cement facility (138 kV), acquired using 100 samples per cycle. The fundamental frequency is 60 Hz. This signal will be used for testing an algorithm that works at 64 samples per cycle.

The fractional alteration in the sampling rate is given by:

$$k = \frac{64}{100} = \frac{16}{25}.$$

Then $L = 16$ and $M = 25$ and the resampled sequence is given by:

$$y = \text{resample}(x, 16, 25)$$

The new sequence is shown in Figure 6.9b.

6.6   Real-Time Sampling Rate Alteration

The interpolator, decimator and fractional structures shown before can be used in a real-time application. However, these are limited to integers or fractional sampling alterations. In some real-time applications the conversion ratio may be any arbitrary number and may be slowly time-varying when the frequency of the system is time-varying. For such conventional situations, the procedures shown above are not very effective and alternative methods must be considered. The general principle for an arbitrary sampling rate conversion (ASRC) is described in references [2,3,4]. Some power system applications are described in references [5,6].

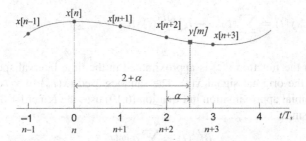

Figure 6.10 Illustration of real-time resampling processing.

In this section the basic principle for real-time and non-fractional sampling rates are introduced. The interpolation method shown below uses spline functions. The spline interpolation is an approximation technique which consists of dividing the interval of interest into several subintervals, interpolating as smoothly as possible subintervals by means of low-order polynomials.

Figure 6.10 illustrates the original samples $x[n]$ and the new samples that need to be estimated, $y[n]$, using the new sampling time T_s'. The example shown in Figure 6.10 illustrates the interpolated sample $y[m]$, requiring the original samples $x[n-1]$ to $x[n+3]$. Note that m has been used as the index of the interpolated sequence $y[m]$ and n as index of the original sequence $x[n]$.

The dynamic resampling (DR) of signal $x[n]$ originally sampled with a frequency F_s consists of the following steps:

1. estimation of the fundamental frequency f_1 from the samples of the input signal $x[n]$;
2. calculation of the resampling moments α;
3. resampling of the signal $x[n]$ at the moment $2+\alpha$ using interpolation in time domain, giving rise to the output signal $y[n]$.

The resampling moments $t_r[i]$ should be spaced in time according to the estimated value of the fundamental frequency $\hat{f}_1$. This should be accomplished in such a manner that the correspondent period $\hat{T}_1$ always contains an integer number N of output samples $y_r[i]$. The output sampling rate F_s' will therefore be equal to $N\hat{f}_1$. The resampling moment $t_r[i]$ is determined from the arbitrary delay $t_d[i]$, $(0 \leq t_d[i] < T_s)$ (where T_s is the original sampling period) which is an offset from the time instant $(n-1)T_s$ of the previous input. The procedure for the calculation of the resampling moments is shown in Figure 6.10, and is described in detail in the next section.

6.6.1 Spline Interpolation

The fundamental idea behind spline interpolation is finding a polynomial of order L to model the continuous functions between two consecutive samples, under the condition that the endpoint values of the polynomial are equal in the two samples and that all derivatives up to order $L-1$ are continuous across the boundary between these two intervals. The polynomial approximation using the spline function is [1,7]:

$$y(t) = \sum_{k=m}^{N+m}(t_{N+k} - t_k)B_k^{(L)}(t)x[t_k], \quad t_m \leq t < t_{N+m}. \tag{6.19}$$

In this equation the function $y(t)$ is approximated in the time interval specified and uses $N+1$ samples of the original signal $x(t)$. The samples used ($x(t_m)$ to $x(t_{N+m})$) are called *knots*. The polynomial approximation in Equation (6.19) uses the Lth order B-spline function that is obtained using:

$$B_k^{(L)}(t) = \sum_{i=k}^{N+k} a_i\phi_i(t) \tag{6.20}$$

where $\phi_i(t)$, referred to as the truncated power function, is a piecewise polynomial of degree L, defined:

$$\phi_i(t) = (t - t_i)_+^L = \begin{cases} 0, & t < t_i \\ (t - t_i)^L, & t \geq t_i. \end{cases} \tag{6.21}$$

The expression $(f(t, t_i))_+$ indicates that the function is zero for $t < t_i$ and equal to $f(t, t_i)$ for $t > t_i$. The coefficients a_i are determined using the fact that Equation (6.20) must be zero for $t > t_{N+m}$ [1]:

$$\sum_{i=m}^{N+m} a_i(t - t_i)^L = 0. \tag{6.22}$$

In signal processing it is common that the knots used in previous equations are equally spaced, which leads to further simplification. If we then assume that the samples are approximating an uniform sampling time, the equation can be rewritten as $x(t_i) = x(iT_s)$ or simply as $x[i]$ and $t_i = iT_s$.

Equation (6.22) for cubic spline ($L = 3$) and choosing $N = 4$ can be written:

$$\sum_{i=m}^{4+m} a_i(t - t_i)^3 = 0 \tag{6.23}$$

$$a_m(t - t_m)^3 + a_{m+1}(t - t_{m+1})^3 + a_{m+2}(t - t_{m+2})^3 + a_{m+3}(t - t_{m+3})^3$$
$$+ a_{m+4}(t - t_{m+4})^3 = 0. \tag{6,24}$$

After algebraic manipulation, if the coefficient of $t^k (k = 0, 1, 2, 3)$ must be zero, the matrix form is obtained:

$$\begin{bmatrix} 1 & 1 & 1 & 1 & 1 \\ t_m & t_{m+1} & t_{m+2} & t_{m+3} & t_{m+4} \\ (t_m)^2 & (t_{m+1})^2 & (t_{m+2})^2 & (t_{m+3})^2 & (t_{m+4})^2 \\ (t_m)^3 & (t_{m+1})^3 & (t_{m+2})^3 & (t_{m+3})^3 & (t_{m+4})^3 \end{bmatrix} \begin{bmatrix} a_m \\ a_{m+1} \\ a_{m+2} \\ a_{m+3} \\ a_{m+4} \end{bmatrix} = \begin{bmatrix} 0 \\ 0 \\ 0 \\ 0 \end{bmatrix}. \tag{6.25}$$

The above equation is an underdetermined system, which means that we can choose any coefficient and find the others as a function of the first. For example, let the coefficient a_{m+4} be the free coefficient; then the others coefficients are:

$$
\begin{bmatrix}
1 & 1 & 1 & 1 \\
t_m & t_{m+1} & t_{m+2} & t_{m+3} \\
(t_m)^2 & (t_{m+1})^2 & (t_{m+2})^2 & (t_{m+3})^2 \\
(t_m)^3 & (t_{m+1})^3 & (t_{m+2})^3 & (t_{m+3})^3
\end{bmatrix}
\begin{bmatrix}
a_m \\
a_{m+1} \\
a_{m+2} \\
a_{m+3}
\end{bmatrix}
= -a_{m+4}
\begin{bmatrix}
1 \\
t_{m+4} \\
(t_{m+4})^2 \\
(t_{m+4})^3
\end{bmatrix}.
\tag{6.26}
$$

By Cramer's rule, we have

$$
a_m = -a_{m+4} \frac{(t_{m+4} - t_{m+1})(t_{m+4} - t_{m+2})(t_{m+4} - t_{m+3})}{(t_m - t_{m+1})(t_m - t_{m+2})(t_m - t_{m+3})}.
\tag{6.27}
$$

By choosing the free parameter to be

$$
a_{m+4} = \frac{1}{(t_{m+4} - t_m)(t_{m+4} - t_{m+1})(t_{m+4} - t_{m+2})(t_{m+4} - t_{m+3})},
\tag{6.28}
$$

we arrive at

$$
a_m = \frac{1}{(t_m - t_{m+1})(t_m - t_{m+2})(t_m - t_{m+3})(t_m - t_{m+4})}.
\tag{6.29}
$$

If $t_i = iT_s$, then

$$
a_m = \frac{1}{24(T_s)^4}.
\tag{6.30}
$$

In a similar manner the final expression for the other coefficients can be found as:

$$
a_{m+1} = -\frac{1}{6(T_s)^4}
\tag{6.31}
$$

$$
a_{m+2} = \frac{1}{4(T_s)^4}
\tag{6.32}
$$

$$
a_{m+3} = -\frac{1}{6(T_s)^4}
\tag{6.33}
$$

and

$$
a_{m+4} = \frac{1}{24(T_s)^4}.
\tag{6.34}
$$

The cubic B-spline is then defined:

$$B_k^{(3)}(t) = \sum_{i=k}^{4+k} a_i(t - t_i)_+^3 \tag{6.35}$$

and the interpolating polynomial is

$$y(t) = \sum_{k=m}^{4+m} (t_{4+k} - t_k)B_k^{(3)}(t)x[t_k], \quad t_m \le t < t_{4+m}. \tag{6.36}$$

In a general case, the coefficients are given by [1]:

$$a_i = \frac{(-1)^{L+1}}{\prod_{k=m,i\ne k}^{N+m}(i - k)(T_s)^{L+1}}, \quad m \le i \le N+m \tag{6.37}$$

and the general Lth B-spline function is of the form:

$$B_k^{(L)}(t) = \sum_{i=k}^{N+k} a_i(t - t_i)_+^L. \tag{6.38}$$

6.6.2 Cubic B-Spline Interpolation

Figure 6.10 depicts an interpolation which has been obtained by using the cubic B-spline. The value of m in Equations (6.35)–(6.37) is $m = -1$ and the interpolated sample is $y[m]$. In the figure it is noted that the interpolation processing of $y[m]$ requires the original samples from $x[n - 1]$ to $x[n + 2]$. The time axis is normalized by T_s. Rewriting Equation (6.36) using the values in Figure 6.10, we have:

$$y[m] = y((2 + \alpha)T_s) = (4T_s) \sum_{k=-1}^{3} B_k^{(3)}((2 + \alpha)T_s)x[t_k] \tag{6.39}$$

and

$$B_k^{(3)}((2 + \alpha)T_s) = T_s^3 \sum_{i=k}^{4+k} a_i(2 + \alpha - i)_+^3. \tag{6.40}$$

Note that the sampling period T_s will be canceled. Furthermore, the fourth power of the sampling appears in a_i coefficients. Equations (6.30)–(6.34) will then become

$$a_{-1} = \frac{1}{24}, a_0 = -\frac{1}{6}, a_1 = \frac{1}{4}, a_2 = -\frac{1}{6} \text{ and } a_3 = \frac{1}{24} \tag{6.41}$$

and

$$y[m] = 4\Big\{ B_{-1}^{(3)}(2+\alpha)x[n-1] + B_0^{(3)}(2+\alpha)x[n] + B_1^{(3)}(2+\alpha)x[n+1]$$
$$+ B_2^{(3)}(2+\alpha)x[n+2] + B_3^{(3)}(2+\alpha)x[n+3] \Big\}. \tag{6.42}$$

Each term of Equation (6.42) is computed by using Equation (6.40). The term T_s can be neglected, because it will ultimately be canceled by computing:

$$B_{-1}^{(3)}(2+\alpha) = \sum_{i=-1}^{3} a_i(2+\alpha-i)_+^3$$

$$B_{-1}^{(3)}(2+\alpha) = a_{-1}(2+\alpha+1)_+^3 + a_0(2+\alpha)_+^3 + a_1(2+\alpha-1)_+^3 \tag{6.43}$$
$$+ a_2(2+\alpha-2)_+^3 + a_3(2+\alpha-3)_+^3.$$

Using Equation (6.21) for each term in Equation (6.43),

$$(2+\alpha-i)_+^3 = \begin{cases} 0, & 2+\alpha < i \\ (2+\alpha-i)^3, & 2+\alpha \geq i \end{cases} \tag{6.44}$$

we arrive at:

$$B_{-1}^{(3)}(2+\alpha) = a_{-1}(\alpha+3)^3 + a_0(\alpha+2)^3 + a_1(\alpha+1)^3 + a_2(\alpha)^3. \tag{6.45}$$

Simplification yields:

$$B_{-1}^{(3)}(2+\alpha) = -\frac{\alpha^3}{6} + \frac{\alpha^2}{2} - \frac{\alpha}{2} + \frac{1}{6}. \tag{6.46}$$

In the same way,

$$B_0^{(3)}(2+\alpha) = \frac{\alpha^3}{2} - \alpha^2 + \frac{2}{3}, \tag{6.47}$$

$$B_1^{(3)}(2+\alpha) = -\frac{\alpha^3}{2} + \frac{\alpha^2}{2} + \frac{\alpha}{2} + \frac{1}{6}, \tag{6.48}$$

$$B_2^{(3)}(2+\alpha) = \frac{\alpha^3}{6} \tag{6.49}$$

and

$$B_3^{(3)}(2+\alpha) = 0. \tag{6.50}$$

The interpolated value is then obtained from Equation (6.39),

$$y[m] = y((2 + \alpha)T_s) = B_{-1}^{(3)}(\alpha)x[n-1] + B_0^{(3)}(\alpha)x[n] + B_1^{(3)}(\alpha)x[n+1] + B_2^{(3)}(\alpha)x[n+2].$$
$$(6.51)$$

In the previous equation the function $B_i^{(3)}(2 + \alpha)$ was simply written $B_i^{(3)}(\alpha)$ by substituting Equations (6.46) and (6.49) into Equation (6.51), yielding:

$$y[m] = \alpha^3 \left(-\frac{1}{6}x[n-1] + \frac{1}{2}x[n] - \frac{1}{2}x[n+1] + \frac{1}{6}x[n+2] \right)$$
$$+ \alpha^2 \left(\frac{1}{2}x[n-1] - x[n] + \frac{1}{2}x[n+1] \right) + \alpha \left(-\frac{1}{2}x[n-1] + \frac{1}{2}x[n+1] \right) \quad (6.52)$$
$$+ \frac{1}{6}x[n-1] + \frac{2}{3}x[n] + \frac{1}{6}x[n+1].$$

In the z domain:

$$Y(z) = \left(\alpha^3 H_0(z) + \alpha^2 H_1(z) + \alpha H_2(z) + H_3(z) \right)X(z) \quad (6.53)$$

where $H_i(z), i = 0, 1, 2, 3$ can be easily identified from Equation (6.52).

Figure 6.11 shows an efficient implementation of Equation (6.53), which is referred to as the Farrow structure [7]. The figure displays a slight alteration from its original structure as proposed in [7], and the feedback branch is controlled by the incorporated value of α. The listing below describes the fractional-rate interpolator, demonstrating how the Farrow structure is used:

1. Initialization: $T_s' = T_s$; $n = 0$; $m = 0$; $\alpha = 0$
2. Update T_s' and compute $\lambda = T_s'/T_s$
3. If $\alpha \le 1$, then calculate $y[m]$ from Equation (6.52);
   if $\alpha = \alpha + \lambda$, $m = m + 1$;
   else if $\alpha = \alpha - 1$, $n = n + 1$ and update filters memory with the new sample $x[n]$
4. Return to step (2).

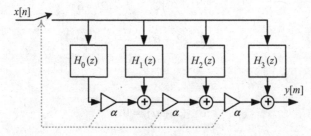

Figure 6.11 Implementation of the fractional-rate interpolator using the Farrow structure.

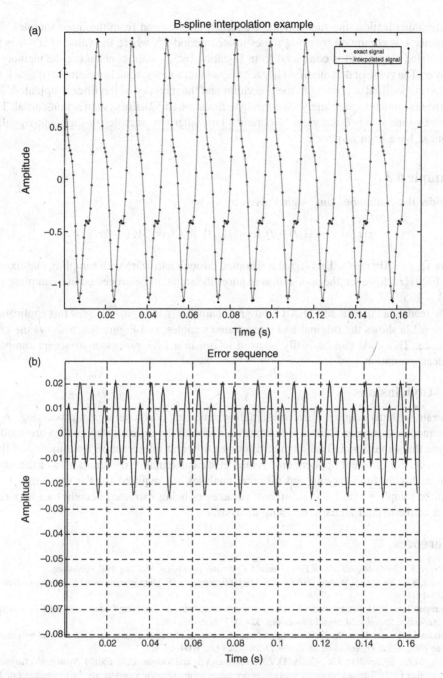

Figure 6.12 B-spline interpolation: (a) exact and interpolated samples; and (b) error sequence.

After initialization, the fundamental frequency is estimated from the input samples. The parameter λ is computed by using the estimated period T'_s, where the value of $\alpha \leq 1$ is the interpolated output and is computed from Equation (6.52) without changing the memory of the filter. The value of α is modified in each step so that a new output is generated. If $\alpha > 1$, the α value is modified according to the algorithm and the memory of the filter is up-dated. The algorithm continues indefinitely. According to Equation (6.52), the system is non-causal. This means that the interpolated value is obtained only after two samples of the input signal, at instant n, have been acquired.

Example 6.3

Consider the continuous-time signal given by

$$x(t) = \cos(\Omega_1 t) + 0.1\cos(2\Omega_1 t) + 0.3\cos(3\Omega_1 t + \pi/3),$$

where $\Omega_1 = 120\pi$ rad/s. This signal is sampled through hardware with sampling rate fixed at $F_s = 1920$ Hz. However, the application requires the signal to be processed at a sampling rate of $F'_s = 1952$.

The reader can use the MATLAB® program named *b_spline_interp.m* for this application. Figure 6.12a shows the original and interpolated samples, and Figure 6.12b shows the error sequence. This code can be easily adapted to run in a DSP processor to accept sampling frequency variation.

6.7 Conclusions

Multirate systems require different sampling rates for adequate signal processing. As a consequence, sampling alteration becomes a necessity as different techniques are used to analyze the same signal. This chapter discusses different interpolation techniques and their advantages for different applications. As the electrical signal of the smart grid of the future becomes highly time-varying and different analyses are required for the same signal, the adequate sampling rate for the different parameters being extracted becomes an important aspect of the overall signal processing analysis.

References

1. Mitra, S.K. (2006) *Digital Signal Processing: A Computer Approach*, McGraw-Hill Publishing.
2. Dooley, S.R., Stewart, R.W. and Durrani, T.S. (1999) Fast on-line B-spline interpolation. *Eletronics Letters*, **35**, 1130–1131.
3. Ramstad, T.A. (1984) Digital methods for conversion between arbitrary sampling frequencies. *IEEE Transactions on Acoustics, Speech and Signal Processing*, **32** (3), 577–591.
4. Borkowski, D. and Bien, A. (2009) Improvement of accuracy of power system spectral analysis by coherent resampling. *IEEE Transactions on Power Delivery*, **24** (3), 1004–1013.
5. Lima, M.A., deCarvalho, J.R., Coury, D.V., Cerqueira, A.S. and Duque, C.A. (2010) A method of dynamic resampling for DFT-based harmonic analysis under time-varying frequency conditions. In *Proceedings of 14th International Conference on Harmonics and Quality of Power (ICHQP)*. IEEE Conference Publications. pp. 1–6.
6. Zolzer, U. (1997) *Digital Audio Signal Processing*, John Wiley & Sons, New York.
7. Cucchi, S., Desinan, F. and Sicuranza, G. (1991) DSP implementation of arbitrary sampling frequency conversion for high quality sound application. In *Proceedings of IEEE International Conference on Acoustics, Speech, and Signal Processing*, Toronto, Canada, pp. 3609–3612.

7

Estimation of Electrical Parameters

7.1 Introduction

In the new scenario of smart grids more signal processing methods for electrical parameters are needed to keep the network under control and operating at the desired quality of the service (QoS) and reliability. For this purpose, electrical parameters are reviewed from a more complex environment where the new frequency components added to the fundamental experience higher variations, affecting measurement-corrupting noise. Furthermore, the unpredictability and non-feasibility of monitoring all system parameters at all locations demands analytical tools for the state estimation of system parameters. In this way, the estimation and further processing of electrical power systems parameters becomes an essential feature of power systems analysis. This chapter presents the theory and the techniques of signal processing procedures for the estimation of electrical parameters in the complex context of smart grids.

7.2 Estimation Theory

Estimation theory deals with the estimation of the quantities of interest from a given finite set of measurements that are noise corrupted. This section introduces some basic concepts of estimation theories and presents the main methods used in power system signal processing. A complete treatment of estimation theory can be found in reference [1].

N measurements are collected from a raw observation dataset, which may contain uncertain values due to sensor inaccuracies, additive noise and so on. If these data are to be used to estimate parameters of interest, the data vector of N samples and the parameter vector of p parameters are given by

$$\mathbf{x} = [x[1] \quad x[2] \ldots x[N-1] \quad x[N]]^t \qquad (7.1)$$

$$\boldsymbol{\theta} = [\theta[1] \quad \theta[2] \ldots \theta[N-1] \quad \theta[p]]^t. \qquad (7.2)$$

Power Systems Signal Processing for Smart Grids, First Edition. Paulo Fernando Ribeiro, Carlos Augusto Duque, Paulo Márcio da Silveira and Augusto Santiago Cerqueira.
© 2014 John Wiley & Sons, Ltd. Published 2014 by John Wiley & Sons, Ltd.
Companion Website: http://www.wiley.com/go/signal_processing/

The estimator $\hat{\boldsymbol{\theta}}$ is the mathematical expression or function by which the parameters can be estimated, where

$$\hat{\boldsymbol{\theta}} = \mathbf{g}(\mathbf{x}). \tag{7.3}$$

Fortunately, a large number of power system estimation problems can be represented by a linear model, relating the samples (measurements) and the parameters by

$$\mathbf{x} = \mathbf{H}\boldsymbol{\theta} + \boldsymbol{\varepsilon} \tag{7.4}$$

where $\mathbf{H}$ is an $N \times p$ matrix referred to as the observation matrix and $\boldsymbol{\varepsilon}$ is a noise vector of dimension $N \times 1$. The objective is to estimate $\boldsymbol{\theta}$, known as the *optimum estimator*, as accurately as possible according to a specific minimization criterion. This optimal estimator can sometimes be difficult to obtain however, due to its high computational requirements or due to the difficulty in identifying the statistical properties of $\boldsymbol{\varepsilon}$ if the chosen criterion should require it. In these situations, *suboptimal estimators* may be preferred.

An optimum estimator can be defined as one that is unbiased and has minimum variance. This estimator is known as the minimum variance unbiased (MVU) estimator, defined:

$$E\left(\hat{\boldsymbol{\theta}}\right) = \boldsymbol{\theta} \qquad \text{unbiased condition} \tag{7.5}$$

and

$$\hat{\boldsymbol{\theta}} = \operatorname*{argmin}_{\hat{\boldsymbol{\theta}}}\{\operatorname{var}(\hat{\boldsymbol{\theta}})\} \qquad \text{minimum variance condition} \tag{7.6}$$

where $E(x)$ is the expectation value of x and $\operatorname{var}(x)$ is the variance of x.

The main questions that arise in estimation theory are: (a) does the MVU estimator exist? and (b) how to find it? Both questions are difficult to answer, due to the fact that in some cases the MVU estimator does not exist and, when it does, there is no known procedure that will always produce it. However an important result in estimation theory is the Cramer-Rao lower bond (CRLB) which states that any unbiased estimator has a variance greater than or equal to the value given by the CRLB. The CRLB is very important in practice because it provides a way to compare the performance of any unbiased estimator. Furthermore, if the performance of a given estimator is equal to the CRLB, the estimator is a MVU estimator. In some specific cases it is possible to identify the estimator at the same time as the computation of the CRLB.

In the Cramer-Rao lower bound, the probability density function (pdf) $p(\mathbf{x}; \boldsymbol{\theta})$, must be known. If the following equality holds [1],

$$\frac{\partial \ln p(\mathbf{x}; \boldsymbol{\theta})}{\partial \boldsymbol{\theta}} = \mathbf{I}(\boldsymbol{\theta})(\mathbf{g}(\mathbf{x}) - \boldsymbol{\theta}) \tag{7.7}$$

then the estimator is

$$\hat{\boldsymbol{\theta}} = \mathbf{g}(\mathbf{x}) \tag{7.8}$$

where $\mathbf{I}(\boldsymbol{\theta})$ is a $p \times p$ matrix dependent only on $\boldsymbol{\theta}$ and defined as the Fisher information matrix. The elements (i,j) of the Fisher information matrix originate from

$$[\mathbf{I}(\boldsymbol{\theta})]_{i,j} = -E\left[\frac{\partial^2 \ln p(\mathbf{x}; \boldsymbol{\theta})}{\partial \theta_i \partial \theta_j}\right]. \tag{7.9}$$

In Equation (7.7), $\mathbf{g}(\mathbf{x})$ is a p-dimensional function of the data $\mathbf{x}$.

Finally, the minimum variance of the parameter $\hat{\theta}_i$ is given by

$$\mathrm{var}\left(\hat{\boldsymbol{\theta}}_i\right) = \left[\mathbf{I}^{-1}(\boldsymbol{\theta})\right]_{i,i}. \tag{7.10}$$

It may be the case that the MVU estimator cannot be found because the pdf of the data is unknown or because is not possible (or easy) to write the equation in the form given by Equation (7.7). In such cases one solution is to assume that the estimator model is linear in data, and the first- and second-order statistics are known (mean and variance). In such a situation we can find the best linear unbiased estimator (BLUE).

By using the general linear model described by Equation (7.4), assuming the noise vector with zero mean and covariance matrix of $\mathbf{C}$ (the pdf of $\mathbf{w}$ is arbitrary), then according to the Gauss-Markov Theorem [1] the BLUE of $\boldsymbol{\theta}$ is

$$\hat{\boldsymbol{\theta}} = \left(\mathbf{H}^t\mathbf{C}^{-1}\mathbf{H}\right)^{-1}\mathbf{H}^t\mathbf{C}^{-1}\mathbf{x} \tag{7.11}$$

and the minimum variance of the parameter θ_i is

$$\mathrm{var}(\hat{\theta}_i) = \left[\left(\mathbf{H}^t\mathbf{C}^{-1}\mathbf{H}\right)^{-1}\right]_{i,i}. \tag{7.12}$$

If the data are Gaussian in nature, then the BLUE is also the MVU estimator.

In some cases the MVU estimator may not exist or is unknown. Under such circumstances, the maximum likelihood estimator (MLE) is an alternative method when the pdf function $p(\mathbf{x}; \boldsymbol{\theta})$ is known. The parameters are found using the MLE by maximizing the pdf function:

$$\hat{\boldsymbol{\theta}} = \underset{\boldsymbol{\theta}}{\mathrm{argmax}}\{p(\mathbf{x}; \hat{\boldsymbol{\theta}})\} \ldots \tag{7.13}$$

If the model is linear and the pdf of $\boldsymbol{\varepsilon}$ is the normal function (Gaussian), $\mathcal{N}(0,C)$, then the MLE is the same as the BLUE, which is equal to the MVU, which holds for Equation (7.11).

7.3 Least-Squares Estimator (LSE)

The MVU, BLUE and MLE estimators require some information about the statistics of the data: the MVU and MLE require the pdf function and the BLUE requires the first- and second-order statistics. The least-squares estimator (LSE) does not need statistical information however, only the signal model. It is widely used in practice due to the ease of its implementation. As such, it deserves a separate section in the estimation theory review.

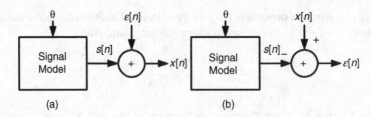

Figure 7.1 Linear model for LSE: (a) observed signal corrupted by noise; (b) the same model highlighting the error signal.

A disadvantage of the LSE is that it does not guarantee optimal performance and its statistics cannot be assessed, unless additional information is known about the probabilistic structure of the data. In the least-squares approach the signal model is described as a function of the unknown parameters:

$$s[n] = g(n; \boldsymbol{\theta}). \qquad (7.14)$$

Figure 7.1 illustrates this method. In Figure 7.1a, the signal is generated by a model that is a function of these parameters. The signal $s[n]$ is deterministic but, due to model inaccuracies and external noises, the observed signal $x[n]$ is a corrupted version of $s[n]$. The error signal is highlighted in Figure 7.1b.

The LSE estimator searches for $\boldsymbol{\theta}$ in order to minimize the error between $s[n]$ and the observed data $x[n]$ using the least-squares error criterion:

$$J(\boldsymbol{\theta}) = \sum_{n=0}^{N-1} (\varepsilon[n])^2 = \sum_{n=0}^{N-1} (x[n] - s[n])^2. \qquad (7.15)$$

The value of $\boldsymbol{\theta}$ that minimizes $J(\boldsymbol{\theta})$ is the LSE estimator. Note that no probabilistic assumptions have been made about the data $x[n]$; the only assumption needed to produce a good unbiased estimator is that the noise $\varepsilon[n]$ is zero-mean. An interesting variation of the LS problem is when a weight vector is used to evaluate Equation (7.15), resulting in a new cost function:

$$J(\boldsymbol{\theta}) = \sum_{n=0}^{N-1} w[n](x[n] - s[n])^2. \qquad (7.16)$$

7.3.1 Linear Least-Squares

Equation (7.14) is a generic relationship between the signal model and the parameters to be estimated. If this relationship is linear, then the solution is simple. In this case the relationship can be represented using the matrix:

$$\mathbf{s} = \mathbf{H}\boldsymbol{\theta} \qquad (7.17)$$

where $\mathbf{s} = [s[0] \quad s[1] \quad \cdots \quad s[N-1]]^t$ is the data vector generated, although there is no direct access to the vector, and $\boldsymbol{\theta}$ is the parameter vector and $\mathbf{H}$ is a known $N \times p$ matrix. When the noise is included we have:

$$\mathbf{x} = \mathbf{H}\boldsymbol{\theta} + \boldsymbol{\varepsilon}. \tag{7.18}$$

The cost function of Equation (7.16) can be written in matrix form as

$$J(\boldsymbol{\theta}) = (\mathbf{x} - \mathbf{H}\boldsymbol{\theta})^t \mathbf{W} (\mathbf{x} - \mathbf{H}\boldsymbol{\theta}) \tag{7.19}$$

where $\mathbf{W}$ is a diagonal matrix of dimension $N \times N$ with diagonal elements $[\mathbf{W}]_{ii} = w_i, (i = 0, 1, \ldots, N-1)$. The desired parameter estimation is found by differentiating Equation (7.19) and equating the result to zero, that is,

$$\left. \frac{\partial J(\boldsymbol{\theta})}{\partial \boldsymbol{\theta}} \right|_{\boldsymbol{\theta} = \hat{\boldsymbol{\theta}}} = \mathbf{0}. \tag{7.20}$$

After deriving the quadratic form of Equation (7.19) and equating it to zero, we have

$$-2\mathbf{H}^t\mathbf{W}\mathbf{x} + 2\mathbf{H}^t\mathbf{W}\mathbf{H}\boldsymbol{\theta} = \mathbf{0} \tag{7.21}$$

and the LSE can be found as

$$\hat{\boldsymbol{\theta}} = (\mathbf{H}^t\mathbf{W}\mathbf{H})^{-1}\mathbf{H}^t\mathbf{W}\mathbf{x}. \tag{7.22}$$

Equation (7.22) is the general case of the weighted LS method. If the weight matrix is the identity matrix, then the standard LS normal equation is obtained:

$$\hat{\boldsymbol{\theta}} = (\mathbf{H}^t\mathbf{H})^{-1}\mathbf{H}^t\mathbf{x}. \tag{7.23}$$

Note that in this equation the matrix $\mathbf{H}$ and the vector $\mathbf{x}$ are known. The following example will illustrate how the amplitude and phase of a sinusoid can be estimated using LSE.

Example 7.1

Find the phase and amplitude of a signal $s[n] = A\cos(\omega_1 n + \phi)$, where ω_1 is a known frequency.

At first glance the problem is non-linear and cannot be solved directly using Equation (7.23). However, the transformation procedure can be used to make the problem linear. For this, the observed signal can be written:

$$x[n] = a\cos(\omega_1 n) + b\sin(\omega_1 n) + \varepsilon[n] \quad (n = 0, 1, \ldots, N-1)$$
$$a = A\cos(\phi)$$
$$b = -A\sin(\phi).$$

The vector to be estimated contains parameters a and b. In vector form we have

$$
\begin{bmatrix} x[0] \\ x[1] \\ \vdots \\ x[N-1] \end{bmatrix} = \begin{bmatrix} 1 & 0 \\ \cos \omega_1 & \sin \omega_1 \\ \vdots & \vdots \\ \cos \omega_1 (N-1) & \sin \omega_1 (N-1) \end{bmatrix} \cdot \begin{bmatrix} a \\ b \end{bmatrix} + \begin{bmatrix} \varepsilon[0] \\ \varepsilon[1] \\ \vdots \\ \varepsilon[N-1] \end{bmatrix}
$$

or

$$
\mathbf{x} = \mathbf{H}\boldsymbol{\theta} + \boldsymbol{\varepsilon}.
$$

The LSE is obtained using Equation (7.23), where the amplitude and phase can be found directly from the parameters as:

$$
A = \sqrt{a^2 + b^2} \tag{7.24}
$$

and

$$
\phi = \tan^{-1}\left(\frac{-b}{a}\right). \tag{7.25}
$$

Consider the frequency ω_1 being substituted by $\omega_k = \dfrac{2\pi}{N}k$, $k = 1, 2, \ldots, (N/2 - 1)$, then the observation matrix becomes:

$$
\mathbf{H} = \begin{bmatrix} \mathbf{h_1} & \mathbf{h_2} \end{bmatrix} \tag{7.26}
$$

where

$$
\mathbf{h_1} = \begin{bmatrix} 1 & \cos \omega_k & \cdots & \cos \omega_k (N-1) \end{bmatrix}^t \tag{7.27}
$$

$$
\mathbf{h_2} = \begin{bmatrix} 0 & \sin \omega_k & \cdots & \sin \omega_k (N-1) \end{bmatrix}^t. \tag{7.28}
$$

The normal Equation (7.23) becomes

$$
\hat{\boldsymbol{\theta}} = \begin{bmatrix} \mathbf{h_1^t.h_1} & \mathbf{h_1^t.h_2} \\ \mathbf{h_2^t.h_1} & \mathbf{h_2^t.h_2} \end{bmatrix}^{-1} \cdot \begin{bmatrix} \mathbf{h_1^t} \\ \mathbf{h_2^t} \end{bmatrix} \cdot \mathbf{x}. \tag{7.29}
$$

It is easy to demonstrate the following identities:

$$
\mathbf{h_1^t} \cdot \mathbf{h_2} = \sum_{n=0}^{N-1} \cos\left(\frac{2\pi}{N}kn\right) \cdot \sin\left(\frac{2\pi}{N}kn\right) = 0 \tag{7.30}
$$

$$
\mathbf{h_2^t} \cdot \mathbf{h_1} = 0 \tag{7.31}
$$

$$\mathbf{h}_1^t \cdot \mathbf{h}_1 = N/2 \tag{7.32}$$

$$\mathbf{h}_2^t \cdot \mathbf{h}_2 = N/2 \tag{7.33}$$

Taking these identities into account, Equation (7.29) becomes

$$\hat{\boldsymbol{\theta}} = \frac{2}{N} \mathbf{H}^t \cdot \mathbf{x} \tag{7.34}$$

or

$$\hat{a} = \frac{2}{N} \sum_{n=0}^{N-1} x[n] \cos\left(\frac{2\pi}{N} kn\right) \tag{7.35}$$

$$\hat{b} = \frac{2}{N} \sum_{n=0}^{N-1} x[n] \sin\left(\frac{2\pi}{N} kn\right). \tag{7.36}$$

Equations (7.35) and (7.36) are identified as the real and imaginary elements of the discrete Fourier transform (DFT) of $x[n]$. These results reveal that the DFT of a signal

$$s[n] = A \cos\left(\frac{2\pi}{N} kn + \phi\right)$$

corresponds to the optimal LSE estimator regarding the estimation of the amplitude and phase. Note this affirmation holds only if the frequency is known and constant, a constraint that cannot be held for weak power systems.

7.4 Frequency Estimation

An important parameter of a power system is its fundamental frequency, due to the fact that it is generally used to indicate the system operation state. Furthermore, it is the basis for the estimation of several other parameters including the amplitude and phase of voltage and current signal. The accuracy of the frequency estimation is therefore of great importance in many applications of power systems such as effective power control, the setting of protective relays for load shedding and restoration, power quality monitoring and generation protection. The trade-off between accuracy and convergence time of estimation is dependent on its application. In control and protection for example, where the estimation time is the dominant factor, some error margin may be acceptable providing it does not compromise the final result. In monitoring applications however, accuracy is the major concern. Since transient and abnormal conditions such as harmonic distortions may occur in a power system, the waveform of voltage and current will not be pure sinusoids. As such, a conventional frequency estimator such as zero-crossing algorithms can produce large errors. Fast and accurate frequency estimators for distorted and time-varying signals are therefore a constant field of research.

It should be highlighted that various numerical algorithms for electrical parameter measurements and estimations are sensitive to frequency variations, for example those based on the discrete Fourier transform (DFT) or the least-mean-squares (LMS) technique that assumes that the system frequency is known in advance and constant (50 or 60 Hz) [2]. On the other hand, frequency estimation algorithms can produce erroneous results due to the variation of the amplitude or phase of the input signal. As such, further processing must be included in the estimator in order to eliminate this false variation. Finally, new challenges are becoming apparent due to the proliferation of dispersed and distributed generation as several estimation methods for electrical parameters are based on the fact that the fundamental frequency is almost constant; in weak power systems the frequency variation is however larger than in conventional and strong power systems. As such, the operational success of several algorithms that are used today in intelligent electronic devices) (IEDs) may require careful evaluation in this new context.

In this section we present a range of frequency estimators from the traditional zero-crossing to more complex estimators. Digital signal structures are presented and possible uses and related limitations discussed.

7.4.1 Frequency Estimation Based on Zero Crossing (IEC61000-4-30)

Traditionally, frequency is estimated by the time between two zero crossings. However, this method is relatively sensitive to distorted signals and further processing must be considered in order to minimize the effect of harmonic, interharmonic and others power quality disturbances. This method is introduced first due to its simplicity and the fact that, after some modifications, it can be a very attractive tool in terms of accuracy and convergence time. A modified version of frequency estimation based on zero crossing, one which is computationally efficient and very accurate for some classes of distorted signals, is presented. The other motivation for exploring zero crossing is that this is the method of choice in IEC 61000-4-30 to estimate power system frequency.

The first step is estimating the time between two or several zero crossings in the same direction. A 'zero crossing' means that the signal changes from a positive to a negative value (or vice versa), as depicted in Figure 7.2. From this figure the frequency can be found by simply taking $\hat{f}_1 = 1/T_{sc}$, where T_{sc} is the time between two consecutives zero crossings. More accurately, N_{zc} zero crossings can be taken in the same direction and the frequency estimated as $\hat{f}_1 = N_{zc}/T_{mc}$, where T_{mc} is the time between N_{zc} zero crossings in the same direction.

In the case of a digital frequency estimator, where the samples of the signal are acquired at the sampling time T_s, the algorithm is needed to identify the zero crossing and the time between them. Figure 7.3 shows a small portion of the sinusoid signal of frequency f_1 and some of its samples. $y_c[n-1]$ and $y_c[n]$ are the sample values when the signal changes from

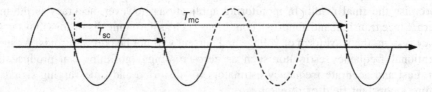

Figure 7.2 Zero-crossing approach.

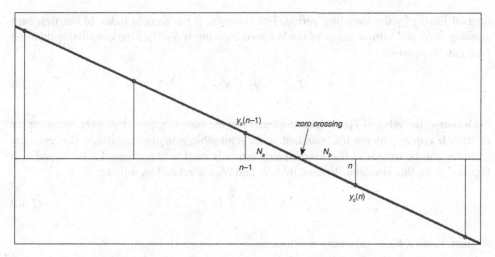

Figure 7.3 Samples in the zero-crossing approach: linear interpolation.

positive to negative. The indices of the samples are $n - 1$ and n, respectively. N_a and N_b are the fractional distance (i.e. N_a and $N_b < 1$) between the zero crossing and the sample $n - 1$ and the sample n, respectively, estimated by linear interpolation as described below.

Let the number of samples between two consecutive zero crossings in the same direction be N, then the estimated frequency is given by:

$$\hat{f}_1 = \frac{F_s}{N} \qquad (7.37)$$

where F_s is the sampling frequency.

Using this method, the accuracy of the estimation is a function of the sampling frequency. The higher the sampling frequency, the better is its estimation. The value of the sampling frequency is however limited by hardware costs and other procedures, so the sampling frequency cannot be increased arbitrarily. The error may also be reduced when the algorithm uses several consecutive zero crossings. Such an example is found in IEC 61000-4-30 [3], where the frequency is computed using the total number of zero crossings taking into account the time interval of 10 s divided by the cumulative duration of the integer cycles. In the case of a 60 Hz system, the number of zero crossings will be 600 if the actual frequency is 60 Hz; the cumulative duration is therefore 10 s. If the frequency is not equal to 60 Hz however, the number of integral cycles will be N_{zc} and the cumulative duration time will be T_{mc}. Then the frequency of the system is computed as:

$$\hat{f}_1 = \frac{N_{zc}}{T_{mc}} \qquad (7.38)$$

There are several strategies for computing T_{mc}. The easiest is by multiplying the total number of samples until the last zero crossing (the zero-crossing counter is stopped at the 10 s interval finish) by the sampling period. For example, if the sample index of the first zero crossing is N_1 and sample index of the last zero crossing is N_2, then the cumulative duration time can be approximated:

$$T_{\text{mc}} = (N_2 - N_1)T_{\text{s}}. \tag{7.39}$$

Of course, the value of T_{mc} is again dependent on the sampling time. However, the accuracy of 10 mHz required by the IEC standard is not achievable using this approach. However, the accuracy can be improved if a linear interpolation is used between samples $n - 1$ and n in Figure 7.3. In this situation, assume that N_a and N_b are related as follows:

$$N_a + N_b = 1. \tag{7.40}$$

Then, using a basic geometry relationship,

$$\frac{y_{\text{c}}[n - 1]}{N_a} = -\frac{y_{\text{c}}[n]}{N_b} \tag{7.41}$$

Substituting Equation (7.40) into Equation (7.41), we find that

$$N_a = \frac{y_{\text{c}}[n - 1]}{y_{\text{c}}[n - 1] - y_{\text{c}}[n]} \tag{7.42}$$

$$N_b = \frac{y_{\text{c}}[n]}{y_{\text{c}}[n] - y_{\text{c}}[n - 1]}. \tag{7.43}$$

Using a linear interpolation to find an approximation to the exact moment that the signal crosses the zero for both the first and the last zero crossing, N_{a1} and N_{a2} are found, respectively, according to Equation (7.42). Inserting these values into Equation (7.39), a better estimation is

$$T_{\text{mc}} = (N_2 + N_{a2} - N_1 - N_{a1})T_{\text{s}}. \tag{7.44}$$

Finally, the frequency estimation is given by Equation (7.38).

Example 7.2

A pure sinusoid signal is used to test the zero-crossing frequency estimation. From the program listed below, it is possible to modify several parameters including system frequency, the frequency of the actual signal, the sampling frequency and so on. The example calls the function zc_freq_iec61000.m (http://www.ufjf.br/pscope-eng/digital-signal-processing-to-smart-grids/) which returns the value of the estimated frequency.

```
%-----------------------------------------------------------------
% Example7_2.m Frequency estimation IEC 6100-4-30 using zero crossing

f1=60;                         % System frequency
Nppc=32;                       % Number of samples per cycle
Nc=602;                        % Number of cycles
Nt=Nc*Nppc;                    % Total of samples
fs=Nppc*f1;     Ts=1/fs;       % Sampling frequency and sampling time
fa=58.32;                      % Real frequency of the test signal
for n=1:100
    i=1:Nt;
    w=0.0071*randn(1,Nt);      % SNR about 40dB
    phas=(0.5*rand(1)-0.5)*pi; % phase varying between -pi and pi
    x=cos(2*pi*fa*i*Ts+phas)+w; %test signal
    f_est=zc_freq_iec61000(x,Nppc,f1);% function for frequency
estimation based on IEC6100-4-3
    err(n)=abs(fa-f_est);          % error
end
plot(err)
%-----------------------------------------------------------------
```

Figure 7.4 shows the estimation error for 100 simulated cases. The phase of the sinusoid falls between $-\pi$ and π. The error of the estimation in this procedure is smaller than 10 mHz, as required by IEC 61000-4-30.

7.4.2 Short-Term Frequency Estimator Based on Zero Crossing

In an application where the frequency needs to be estimated as quickly as possible, the zero-crossing method can still be used although it will be necessary to pre-filter the input signal

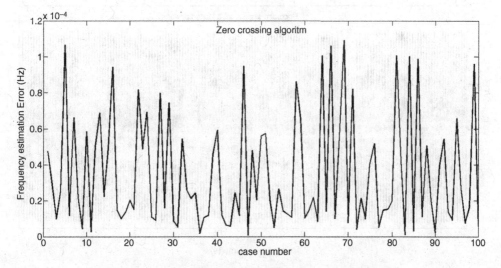

Figure 7.4 Frequency estimation error of the IEC-based method.

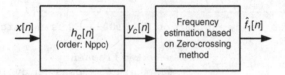

Figure 7.5 Short-term frequency estimation based on filtering and zero-crossing method.

using a low-pass filter to reduce the influence of the harmonic and interharmonics. Although the IEC standard uses a long-term estimator, it also requires the use of pre-filtering to attenuate the influence of unwanted components. The first candidate to be used to accomplish this task is the cosine filter presented in Chapter 4. This FIR filter is able to eliminate harmonic components and attenuates high-order interharmonics [4]. In this section, some advantages of this method for short-term estimation are explored.

Figure 7.5 shows the block diagram of the proposed method. The input signal is filtered using the cosine FIR filter with an impulse response of $h_c[n]$, where the filter output signal becomes the input signal for the frequency estimator. The frequency estimator finds the time between two consecutive zero crossings in the same direction, T_{sc}, using the interpolation method described at Section 7.4.1. The estimated frequency is then given by $\hat{f}_1 = 1/T_{sc}$.

Example 7.3

The MATLAB® file example7_3.m (http://www.ufjf.br/pscope-eng/digital-signal-process-ing-to-smart-grids/) implements the short-term frequency estimator based on zero crossings. The input signal is loaded from the file dados7_1_a.mat and the input variable v0 is saved in the file. Figure 7.6 shows the signal v0 which corresponds to a real signal recorded during a fault in a real power system. The sampling rate is $32 \times 60 = 1920$ Hz. Figure 7.7 shows the output of the frequency estimator.

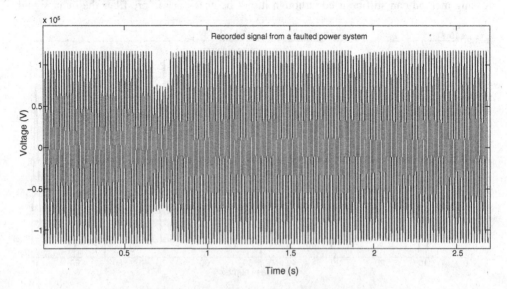

Figure 7.6 Real signal recorded from a faulted power system.

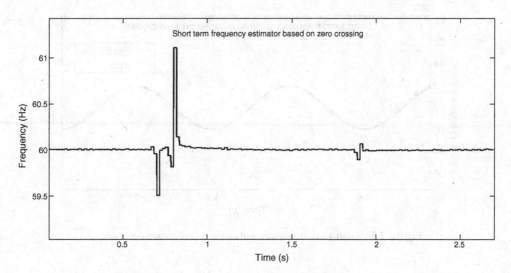

Figure 7.7 Frequency estimation from real case.

As can be observed from Figure 7.7 the frequency estimator is sensitive to sag variations. Variations such as that highlighted in the plot can be caused by a phase variation, common during a sag. The accuracy of the algorithm can be increased using a MAF (moving average filter) of M periods. As discussed in Chapter 4 the MAF smoothes the input signal and eliminates abrupt changes in the plot. The reader is invited to change the order of the MAF in example7_3.m (http://www.ufjf.br/pscope-eng/digital-signal-processing-to-smart-grids/). The reader is also invited to test the algorithm with other signals.

Example 7.3 illustrates an important characteristic that any frequency estimator must have: the capacity to track frequency changes. Step, ramp and sinusoidal changes are commonly used to test the tracking ability of the estimator. The reader can create his own test signal, save it in *.mat format and then paste it into file example 7_3.m.

Example 7.4

This example shows the performance of the short-term zero crossing estimator when the frequency of the input signal varies in a sinusoidal form. The input signal is defined:

$$v_0(t) = \cos(\phi(t)) \tag{7.45}$$

$$\theta(t) = 2\pi \int_0^t f(\tau)d\tau \tag{7.46}$$

and

$$f(t) = f_1 + A_m \cos(2\pi f_m t). \tag{7.47}$$

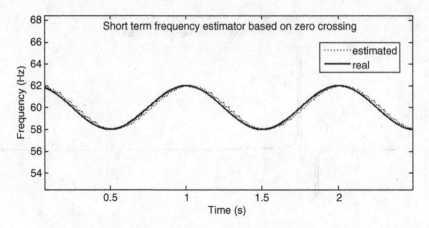

Figure 7.8 Short-term frequency estimator based on zero crossing.

In the example $f_1 = 60$ Hz, $A_m = 2$ and $f_m = 1$ Hz. The data generated by Equation (7.45) are saved in the file dados7_2_b.mat (http://www.ufjf.br/pscope-eng/digital-signal-processing-to-smart-grids/). When loaded, it brings to the workspace the variable v0 and the real frequency given by Equation (7.47). Figure 7.8 shows the real and estimated frequencies. However, this ability to track the frequency is not possible without a delay. In the zero-crossing method the frequency is tracked in one cycle of delay.

The instantaneous error in this example reaches the maximum of 0.4 Hz. This error can be reduced to about 0.25 Hz if the zero-crossing algorithm is edited to obtain the time interval between two consecutive zero crossings instead of two zero crossings in the same direction. In this case, the tracking frequency delay would be changed to a half cycle.

7.4.3 Frequency Estimation Based on Phasor Rotation

Several methods of frequency estimation are based on the definition of the instantaneous frequency as an angular velocity of the rotating voltage phasor. The phasor of the fundamental component is commonly calculated using the DFT algorithm, presented in [5]. New variants are introduced in the following in order to overcome the new challenges that arise due to the presence of high distortion signals in the power system.

Let $x[n]$ be the sample of a real signal of fundamental frequency f_1, sampled using N points per cycle where N is an integer and $F_s = Nf_1$. Let L be the sample index of the second window as depicted in Figure 7.9. The signal is assumed to be a pure cosine function with off-nominal frequency given by

$$x[n] = \cos[2\pi(f_1 + \Delta f)T_s n + \theta_1]. \tag{7.48}$$

The difference (in radians) between the phase generated by the DFT of the second window and the first window is a function of the frequency deviation (remember that the recursive

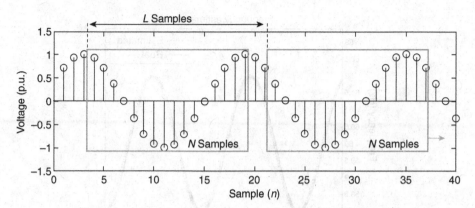

Figure 7.9 Consecutive data windows used to compute the phase difference.

DFT algorithm implemented in Chapter 4 generates the phase referred to the zero point) as follows:

$$\Delta f[n] = \frac{Nf_1}{2\pi L}(\angle X[1,n] - \angle X[1,n-L]). \tag{7.49}$$

If the recursive DFT described in Chapter 4 is used as a phasor estimator, $\Delta f[n]$ can exhibit oscillation due to the presence of noise. The average of the estimator output can then be used as an estimation of the frequency variation. Figure 7.10 depicts the block diagram to estimate the deviation frequency of the signal. The MAF is used to obtain the average of the frequency deviation output. The size of L and the order of MAF (M) can be chosen as $L = N/4$ and $M = N$. The reader can explore the performance of this method in the following example.

Example 7.5

This example demonstrates the performance of frequency estimator based on the phasor rotating approach. The MATLAB$^{®}$ program named dft_freq_estimation.m is used to generate the estimation presented in Figure 7.11. The data file that contains the frequency variation is saved in file dados7_1_b.mat (http://www.ufjf.br/pscope-eng/digital-signal-processing-to-smart-grids/). This file contains both the signal and the reference frequency used to compare the estimator performance.

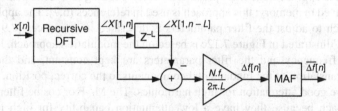

Figure 7.10 Block diagram for frequency estimation.

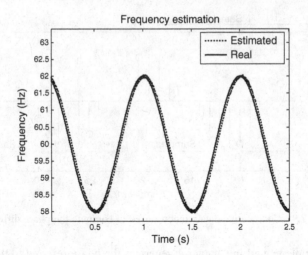

Figure 7.11   Frequency estimation based on phasor rotation.

7.4.4 Varying the DFT Window Size

The objective of frequency estimation is twofold: first, direct measurement of the system frequency for protection and monitoring; second, to improve the accuracy of the estimation of other electrical parameter estimators that are frequency dependent and assume a synchronous sampling rate. For example, the estimators for phase, magnitude and RMS are in general frequency dependent and need a synchronous sampling rate. If these conditions fail, errors occur in the estimation.

Furthermore, the frequency estimator can itself lose acuracy if the input signal is distorted with harmonics and the frequency is off-nominal, since the DFT filter can no longer eliminate the harmonic components. Figure 7.12 depicts three solutions for improving the accuracy of the estimator: (a) sampling frequency adjustment; (b) window size adjustment; and (c) modulation approaching. As the frequency is tracked the algorithm can adjust the sampling frequency in order to match an integer number of samples per cycle. Solution (a) is seldom used for practical reasons; for example, the sharing of recorded data with non-constant sampling rates is difficult when reproducing the original data. The second and third structures keep the sampling frequency constant. In solution (b) the estimated frequency is used to adjust the length of the window and consequently the parameters of the preprocessing filter (generally a DFT filter).

Filter parameters can be updated from a look-up table or by tuning the filter coefficients. In the look-up table approach, the filter parameters corresponding to a set of possible frequency variations is saved in memory; this approach is used in references [6,7]. The application of the tuning approach to adjust the filter parameters is described in references [8,9].

The method illustrated in Figure 7.12c is based on the modulation approach. In this method the sampling frequency and the filter parameters are kept constant, and the modulation provides the translation of the fundamental components to the correct position. However, the filter must have good attenuation for high harmonics. The MAF or cosine filter are not good for this approach because they have a low attenuation capability for high harmonics. In reference [10] an adaptive method is discussed to change the notch frequency of a MAF-based

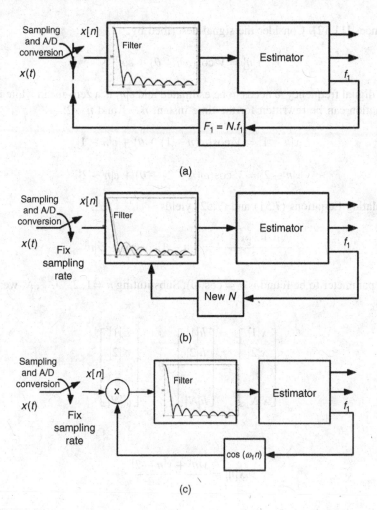

Figure 7.12 Improvement in frequency estimation: (a) sampling frequency adjustment; (b) window size adjustment; and (c) the modulation approach.

filter without changing its order. However, these methods have to deal with fake frequency changing. As observed, disturbances such as sags can affect the frequency estimation, producing erroneous results during the transient. If the algorithm feeds back this variation trying to improve the estimation, the final results will be poor. Care must be taken every time the loop is closed. It is important to be sure that the frequency is real and not a false variation. Remember that frequency estimators are not able to track the frequency variation without some delay, and this delay can worsen the final results.

7.4.5 Frequency Estimation Based on LSE

Estimation of the power frequency based on estimation theory is typically a non-linear process. However, transformations procedures can model the problem as a linear estimation;

see references [11,12]. Consider the signal described by:

$$v[n] = V \cos(\omega n + \theta) + \varepsilon[n] \tag{7.50}$$

where the digital frequency ω needs to be estimated and $\varepsilon[n]$ is a zero-mean white noise. The above equation can be rewritten for the time instant $n-1$ and $n-2$:

$$v[n-1] = V \cos(\omega(n-1) + \theta) + \varepsilon[n-1] \tag{7.51}$$

$$v[n-2] = V \cos(\omega(n-2) + \theta) + \varepsilon[n-2]. \tag{7.52}$$

Manipulating Equations (7.51) and (7.52) yields

$$\frac{v[n] + v[n-2]}{2} = v[n-1]\cos(\omega) + \varepsilon'[n] \tag{7.53}$$

where the parameter to be found is $\theta = \cos(\omega)$. Substituting $n = 1, 2, \ldots, N$, we obtain the equations:

$$\begin{bmatrix} x[1] \\ x[2] \\ \vdots \\ x[N] \end{bmatrix} = \begin{bmatrix} h[1] \\ h[2] \\ \vdots \\ h[N] \end{bmatrix} \cdot \theta + \begin{bmatrix} \varepsilon'[1] \\ \varepsilon'[2] \\ \vdots \\ \varepsilon'[N] \end{bmatrix} \tag{7.54}$$

where

$$x[n] = \frac{v[n] + v[n-2]}{2} \tag{7.55}$$

and

$$h[n] = v[n-1]. \tag{7.56}$$

In matrix form, $\mathbf{x} = \mathbf{H}\theta + \boldsymbol{\varepsilon}$. Its solution is given by the normal equation of LSE. This procedure is used in reference [11] with some variance. In the reference, the author uses a forgetting factor φ so that Equation (7.54) becomes:

$$\begin{bmatrix} \varphi^{N-1}x[1] \\ \varphi^{N-2}x[2] \\ \vdots \\ x[N] \end{bmatrix} = \begin{bmatrix} \varphi^{N-1}h[1] \\ \varphi^{N-2}h[2] \\ \vdots \\ h[N] \end{bmatrix} \cdot \theta + \begin{bmatrix} \varphi^{N-1}\varepsilon'[1] \\ \varphi^{N-2}\varepsilon'[2] \\ \vdots \\ \varepsilon'[N] \end{bmatrix}. \tag{7.57}$$

After mathematical manipulation the LSE estimator yields

$$\hat{\theta}[N] = \frac{\sum_{i=1}^{N} \varphi^{2(N-1)} h[i] x[i]}{\sum_{i=1}^{N} \varphi^{2(N-1)} h[i]^2}. \tag{7.58}$$

7.4.6 IIR Notch Filter

The adaptive notch filter (ANF) has been used in many signal processing applications, mostly as a support for frequency measurement in high-noise environments. Even though a finite impulse response (FIR) ANF can be used for this, it is not sufficiently discriminative in frequency for a noisy ambient. Its magnitude response is not sufficiently steep in the notch frequency vicinity. Placing poles near the notch frequencies enhances the magnitude steepness as desired, but this also requires infinite impulse response (IIR) filters. An important issue related to ANF algorithms concerns the location of these critical poles. In most methods, the pole radii are brought to almost unity in order to optimize selectivity. It is however possible to show that the pole sensitivities to the digital filter coefficients increase tremendously as the pole approaches the unit circumference.

7.4.7 Small Coefficient and/or Small Arithmetic Errors

Small coefficient and/or small arithmetic errors may be caused by noise amplification to instability. Avoiding this situation requires enhanced arithmetic facilities, such as fixed-point double precision and/or the use of especially robust structures.

A number of notch filter structures have been proposed in references [13–16]. The constant bandwidth biquad [17] is commonly used, which is controlled by two coefficients which yield the pole angle and radius for each adaptive pole. One is kept constant, preserving the pole radius, while the other is adapted in order to place the notch frequency canceling the sinusoid poles. Multiple sinusoids may be detected by a cascade of second-order notch filters.

The algorithm presented in reference [17] is:

$$k_1[n + 1] = k_1[n] - \mu \frac{e[n] s'[n]}{\|s'[n]\|^2} \tag{7.59}$$

where $s'[n]$ is the sensitivity output (the derivative of the input regarding k_1), $e[n]$ is the notch output, μ is a gain factor ($0 < \mu < 1$) and $\|s'[n]\|^2$ is the power of the sensitivity filter. The frequency is obtained using

$$k_2 = 1 - r^2 \tag{7.60}$$

where r is the radius of the notch pole, and

$$\omega = 2 \arcsin\left(\frac{k_1}{2\sqrt{1 - k_2/2}}\right). \tag{7.61}$$

Figure 7.13 depicts the performance of the frequency estimation based on the adaptive notch filter. The MATLAB$^{®}$ program named freq_est_notch.m was used to generate Figure 7.13a and b. In Figure 7.13a the signal presents a step change in frequency from 60 Hz to 62 Hz. The signal is corrupted by noise such that the signal-to-noise ratio is SNR $= 37$ dB. The algorithm takes about 10 cycles to reach the new frequency value.

Figure 7.13b is the frequency estimation of the real recorded signal of Example 7.3. It can be observed that the estimation error due to the sag in the signal is about 2% in the case presented in Figure 7.7 compared to 0.2% in Figure 7.13b, highlighting that the frequency estimator based on the adaptive notch is less sensitive to sag variation. This method is however highly influenced by the harmonic presence and preprocessing is required to reduce the harmonic influence.

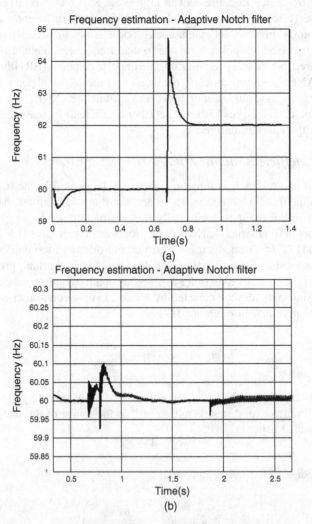

Figure 7.13 Frequency estimator based on adaptive notch filter: (a) sinusoid signal with step frequency variation; and (b) real signal recorded from a faulted signal.

7.5 Phasor Estimation

7.5.1 Introduction

The DFT is the most common method used for phasor estimation. It has often been applied in power system protection, monitoring and operation. It is sometimes used in the form of the more efficient fast Fourier transform (FFT) algorithm or based on its recursive formulation [18,19]. This recursive version constitutes the core of a large number of IEDs.

Although being simple, care must be taken when implementing the DFT algorithm in order to avoid inconsistent results due to voltage and current disturbances. The IEDs are mostly affected by asynchronous sampling by the existence of interharmonics and DC components. Asynchronous sampling occurs when the ratio of the sampling frequency to the fundamental component frequency is a fractional number. In the context of a protective power system, the DC component is not really an input signal DC component (which is usually rejected by the input coupling transformer) but instead a slowly decaying exponential function that appears as part of the faulty line current.

Equations (7.62) and (7.63) yield the real $Y_c[n]$ and imaginary $Y_s[n]$ parts of the estimated phasor. Applying the DFT approach to the input vector

$$\mathbf{x}_n = [x[n-N+1] \quad \cdots \quad x[n]]^t$$

$$Y_c[n] = \frac{2}{N}\sum_{k=0}^{N-1} x_n[k]\cos(2\pi k f_1/F_s) \tag{7.62}$$

and

$$Y_s[n] = \frac{2}{N}\sum_{k=0}^{N-1} x_n[k]\sin(2\pi k f_1/F_s) \tag{7.63}$$

where $x_n[k]$ represents the kth element in vector $\mathbf{x}_n$.

A possible interpretation of these two equations is a demodulation process followed by low-pass filtering, as depicted in Figure 7.14. In this figure the MAF block represents an Mth-order moving average filter, of magnitude as depicted in Figure 7.15.

If the MAF input signal is composed of a DC plus the fundamental component, for instance 60 Hz and its harmonics, the DC component will not be rejected by the zeros gain regularly placed at 60 Hz, 120 Hz and so on. The demodulation interpretation for synchronous sampling and $f_1 = 60$ Hz moves the fundamental frequency component of $x[n]$ to DC. This occurs while the other shifted components still coincide with the MAF zero gains and are rejected as such. In the case of asynchronous sampling however, the mapping of the fundamental to DC fails

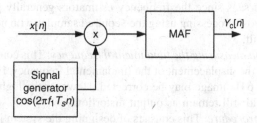

Figure 7.14 Demodulation interpretation of the DFT real part.

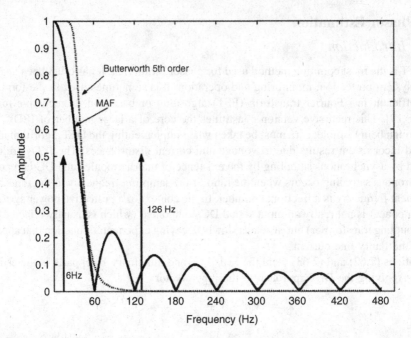

Figure 7.15 Frequency component after demodulation and frequency response of the MAF and the 5th-order Butterworth filter.

and a corresponding non-zero low-frequency image appears. This situation is illustrated in Figure 7.15 for $x[n]$ representing a 66 Hz sinusoid, where a non-practical frequency deviation is used only to illustrate the process. Observe that the sinusoid after demodulation produces a 6 Hz low-frequency image attenuated by the MAF. However, it also shifts the other image in 126 Hz out of the MAF zeros. Similarly, the output distortion further increases if $x[n]$ also contains harmonics.

In Section 7.4.4 three procedures that aid the understanding of problems related to asynchronous frequency in the context of frequency estimation are provided. For phasor estimation, the procedures are slightly different as follows.

- *Adjust sampling frequency according to the fundamental frequency estimation*: This recovers the correct functioning of the process (as in the moving average filtering) but requires continuous tracking of the fundamental frequency as it makes frequent adjustments in the sampling frequency. This makes it more difficult to deal with transient signals such as spikes and sags since the frequency estimators generally give incorrect results during transients. Post-processing using the acquired samples with non-constant sampling rates is also difficult.
- *Correction based on measuring the fundamental frequency*: This consists of estimating the output error due to the displacement of the fundamental frequency. For instance, the MAF attenuation for the 6 Hz image may be corrected. However the higher-frequency components leakage would still remain as output distortions.
- *Synchronism free procedure*: This consists of designing the system in such a way as to be independent of the fundamental frequency, for example by substituting the MAF with a

5th-order Butterworth filter with a magnitude response (also displayed in Figure 7.15) sufficiently flat to accept large fundamental frequency fluctuations, while rejecting higher frequencies.

As observed, the estimation of a signal of electrical nature usually requires identification of the fundamental and eventually its harmonics. Some of these methods are discussed in the sequence of this work, but it is worth pointing out the DFT sensitivity with respect to fundamental synchronisms. Although the DFT is almost insensitive to small synchronism errors in the fundamental frequency, this error is scaled by the harmonic order which can lead to incorrect estimations of the high-frequency harmonic. As such, when interested in estimating the magnitude and phase of harmonics, care is needed regarding the synchronism.

7.5.2 The PLL Structure

Contrary to the adaptive notching filter described in Section 7.4.6, the detection of a sinusoid (characterized by its rejection from the notch output signal) by phase-locked loop (PLL) acts simply by comparing the unknown sinusoid characteristics with those of an idealized sinusoid. For instance if the sinusoid frequency is ω_x, delaying it by m samples should result in a extra phase of $-m(2\pi\omega_x)$. The PLL acts in such a way in order to continuously match the expected sinusoid frequency with the corresponding extra phase. When such a phase-controlled situation occurs, it is said to be locked. The PLL adaptation process is based on control theory, and both the convergence and the steady-state responses are controlled by internal gains.

PLL techniques have supported many power system applications such as peak voltage detection, harmonic detection and analysis, amplitude demodulation, reactive current extraction and so on. The main objective of this chapter concerns the PLL connections with parameter estimation and the processing structures that have been designed to work adequately in the presence of power quality (PQ) disturbances.

There are many different methods of describing the PLL [20–22]. However, most can be described by a set of non-linear differential equations similar in form to:

$$A'(t) = \mu_1 e(t)\sin(\phi(t))$$
$$\Omega'(t) = \mu_2 e(t)A(t)\cos(\phi(t)) \tag{7.64}$$
$$\phi'(t) = \mu_3 e(t)A(t)\cos(\phi(t)) + \Omega(t)$$

$$y(t) = A(t)\sin(\phi(t))$$
$$e(t) = u(t) - y(t) \tag{7.65}$$

in which $u(t)$ and $y(t)$ are the algorithm input and output signal, respectively. The state variables A, ϕ and Ω are correspondingly the estimation of amplitude, the phase and frequency of the input signal $u(t)$. Furthermore, the primes in Equations (7.64) indicate a derivative operation. The parameters μ_1, μ_2 and μ_3 are positive numbers that determine the behavior of the system in terms of convergence, speed and accuracy. These equations are obtained by the minimization of some objective functions. The LSE between the input and out signals $u(t)$ and $y(t)$ are typically minimized using the gradient descent method [23].

It has been demonstrated [23] that the above set of differential equations possesses a unique asymptotically stable periodic orbit that lies in the neighborhood of the orbit associated with the desired component of $u(t)$, if $u(t)$ is a single sinusoid corrupted by noise. In other words, this dynamic system works as an adaptive notch filter.

Karimi-Ghartemani and Iravani [23] used a simplification of Equation (7.64) which incorporates $A(t)$ in the respective constants μ_2 and μ_3. Considering the convergence of this parameter, this is a consistent procedure after the transient algorithm. Grouping the constants in this way yields the set of differential equations:

$$\frac{dA(t)}{dt} = \mu_1 e(t)\sin(\phi(t))$$

$$\frac{d\omega(t)}{dt} = \mu_2 e(t)\cos(\phi(t)) \qquad (7.66)$$

$$\frac{d\phi(t)}{dt} = \omega(t) + \mu_3 e(t)\cos(\phi(t)).$$

The discrete-time model can be obtained from Equation (7.66) using a first-order (Euler backward) approximation for derivatives as suggested in references [21,20], yielding the final equations:

$$A[n+1] = A[n] + \mu_1 T_s e[n]\sin(\phi[n])$$

$$\omega[n+1] = \omega[n] + \mu_2 T_s e[n]\cos(\phi[n]) \qquad (7.67)$$

$$\phi[n+1] = \phi[n] + T_s\omega[n] + \mu_3 T_s e[n]\cos(\phi[n]).$$

The output and error signals are:

$$y[n] = A[n]\sin(\phi[n])$$

$$e[n] = u[n] - y[n]. \qquad (7.68)$$

The results presented in Equation (7.67) are very attractive since the tracking capability is fast (3–5 cycles) and the accuracy is very good. However, these characteristics are degraded when high-energy harmonic components are present at the input signal and when the structure is implemented in a fixed-point processor with word length of 16 bits [24].

Note that, unlike the DFT, this method provides an equation for simultaneously calculating the amplitude, phase and frequency, allowing the tracking of this parameter when it is time varying. Given an input signal $x[n]$, the method basically consists of applying Equations (7.67) and (7.68). The initial values of $A[0]$, $\omega[0]$ and $\phi[0]$ must be specified. Care is required when choosing the initial value of frequencies. Generally, this parameter should be initiated with a value close to the component that needs to be extracted.

The performance of the algorithm is dependent on the composition of the input signal and the parameters μ_1, μ_2 and μ_3. Figure 7.16 shows the amplitude and frequency estimation of a pure 60 Hz sinusoid signal and another with harmonic content. Each harmonic amplitude is equal to the inverse of their order. The sampling rate is $F_s = 64 \times f_1$ and the constants used in the simulation are $\mu_1 = 50$, $\mu_2 = 1500$ and $\mu_3 = 50$. The presence of harmonics increases the error, even though the overall error is very low. For this set of parameters the convergence time is very low; however, it takes 12 cycles to arrive at the final value (not shown in figure). A strategy for the reduction of the convergence time leads to a greater estimation error, so the designer must adjust the parameter according to the application.

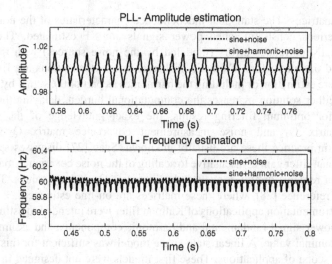

Figure 7.16 Effect of harmonics in the estimation of the magnitude of fundamental component.

To deal with the problems of high-energy harmonic/interharmonic components and keeping a short convergence time some variants, including a internal low-pass filter inside the PLL structure, can be used. Another possibility is pre-filtering the input signal with a pass-band filter centered at the desired component. This approach is described in reference [22].

7.5.3 Kalman Filter Estimation

Kalman filtering has been extensively applied to many different areas such as target tracking, adaptive control and radar. In this section we are concerned with its application to power system signal frequency and phasor estimation, although it is also the basis of a large variety of power system methods. The applicability of the Kalman filter (KF) to real-time signal processing problems is generally limited by its significant computational burden, as for the related recursive least-squares adaptive algorithm. The KF and other complex algorithms are however being absorbed by the digital technology boom, for example the continuous reductions in costs associated with the increased velocity/speed of the modern digital signal processor and virtual instrumentation (VI) allow the use of the KF in real-time applications [25].

A distinctive feature of Kalman filters is in relation to their mathematical formulation in terms of state-space linear or non-linear models. In the non-linear case, use of the KF can be extended through the linearization procedure referred to as the extended Kalman filter (EKF) [26], where the general state and output space equations are:

$$\begin{aligned} \mathbf{x}[n+1] &= \mathbf{f}(n, \mathbf{x}[n]) + \mathbf{v}_1[n] \\ \mathbf{y}[n] &= \mathbf{g}(n, x[n]) + \mathbf{v}_2[n], \end{aligned} \tag{7.69}$$

since $\mathbf{v}_1[n]$ and $\mathbf{v}_2[n]$ are zero-mean white noise processes, uncorrelated to each other and to the state and output vectors. Their correlation matrices are Q_1 and Q_2, respectively. The operators $\mathbf{f}(n, \mathbf{x}[n])$ and $\mathbf{g}(n, \mathbf{x}[n])$ denote possible non-linear time-variant state transition and measurement matrices. The statistical independence characteristics of the noise parameters constitute a serious constraint when power signals are KF estimated. The presence of harmonics usually correlates errors (modeled by the noise sources) with themselves and to the state and output vectors. To avoid such a situation, the options are filtering out the harmonics before estimation or including the harmonics as part of the signal by increasing the model order. Either solution increases the computational burden, delaying the final results. Furthermore, for optimum filtering results, the exact knowledge of the process noise covariance matrix Q_1 and noise measurement covariance matrix Q_2 are required. Unfortunately, in practice these are usually unknown. Saab [27] discusses special cases in which the Kalman filter gain is insensitive to scaling of the noise covariance matrix, while the state estimation remains optimum even under incorrect Q_1 and Q_2 values. This procedure contrasts with reference [28], where these matrices are on-line estimated.

The first instrumentation applications of Kalman filter were intended to estimate phase and amplitude of power signal fundamental and harmonic components, and assumed the system frequency of nominal value. A linear state-space model was sufficient for this objective, but this limited its scope of applications. These first models were not designed in order to work well under protection or power quality disturbance conditions, especially in a frequency-varying scenario.

On the other hand, Kalman filters have often been applied to frequency estimations. In reference [28], a KF was used in the process of estimating power system frequency deviation. The proposed method is a two-stage algorithm. In the first stage the frequency deviation is estimated using an EKF, and the result of the first stage feeds the second stage where a linear KF is used in order to estimate the voltage phasor amplitude and phase. The process noise covariance matrix Q_1 and measurement covariance matrix Q_2 are on-line estimated. Although this works properly, the tracking capability of this approach is too slow, converging in close to 0.2 s (12 cycles).

In reference [29] a non-linear state-space model is proposed for estimating a complex sinusoidal signal and its parameters (frequency, amplitude and phase) which are corrupted by white noise measurements. This filter was derived by applying an extended complex Kalman filter (ECKF) to a non-linear stochastic system. Its state variables are a function of the signal fundamental frequency and of the signal itself. Simulations demonstrate the proposed non-linear filter to be effective as a method for estimating a single complex sinusoid and its frequency under a low SNR. The effect of the assumed initial conditions on the estimated frequency is also discussed [29].

Reference [30] describes another ECKF approach for measurement of power system frequency in the presence of several PQ disturbances. The proposed non-linear model has three state variables and estimates the frequency, amplitude and phase of the fundamental component. Unfortunately, this requires a high computational effort due to the necessary matrix inversions. The final results indicate a high convergence rate (typically converging in 1 cycle) and high noise immunity (13 dB SNR is considered in some examples).

As happens with other adaptive filters, the process gain is automatically reduced as the estimation approaches the target. In this situation, the adaptive algorithm loses its tracking ability, requiring some kind of remedial action. Resetting vectors or matrices may be one of them, as discussed in reference [30].

Reference [25] describes the use of another EKF for the measurement of power system frequency. The authors perform a comparison of this estimation method to other methods, under power disturbance conditions. The proposed method feasibility tests indicate convergence in less than 1.5 cycles, under at most 20.12 dB SNR.

Paper [31] deals with a simple voltage estimation technique for single- or three-phase systems, using a complex KF in linear form. The model uses a single-state variable leading to low-computational-effort results. The model frequency response looks similar to a band-pass filter with the central band-pass frequency equal to the nominal frequency. This is sufficient for low-energy harmonic components, but no strong harmonic contamination is discussed in the paper. Being almost a band-pass model, the method remains sensitive to frequency deviations; in order to be used with the phasor estimation algorithm, the frequency estimation updating technique is required. A single linear model KF is used in reference [31] to estimate a single sinusoid with 40 dB SNR.

7.5.4 Example of Phasor Estimation using Kalman Filter

The first step in the use of Kalman filters for estimation of electrical power systems parameters is the mathematical description of the state equation from which the parameters can be identified. The papers referenced in the previous section show how the authors proposed their state equations to estimate the desired parameters. A single model of a Kalman filter is defined below. The purpose here is to use the KF in a linear model with no frequency variation (at least no significant variation) and estimate their magnitude and phase. The signal described in the compact Fourier trigonometric series is defined [32]:

$$z[n] = \sum_{h=1}^{K} s_{h,c}[n] + v_2[n] \tag{7.70}$$

where

$$s_{h,c}[n] = a_h \cos(n\omega_h + \phi_h), \tag{7.71}$$

h is the harmonic component, K is the total harmonic component considered in the model, a_h, ω_h and ϕ_h are the amplitude, frequency and phase of the respective harmonic, respectively, and $v_2[n]$ is the noise.

Note that

$$s_{h,s}[n+1] = a_h \cos(n\omega_h + \phi_h)\cos\omega_h - a_h \sin(n\omega_h + \phi_h)\sin\omega_h \tag{7.72}$$

and if we set

$$s_{h,s}[n] = a_h \sin(n\omega_h + \phi_h) \tag{7.73}$$

it is easy to verify that

$$s_{h,c}[n+1] = s_{h,c}[n]\cos\omega_h - s_{h,s}[n]\sin\omega_h \tag{7.74}$$

$$s_{h,s}[n+1] = s_{h,c}[n]\sin\omega_h + s_{h,s}[n]\cos\omega_h. \tag{7.75}$$

If the states are chosen as

$$\mathbf{s} = \begin{bmatrix} s_{1,c}[n] & s_{1,s}[n] & \cdots & s_{K,c}[n] & s_{K,s}[n] \end{bmatrix}^{t},\tag{7.76}$$

then the state equation is

$$\mathbf{s}[n+1] = \mathbf{A}\,\mathbf{s}[n] + \mathbf{v}_{1}[n]\tag{7.77}$$

where

$$\mathbf{A} = \begin{bmatrix} \cos(\omega_{1}) & -\sin(\omega_{1}) & 0 & \cdots & 0 & 0 & 0 \\ \sin(\omega_{1}) & \cos(\omega_{1}) & 0 & \cdots & 0 & 0 & 0 \\ \vdots & \vdots & \vdots & \ddots & \vdots & \vdots & \vdots \\ 0 & 0 & 0 & \cdots & 0 & \cos(\omega_{k}) & -\sin(\omega_{k}) \\ 0 & 0 & 0 & \cdots & 0 & \sin(\omega_{k}) & \cos(\omega_{k}) \end{bmatrix}.\tag{7.78}$$

Equation (7.70) can then be written:

$$z[n] = \mathbf{h} \cdot \mathbf{s[n]}\tag{7.79}$$

where

$$\boldsymbol{h} = \begin{bmatrix} 1 & 0 & \cdots & 1 & 0 \end{bmatrix}\tag{7.80}$$

Note that both $\mathbf{A}$ and $\mathbf{h}$ are constant matrices that are functions of only the fundamental and harmonic frequencies.

Finally, the estimation of the amplitude and phase are:

$$a_{k}[n] = \sqrt{s_{k,c}^{2}[n] + s_{k,s}^{2}[n]}\tag{7.81}$$

and

$$\phi_{h}[n] = \tan^{-1}\left(\frac{s_{h,s}[n]}{s_{h,c}[n]}\right) - n\omega_{h}.\tag{7.82}$$

Figure 7.17 shows the performance of a KF in tracking the amplitude of the fundamental and harmonics (3, 5 and 7 harmonics). The amplitudes of the harmonics are inverse in their order and the convergence time is about 1.25 cycles.

7.6 Phasor Estimation in Presence of DC Component

Estimation of the magnitude and phase angle of the current and voltage of a fundamental component in the power system is the essential task in many power system applications such as protection, operation and power quality analysis. It is vital that the phasor estimator can

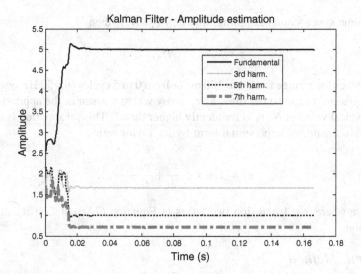

Figure 7.17 Kalman filter applied to amplitude estimation of harmonics.

assess the component of interest, rejecting the unwanted components such as harmonics, interharmonics (subharmonics), DC offset and noise.

Several studies have been conducted in order to derive estimators that are immune to the influence of harmonics, DC offset and noise [33–36]. This type of signal occurs during the faults in conventional transmission lines where there is no compensation. During a fault in the compensated transmission line [35] the signals of voltage and current are composed of the fundamental component, harmonics, DC offsets, subharmonics and noise. In this case, the presence of a subhamonic makes the estimation of the fundamental component even more complex. In distribution systems, the presence of large non-linear loads can introduce interharmonics close to the fundamental components. These can lead conventional estimators to produce inaccurate estimations.

This section introduces some procedures used for the estimation of the fundamental phasor in the presence of a decaying DC. While there are several methods for estimating the phasor in the presence of a decaying DC, there are only a few that address the issue of estimation in the presence of interharmonic and subharmonic components.

7.6.1 Mathematical Model for the Signal in Presence of DC Decaying

A signal with DC decaying can be represented

$$x(t) = A_0 e^{-t/\tau} + A_1 \cos(\Omega_1 t + \theta_1) + h(t), \tag{7.83}$$

where A_0 and τ are the magnitude and the time constant of the decaying DC component, respectively. A_1 and θ_1 are the fundamental parameters to be estimated and $h(t)$ corresponds to the sum of its harmonic components. The discrete form of Equation (7.83) is

$$x[n] = A_0 e^{-n/\tau_1} + A_1 \cos(\omega_1 n + \theta_1) + h(n)$$

$$\tau_1 = \tau F_s. \tag{7.84}$$

If we assume synchronous sampling, that is, $F_s = Nf_1$, then

$$\tau_1 = \tau Nf_1. \tag{7.85}$$

Typical values for τ range from 0 to 100 ms or from 0 to 5 cycles (for 50 Hz systems). From the above equation, τ_1 can be rewritten as $\tau_1 = \alpha N$ where α assumes the approximate values 0–5. For a typical value of N, τ_1 is frequently higher than 1. This information is useful when trying to approximate the exponential term by its Taylor series:

$$e^x = 1 + x + \frac{x^2}{2} + \frac{x^3}{6} + \cdots \tag{7.86}$$

In the case where $x < 1$, the first two terms in Equation (7.86) are sufficient for a good approximation.

7.6.2 Mimic Method

In the mimic method [33] a simple digital filter is used to remove the decaying DC as depicted by Figure 7.18. The transfer function of the mimic filter, in the Z domain is:

$$H_m(z) = k(a + bz^{-1}) \tag{7.87}$$

where a and b are chosen in order to eliminate the exponential term of Equation (7.84) and the k term is chosen to guarantee unitary gain at fundamental frequency.

Taking into account only the DC term of Equation (7.84), the output of the mimic filter is:

$$y[n] = k\left(a + be^{1/\tau_1}\right)e^{-n/\tau_1}. \tag{7.88}$$

Using the fact that τ_1 is mostly higher than 1, and using the first two terms of the Taylor series, we have

$$e^{1/\tau_1} \cong 1 + \frac{1}{\tau_1}. \tag{7.89}$$

Letting $a = 1 + \tau_1$, b can be easily found as $b = -\tau_1$ and the DC term will be filtered. The mimic filter will then be

$$H_m(z) = k\left((1 + \tau_1) - \tau_1 z^{-1}\right) \tag{7.90}$$

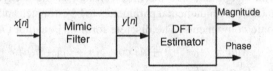

Figure 7.18 Mimic method block diagram.

The constant k is determined so that $|H_m(e^{j\omega_1})| = 1$, or

$$k = \frac{1}{\sqrt{(1 + \tau_1 - \tau_1 \cos \omega_1)^2 + (\tau_1 \sin \omega_1)^2}}. \tag{7.91}$$

The mimic method is widely used in practice due to its simplicity. The drawback of the mimic method is that the time constant needs to be known in advance, which is unrealistic. Power quality studies use the mimic method as a practical range of time constant variation, and the error is considerably smaller than for the DFT without correction.

7.6.3 Least-Squares Estimator

The linear least-squares estimator can be used to estimate the fundamental component, as well as the DC decaying component. The signal model described by Equation (7.84) can be rewritten:

$$x(t) = A_0 e^{-t/\tau} + \sum_{h=1}^{K} A_h \cos(h\Omega_1 t + \theta_h). \tag{7.92}$$

As before, the linear model of the signal can be obtained by using the first two terms of the Taylor series. It also can be deduced by expanding the cosine function after the signal digitalization:

$$x[n] = A_0 - \frac{A_0}{\tau} nT_s + \sum_{h=1}^{K} Y_c[h]\sin(h\omega_1 n) + \sum_{h=1}^{K} Y_s[h]\cos(h\omega_1 n). \tag{7.93}$$

If a number K of harmonics should exist in the signal, then the number of parameters to be estimated is $2K + 2$:

$$\boldsymbol{\theta} = [A_0 \quad -A_0 T_s/\tau \quad Y_c[1] \quad Y_s[1] \quad \cdots \quad Y_c[K] \quad Y_s[K]]^t \tag{7.94}$$

The number of samples of the input signal must be higher than $2K + 2$. Then, taking L samples of the input signal, for $n = 0, 1, 2, \ldots L - 1$, the matrix form can be written:

$$\mathbf{x} = \mathbf{H} \cdot \boldsymbol{\theta} \tag{7.95}$$

where

$$\mathbf{H} = \begin{bmatrix} 1 & 0 & 0 & 1 & \cdots & 0 & 1 \\ 1 & 1 & \sin(\omega_1) & \cos(\omega_1) & \cdots & \sin(K\omega_1) & \cos(K\omega_1) \\ \vdots & & & \vdots & & & \\ 1 & L-1 & \sin((L-2)\omega_1) & \cos((L-2)\omega_1) & \cdots & \sin(K(L-2)\omega_1) & \cos(K(L-2)\omega_1) \\ 1 & L-1 & \sin((L-1)\omega_1) & \cos((L-1)\omega_1) & \cdots & \sin(K(L-1)\omega_1) & \sin(K(L-1)\omega_1) \end{bmatrix} \tag{7.96}$$

and

$$\mathbf{x} = [x[0] \quad x[2] \quad \cdots \quad x[L-1]]^t. \tag{7.97}$$

The LSE is obtained using the pseudo-inverse

$$\boldsymbol{\theta} = \mathbf{H}^+ \mathbf{x} \tag{7.98}$$

$$\mathbf{H}^+ = (\mathbf{H}'\mathbf{H})^{-1}\mathbf{H}. \tag{7.99}$$

If the fundamental phasor is the only parameter of interest, we only need to determine the third and fourth lines of the pseudo-inverse matrix.

7.6.4 Improved DTFT Estimation Method

As described above, many algorithms have been developed for phasor estimation in the presence of DC offset. These algorithms have some limitations however with regards to previous knowledge of the time constant and noise presence, which is a complex issue at low SNR. This section therefore introduces a phasor estimation algorithm [37] in the presence of a DC offset with high noise immunity, which estimates the fundamental component in just 0.675 cycles. A comparison with other methods in the literature is also made.

Figure 7.19 shows a block diagram of the proposed method. In the left section of this figure the input current (or voltage) signal is processed through the recursive DFT algorithm. On the right, the input signal is modulated by a special window generating the signal $i_{w0}[n]$. The

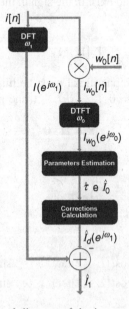

Figure 7.19 General diagram of the improved DTFT estimation.

DTFT of the modulated signal is performed at frequency ω_0 generating the complex signal $I_{w0}(e^{j\omega_0})$. This signal carries only information of the exponential component after special window removes the fundamental and odd harmonic components. The I_0 and τ are estimated from $I_{w_0}(e^{j\omega_0})$. These values are then used to correct the phasor $I(e^{j\omega_1})$, where ω_1 is the power system frequency (60 or 50 Hz).

To understand how the window eliminates the fundamental and harmonic components, we inspect the modulated signal $i_w[n]$:

$$i_w[n] = i[n] \cdot w[n] = i_{0w}[n] + i_{hw}[n], \quad h = 1, 2, \ldots, K \tag{7.100}$$

where the terms $i_{0w}[n]$ and $i_{hw}[n]$ are the modulated version of the DC offset and harmonics, including the fundamental terms, respectively. The DTFT of the harmonics terms $i_{hw}[n]$ are:

$$\text{DTFT}(i_{hw}[n]) = \frac{I_1}{2} e^{j\theta_1} W\left(e^{j(\omega-\omega_1)}\right) + \frac{I_1}{2} e^{-j\theta_1} W\left(e^{j(\omega+\omega_1)}\right) + \frac{I_3}{2} e^{j\theta_3} W\left(e^{j(\omega-\omega_3)}\right)$$
$$+ \frac{I_3}{2} e^{-j\theta_3} W\left(e^{j(\omega+\omega_3)}\right) + (\cdots) + \frac{I_K}{2} e^{j\theta_K} W\left(e^{j(\omega-\omega_K)}\right) + \frac{I_p}{2} e^{-j\theta_K} W\left(e^{j(\omega+\omega_K)}\right) \tag{7.101}$$

where $W(e^{j\omega})$ is the DTFT of the window $w[n]$ and I_h and θ_h ($h = 1, 2, \ldots, K$) are the amplitude and phase, respectively, of each harmonic component of the current signal.

If all terms of Equation (7.101) are canceled for a suitable frequency ω_0, the DTFT of Equation (7.100) will contain only the influence of the DC offset term in that frequency. The goal is therefore to choose a window with a spectrum that has zeros at the frequencies $\omega_0 \pm \omega_h$, that is, $W(e^{j(\omega_0-\omega_h)}) = W(e^{j(\omega_0+\omega_h)}) = 0$, $h = 1, \ldots, K$.

If $w_0[n]$ is the window that contains zeros in the frequencies $\omega_0 \pm \omega_h$ and $i_{w_0}[n]$, the faulted signal filtered by the window, the DTFT of $i_{w_0}[n]$, $I_{w_0}(e^{j\omega})$ calculated at the frequency ω_0, will contain only the DC offset term.

7.6.4.1 Window Synthesis

The window of length J is found by the solution of the following matrix equation:

$$\mathbf{w} = \mathbf{A}^{-1} \cdot \mathbf{d} \tag{7.102}$$

where $\mathbf{w}_{1 \times J}$ is the window vector, $\mathbf{d}_{1 \times J}$ is the magnitude response of the window at the set of frequencies and $\mathbf{A}_{J \times J}^{-1}$ is the inverse of the observation matrix. We therefore have

$$\mathbf{w} = [w_0[0] \quad w_0[1] \quad \cdots w_0[J-1]]^t \tag{7.103}$$

$$\mathbf{A} = \begin{bmatrix} 1 & e^{-j\Delta_1} & e^{-j2\Delta_1} & \cdots & e^{-j(J-1)\Delta_1} \\ 1 & e^{-j\Delta_2} & e^{-j2\Delta_2} & \cdots & e^{-j(J-1)\Delta_2} \\ 1 & e^{-j\Delta_3} & e^{-j2\Delta_3} & \cdots & e^{-j(J-1)\Delta_3} \\ \vdots & \vdots & \vdots & \ddots & \vdots \\ 1 & e^{-j\Delta_J} & e^{-j2\Delta_J} & \cdots & e^{-j(J-1)\Delta_J} \end{bmatrix}. \tag{7.104}$$

The frequencies Δ_j are defined:

$$\Delta = \begin{bmatrix} \Delta_1 \\ \Delta_2 \\ \Delta_3 \\ \Delta_4 \\ \Delta_5 \\ \vdots \\ \Delta_{J-1} \\ \Delta_J \end{bmatrix} = \begin{bmatrix} 0 \\ \omega_0 - \omega_1 \\ \omega_0 + \omega_1 \\ \omega_0 - 3\omega_1 \\ \omega_0 + 3\omega_1 \\ \vdots \\ \omega_0 - (J-1)\omega_1 \\ \omega_0 + (J-1)\omega_1 \end{bmatrix}. \tag{7.105}$$

For example, if $f_0 = 53$ Hz and $J = 36$ samples (equivalent to 0.675 cycles of the fundamental component) the window will have zeroes on the following frequencies: $-2047, \cdots, -247, -127, -7, 113, 233, 353, \cdots, 2153$ Hz. Since the window has no symmetry in the frequency domain, its coefficients are complex numbers. The window $w_0[n]$ can then be rewritten as $w_0[n] = w_{0R}[n] + jw_{0I}[n]$ and consequently,

$$i_{w_{0R}}[n] = i[n] \cdot w_{0R}[n]$$
$$i_{w_{0I}}[n] = i[n] \cdot w_{0I}[n] \tag{7.106}$$

The DTFT of $i_{w_0}[n] = i_{w_{0R}}[n] + ji_{w_{0I}}[n]$ at frequency ω_0 is

$$I_{w_0}(e^{j\omega_0}) = I_{w_{0R}}(e^{j\omega_0}) + jI_{w_{0I}}(e^{j\omega_0}) = \sum_{n=0}^{J-1} a[n]\, i[n] + j\sum_{n=0}^{J-1} b[n]\, i[n] \tag{7.107}$$

where

$$a[n] = w_{0R}[n]\cos(\omega_0 n) + w_{0I}[n]\sin(\omega_0 n) \tag{7.108}$$

$$b[n] = -w_{0R}[n]\sin(\omega_0 n) + w_{0I}[n]\cos(\omega_0 n). \tag{7.109}$$

These equations can be implemented as a filter and are illustrated in Figure 7.20.

Figure 7.21 shows the impulse response of the coefficients $a[k]$ and $b[k]$, $k = 0, 1, \ldots,$ $J - 1$. The real and imaginary sections of the window in this case are $J = 36$. Note the window length that represents a half cycle plus 4 samples, if the sampling rate corresponds to 32 samples per cycle. The extra four samples are necessary in order to guarantee noise attenuation. The magnitude response of $W_0(e^{j\omega})$ is presented in Figure 7.22. Note that there are several zero coefficients in the window impulse response and the non-zero coefficients are equal; this considerably reduces the computational effort.

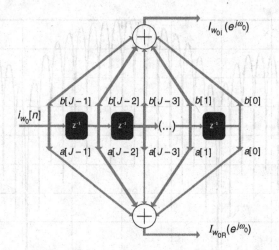

Figure 7.20   Filter implementation.

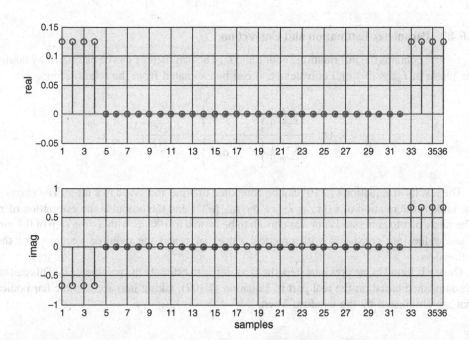

Figure 7.21   Impulse response of the real and imaginary part of the window $J = 36$.

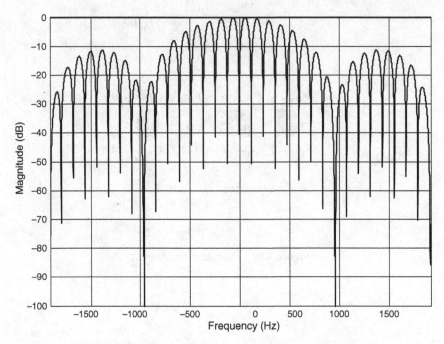

Figure 7.22 The magnitude response of the window $w_0[n]$.

7.6.4.2 Parameter Estimation and Correction

$I_{w_0}(e^{(j\omega_0)})$ contains the information about τ and I_0. The parameter τ can be estimated by taking the phase of $I_{w_0}(e^{(j\omega_0)})$ or, equivalently, it can be estimated from the relationship:

$$k_1(\tau) = \frac{\sum_{n=0}^{J-1}\left\{e^{-\frac{n\Delta t}{\tau}}b[n]\right\}}{\sum_{n=0}^{J-1}\left\{e^{-\frac{n\Delta t}{\tau}}a[n]\right\}} = \frac{I_{w_{0I}}(e^{j\omega_0})}{I_{w_{0R}}(e^{j\omega_0})}. \tag{7.110}$$

The results of Equation (7.110) can be computed off-line and saved in a table. The entry of the table is the relation of $k_1(\tau) = I_{w_{0I}}(e^{j\omega_0})/I_{w_{0R}}(e^{j\omega_0})$ and the output is the estimation of τ. The range of τ to generate $k_1(\tau)$ was chosen to be from 0 to 100 ms with a time step of 0.1 ms. These values are consistent with practical values of τ and the required accuracy for the estimation.

Once τ is found in the previous step, the next step is to estimate the parameter I_0. This can be accomplished based on the real part of Equation (7.107), taking into account the harmonics that are eliminated by the window. Then,

$$I_{w_{0R}}(e^{j\omega_0}) = I_0 \sum_{n=0}^{J-1} a[n]\, e^{-\frac{n\Delta t}{\tau}} \tag{7.111}$$

or simply,

$$\hat{I}_0 \doteq \frac{I_{WOR}(e^{j\omega_0})}{\sum\limits_{n=0}^{J-1} a[n] e^{-\frac{n\Delta t}{\tau}}}. \tag{7.112}$$

Again, for a real-time operation the table $k_2(\tau)$ can be computed off-line, defined

$$k_2(\tau) \doteq \frac{1}{\sum\limits_{n=0}^{J-1} a[n] e^{-\frac{n\Delta t}{\tau}}}. \tag{7.113}$$

It can be seen that the $k_2(\tau)$ table has the same values is independent of the system in which the algorithm is used, since it depends only on $a[n]$ and the sampling frequency. This in turn is related to the coefficients of the window used and the decaying time constant. This means that the estimation of I_0 will not be influenced by other variables such as the amplitude value of the pre-fault current component.

The calculation of corrections is also based on off-line computed table k_3, which relates the value of τ to the influence of their decaying and to the fundamental component. The equation that quantifies this influence is:

$$k_3(\tau) = \sum\limits_{n=0}^{N-1} e^{-\frac{n\Delta t}{\tau}} e^{-j\omega_1}. \tag{7.144}$$

Note that the term $k_3(\tau)$ is the DFT of the pure exponential term with $I_0 = 1$. Then the term to be used in the correction processes is:

$$\hat{I}_{DC}(e^{j\omega_1}) = \hat{I}_0 k_3(\hat{\tau}) = \hat{I}_{DC,R}(e^{j\omega_1}) - j\hat{I}_{DC,I}(e^{j\omega_1}) \tag{7.115}$$

where $\hat{I}_{DC,R}(e^{j\omega_1})$ and $\hat{I}_{DC,I}(e^{j\omega_1})$ are the estimation of the real and imaginary parts, respectively, of the DC exponential in the transform domain.

Finally we arrive at the equation for phasor estimation:

$$\hat{I}_{1R}(e^{j\omega_1}) = \hat{I}_R(e^{j\omega_1}) - \hat{I}_{DC,R}(e^{j\omega_1}) \tag{7.116}$$

$$\hat{I}_{1I}(e^{j\omega_1}) = \hat{I}_I(e^{j\omega_1}) - \hat{I}_{DC,I}(e^{j\omega_1}) \tag{7.117}$$

where $\hat{I}_{1R}(e^{j\omega_1})$ and $\hat{I}_{1I}(e^{j\omega_1})$ are the estimations of the real and the imaginary part of the fundamental phasor; $\hat{I}_R(e^{j\omega_1})$ and $\hat{I}_I(e^{j\omega_1})$ are the DFT of the input signal; and $\hat{I}_{DC,R}(e^{j\omega_1})$ and $\hat{I}_{DC,I}(e^{j\omega_1})$ are the correction to be applied to the DFT estimation.

7.6.4.3 Comparison

To analyze and compare previous and other methods in literature the mean-square error (MSE), noise immunity and computational effort are usually used. The MSE is computed

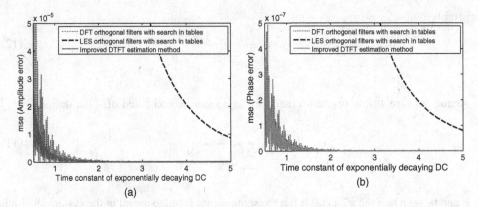

Figure 7.23 Comparison of the improved DTFT estimation method with two other methods presented in literature: (a) MSE error for amplitude estimation; and (b) MSE error for phase estimation.

using the estimated value range from a half cycle after the inception of the fault up to the final estimated value inside 1% of the actual value.

For comparison, a signal based on Equation (7.92) was used with noise addition, 30 cycles and 64 samples per cycle. Odd harmonics of order $h = [1, \quad 3, \quad 5, \quad 7]$, $I_0 = 0$ and $I_1 = 1$ with no fault, $I_0 = I_1 = 10$ during the fault, $I_h = (0.9/h)I_1$ for $h > 1$ where τ ranges from 0.5 to 5 cycles. At first no noise was considered; as such, the performance of each method could be compared by taking into account the DC exponential decaying.

Figure 7.23a compares the amplitude estimation of the improved DTFT method with the methods proposed in [38]. It can be seen that the improved DTFT estimation method performs better than the others. Furthermore, it is not influenced by the time constant values. The same can be seen in Figure 7.23b for phase estimation.

Figure 7.24 shows the comparison of the proposed method with the other methods taking into account noise immunity. In this simulation the signals used encompassed 400 cycles, without DC offset and with a SNR ranging from 25 to 40 dB. It can be seen that the proposed

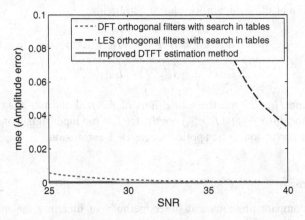

Figure 7.24 Comparison of the noise immunity.

method shows the best performance when the signal is corrupted by an additive white Gaussian noise.

7.6.4.4 Fixed-Point Implementation

The improved DTFT estimation method was implemented in fixed-point arithmetic, using 32-bit long variables, in order to be synthesized in a field-programmable gate array (FPGA) platform. The algorithm was then written in synthesizable Verilog language. It is worth mentioning that the application does not need a high clock frequency, eliminating timing issues. To evaluate the efficiency of the FPGA implemented algorithm, two faulted signals were simulated in the real-time digital simulator (RTDS) [39].

Figure 7.25a and b present a comparison of the results. In this comparison the objective was mainly to verify the accuracy of the fixed-point implementation with the floating-point implementation. For this reason, a comparison with the other previous methods is not presented in the figure. However, the pure DFT estimator was presented in the figure to better illustrate the performance of the improved DTFT estimation method. As can be observed, the fixed-point implementation presents reasonable accuracy when compared to the floating-point method.

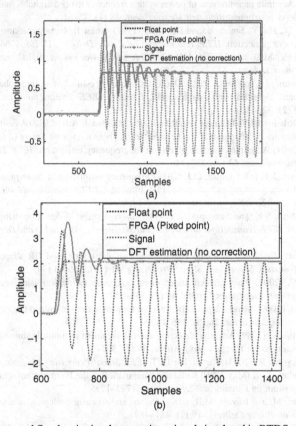

Figure 7.25 Floating- and fixed-point implementation: signal simulated in RTDS with time constant of (a) 33 ms; and (b) 80 ms.

7.7 Conclusions

This chapter describes several techniques for the estimation of electrical parameters in power system grids such as frequency and phasor estimation, based on different signal processing techniques. The smart-grid context has produced conditions in which traditional and advanced signal processing tools are of paramount importance for the operation and control of electrical networks.

References

1. Kay, S.M. (1993) *Fundamentals of Statistical Signal Processing, Estimation Theory*, Prentice Hall, London.
2. Wang, F. and Bollen, M. (2004) Frequency response characteristics and errorestimation in RMS measurement. *IEEE Transactions on Power Delivery*, **19** (4), 1569–1578.
3. IEC 61000-4-30 (2003) Electromagnetic compatibility (EMC), Part 4, Section 30: Power quality measurement. International Electrotechnical Commission, Geneva, Switzerland.
4. Duric, M.B. and Durisic, Z.R. (2005) Frequency measurement in power network in the presence of harmonics using fourier and zero crossing technique. In Proceedings of Power Tech, St Petersburg, Russia. IEEE, 1–6.
5. Phadke, A.G., Thorp, J.S. and Adamiak, M.G. (1983) A new measurement technique for tracking voltage phasors, local system frequency and rate of change of frequency. *IEEE Transactions on Power Apparatus Systems*, **102** (5), 1025–1033.
6. Sidhu, T.S. (1999) Accurate measurement of power system frequency using a digital signal processing technique. *IEEE Transactions on Instrumentation and Measurement*, **48** (1), 75–81.
7. Hart, D., Novosel, D., Hu, Y., Smith, B. and Egolf, M. (1997) A new frequency tracking and phasor estimation algorithm for generator protection. *IEEE Transactions on Power Delivery*, **12** (3), 1064–1073.
8. Tomić, J.J., Kušljević, M.D. and Vujičić, V.V. (2007) A new power system digital harmonic analyzer. *IEEE Transactions on Power Delivery*, **22** (2), 772–780.
9. Kušljević, M.D., Tomić, J.J. and Jovanović, L.D. (2010) Frequency estimation of three-phase power system using weighted-least-square algorithm and adaptive FIR filtering. *IEEE Transactions on Instrumentation and Measurements*, **59** (2), 322–329.
10. Marques, C.A.G., Ribeiro, M.V., Duque, C.A., Ribeiro, P.F., daSilva, E.A.B. (2012) A controlled filtering method for estimating harmonics of off-nominal frequencies. *IEEE Transactions on Smart Grids*, **3** (1), 38–39.
11. Kusljevic, M.D. (2004) A simple recursive algorithm for frequency estimation. *IEEE Transactions on Instrumentation and Measurements*, **53** (2), 335–340.
12. Miodrag, D.K., Josif, J.T. and Ljubisa, D.J. (2010) Frequency estimation of three-phase power system using weighted-least-square algorithm and adaptive FIR filtering. *IEEE Transactions on Instrumentation and Measurements*, **59** (2), 322–329.
13. Petraglia, M.R., Mitra, S.K. and Szczupak, J. (1994) Adaptive sinusoid detection using IIR notch filters and multirate techniques. *IEEE Transactions on Circuits and Systems Part II: Analog and Digital Signal Processing*, 709–717.
14. Rao, B.B. and Peng, R. (1988) Tracking characteristics of the constrained IIR adaptive notch filter. *IEEE Transactions on Acoustics, Speech and Signal Processing*, **41** (11), 1466–1479.
15. Zhou, J. and Li, G. (2004) Plain gradient based direct frequency estimation using second-order constrained adaptive IIR notch filter. *Electronics Letters*, **40** (5), 351–352.
16. Tan, L. and Jiang, J. (2009) Novel adaptive IIR filter for frequency estimation and tracking (DSP Tips&Tricks). *IEEE Signal Processing Magazine*, **26** (6), 186–189.
17. Kwan, T. and Martin, K. (1989) Adaptive detection and enhancement of multiple sinusoids using a cascade IIR filters. *IEEE Transactions on Circuits and Systems*, **36** (7), 937–947.
18. Jacobsen, E. and Lyons, R. (2003) The sliding DFT. *IEEE Signal Processing Magazine*, **20** (2), 74–80.
19. Hartley, R. and Welles, K. II (May 1990) Recursive computation of the Fourier transform. Proceedings of IEEE International Symposium on Circuits and Systems. pp. 1792–1795.
20. Karami-Ghartemani, M. and Iravani, M.R. (2004) Robust and frequency-adaptive measurement of peak value. *IEEE Transactions on Power Delivery*, **19** (2), 481–489.
21. Ziarani, A.K. and Konrad, A. (2004) A method of extraction of nonstationary sinusoids. *Signal Processing*, **84** (8), 1323–1346.

22. Carvalho, J.R., Duque, C.A., Ribeiro, M.V., Cerqueira, A.S., Baldwin, T.L. and Ribeiro, P.F. (2009) A PLL-based multirate structure for time- varying power systems harmonic/interharmonic estimation. *IEEE Transactions on Power Delivery*, **24** (4), 1789–1800.
23. Karimi-Ghartemani, M. and Iravani, R.M. (2003) Periodic orbit analysis of two dynamical systems for electrical engineering applications. *Journal of Engineering and Mathematics*, **45** (2), 135–154.
24. Gomes, P.H., Ribeiro, M.V., Duque, C.A. and Cerqueira, A.S. (2006) An enhanced and robust QPLL technique for power system applications. Proceedings of IEEE 12th International Conference on Harmonic and Quality of Power, pp. 1–6.
25. Routray, A., Pradhan, A.K. and Rao, K.P. (2002) A novel Kalman filter for frequency estimation of distorted signals in power systems. *IEEE Transactions on Instrumentation and Measurements*, **51** (3), 469–479.
26. Haykin, S. (1996) *Adaptive Filter Theory*, Upper Saddle River, Prentice Hall.
27. Saab, S.S. (1995) Discrete-time Kalman filter under incorrect noise covariances. In Proceedings of the American Control Conference. pp. 1152–1156.
28. Girgis, A.A. and Peterson, W.L. (1990) Adaptive estimation of power system frequency deviation and its rate of change for calculating sudden power system overloads. *IEEE Transactions on Power Delivery*, **5** (2), 585–594.
29. Nishiyama, K. (1997) A nonlinear filter for estimating a sinusoidal signal and its parameters in white noise: On the case of a single sinusoid. *IEEE Transactions on Signal Processing*, **45**, 970–981.
30. Dash, P.K., Jena, R.K., Panda, G. and Routray, A. (2000) An extended complex Kalman filter for frequency measurement of distorted signals. *IEEE Transactions on Instrumentation and Measurement*, **49** (4), 746–753.
31. Pradhan, A.K., Routray, A. and Sethi, D. (2004) Voltage phasor estimation using complex linear Kalman filter. Proceedings of 8th IEE International Conference on Developments in Power System Protection, pp. 24–27.
32. Bollen, M.H.J. and Gu, I.Y.H. (2006) *Signal Processing of Power Quality Disturbances*, Wiley & Sons.
33. Benmouyal, G. (1995) Removal of dc-offset in current waveforms using digital mimic filtering. *IEEE Transactions on Power Delivery*, **10** (2), 621–630.
34. Guo, Y., Kezunovic, M. and Chen, D. (2003) Simplified algorithms for removal of the effect of exponentially decaying dc-offset on the Fourier algorithm. *IEEE Transactions on Power Delivery*, **18** (3), 711–717.
35. Yu, C.-S. (2006) A reiterative dft to damp decaying dc and subsynchronous frequency components in fault current. *IEEE Transactions on Power Delivery*, **21** (4), 1862–1870.
36. Sidhu, Z.T. and Balamourougan, V. (2005) A new half-cycle phasor estimation algorithm. *IEEE Transactions on Power Delivery*, **20** (2), 1299–1305.
37. Diniz, A.O., Silva, L.R.M., Martins, C.H., Aleixo, R.R., Duque, C.A. and Cerqueira, A.S. (2012) An improved DFT based method for phasor estimation in fault scenarios. IEEE PES General Meeting. pp. 1–12.
38. Sidhu, T., Zhang, X., Albasri, F. and Sachdev, M.S. (2003) Discrete-fourier-transform-based technique for removal of decaying dc offset from phasor estimates. IEE Proceedings on Generation, Transmission and Distribution.
39. INC (2011) *User Guide to Real Time Digital Simulator*. RTDS Technologies.

8

Spectral Estimation

8.1 Introduction

An important application of digital signal processing is determining the frequency content of a current or voltage signal. This is more commonly known as spectral analysis or spectrum estimation.

The word spectrum was introduced by Newton in relation to his studies of the decomposition of white light into a band of colors when passing through a glass prism. The concept of frequency has evolved from planetary motions in Ancient Greece to the Lagrange rational functions to identify periods (late 18th century), to Fourier analysis and Thomson's harmonic analyzer in the 19th century, followed by Schuster's periodogram in 1897, Einstein's smoothed periodogram (1913), the concept of spectrum (1940), Cooley and Tukey's fast Fourier transform (FFT), Hibert-Huang's empirical mode decomposition (EMD) and, recently, to the time-varying harmonic analyzer.

Applications of the spectral estimation in power systems can be found in power quality analysis, protection and control systems. Spectrum estimation was previously used to estimate the harmonic component of a stationary signal. It was assumed that its nominal frequency was unchanged, and that no further components were present except for harmonics and high signal-to-noise ratios (SNRs). More recently however researchers are focusing on the spectrum analysis of non-stationary signals with a time-varying frequency and interharmonics. The required techniques for such analysis can be divided in non-parametric and parametric spectrum estimations.

In this chapter, the non-parametric and parametric spectrum estimation for the stationary scenario is revisited. The assessment of interharmonics and their possible source of generation are considered. We also discuss the IEC standard 61000-4-7.

8.2 Spectrum Estimation

As presented in Chapter 4, the spectrum of a sampled signal can be calculated using the discrete Fourier transform (DFT). The DFT components are harmonically related to the first harmonic frequency given by:

$$\Delta f = 1/NT_s \tag{8.1}$$

Power Systems Signal Processing for Smart Grids, First Edition. Paulo Fernando Ribeiro, Carlos Augusto Duque, Paulo Márcio da Silveira and Augusto Santiago Cerqueira.
© 2014 John Wiley & Sons, Ltd. Published 2014 by John Wiley & Sons, Ltd.
Companion Website: http://www.wiley.com/go/signal_processing/

where N is the size of the window, in number of samples, T_s is the sampling period and Δf represents the frequency resolution (in Hz), also called the DFT bin. As observed in Equation (8.1), the resolution can be improved by increasing N. If the sampling rate is chosen so that there is an integer number of samples in M periods of the fundamental period of the signal (supposing periodic signal), that is,

$$F_s = \frac{Nf_1}{M} \tag{8.2}$$

then,

$$\Delta f = f_1/M. \tag{8.3}$$

For example, for $f_1 = 60$ Hz, the IEC 61000-4-7 standard [1] recommends a rectangular window containing $M = 12$ integer cycles of the fundamental component ($M = 10$ for 50 Hz), so the frequency resolution is $\Delta f = 5$ Hz. Moreover, the number of samples to be processed by the FFT algorithm should preferably be a power of 2, in order to achieve a more efficient algorithm. If $N = 2^L$ the sampling frequency is given by:

$$F_s = \frac{2^L f_1}{M} = 2^L 5 \text{ Hz}. \tag{8.4}$$

According to IEC 61000-4-7, the number of harmonics to be analyzed must be up to and including the 50th harmonic. If L is chosen as $L = 10$, then

$$F_s = 2^{10} 5 = 5120 \text{ Hz}. \tag{8.5}$$

The maximum frequency that can be observed by the Nyquist Theorem is $F_s/2$. This corresponds to the 42nd harmonic in the case of 60 Hz and 51st for the case of 50 Hz. In this way, the required L for 60 Hz must be higher than 10. If $L = 11$ is used, then the number of samples is more than the necessary to obtain the 50th harmonic and a FFT algorithm that uses a sample number which is not a power of 2 can be more efficient [2].

Note that IEC 61000-4-7 requires an integer number of samples in M cycles and the number of points per cycle ($N_{ppc} = N/M$) does not need to be integer for the standard. For other applications however the constraint of N_{ppc} being integer is held.

The resolution in the frequency domain defines the spacing of the spectral line. If the signal contains a frequency component that falls between two adjacent bins, it cannot be represented and its energy will spread to the neighboring bins, distorting the nearby amplitudes and phases. In this scenario the FFT or DFT of that dataset is not a true spectrum of the process from which the data were obtained, because the FFT analysis is performed using a truncated version of its original signal. Mathematically, the dataset is obtained by the product of the original signal $x[n]$ with a window function $w[n]$, that is,

$$x_w[n] = w[n]x[n]. \tag{8.6}$$

Equation (8.6) is simple example to illustrate the difference between the spectrum of $x[n]$ and $x_w[n]$.

Example 8.1   Spectral Leakage

Let the discrete-time signal be given by $x[n] = A\cos(2\pi f_1 nT_s + \varphi)$, where $f_1 = 55$ Hz. This means that the fundamental system frequency is the off-nominal 5 Hz of deviation. This is of course a non-practical deviation, but can be used to emphasize the spectral leakage. The sampling frequency is chosen to be $F_s = 64 \times 60$ Hz, where the sampling rate is said to be *non-coherent* or simply *asynchronous*. Figure 8.1 shows the spectral magnitude for an observation window of exactly 1 cycle (64 samples, $M = 1$ and $N = 64$), $A = 1$ and $\varphi = 0$. This plot was generated using the FFT MATLAB® command: $X = \text{fft}(x)$.

In this example the resolution according to Equation (8.3) is 60 Hz where $\mathbf{X}$ is a vector of size equal to x. The second term of the $\mathbf{X}$ vector corresponds to the fundamental frequency. However, as can be observed from the figure, there are several components of magnitude different from zero. These components do not exist in the real signal and are the result of the leakage effect. The leakage occurs every time the input signal does not have an integral number of cycles over its window length. Note that as well as the addition of fake components in the spectrum, the amplitude of the real signal is modified as well.

8.2.1   Understanding Spectral Leakage

In order to understand analytically how a spectral leakage develops and how to reduce it, we first recall that the DFT of a finite signal is equivalent to the discrete-time Fourier transform (DTFT) of a typically infinite length signal multiplied by a finite length window:

$$x_w[n] = x[n]w[n] \tag{8.7}$$

where $x_w[n]$ is a signal of length N, $x[n]$ is possibly of infinite duration and $w[n]$ is a window of length N. In Example 8.1, $x[n] = A\cos(2\pi f_1 nT_s + \varphi)$ and $w[n]$ is a rectangular window. The

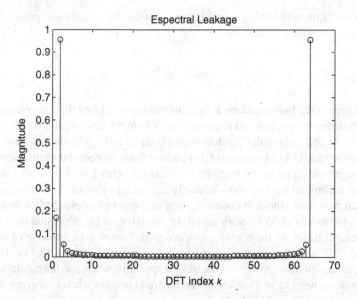

Figure 8.1   Spectral leakage due to non-coherent sampling.

DFT (or FFT) of $x_w[n]$ is

$$X_w[k] = X_w(e^{j\omega})\big|_{\omega_k}, \quad \omega_k = \frac{2\pi}{N}k \text{ and } k = 0, 1, \ldots, N-1. \tag{8.8}$$

Equation (8.8) states that the DFT of a finite signal is obtained from the DTFT of the same signal, sampled at frequencies ω_k. Remember that the DTFT is a continuous function of ω while the DFT is a sampled version of DTFT. If the product in time is a convolution in frequency (see Table 4.6, property 8), then

$$X_w[k] = X_w(e^{j\omega})\big|_{\omega_k} = X(e^{j\omega}) * W(e^{j\omega})\big|_{\omega_k}. \tag{8.9}$$

In Example 8.1, assuming that $-\pi < \omega \leq \pi$ and $\omega_1 < \pi$ rad, the following pairs are used:

$$\cos(\omega_1 n) \leftrightarrow \pi\{\delta(\omega - \omega_1) + \delta(\omega + \omega_1)\}$$

$$w[n] \leftrightarrow \frac{1}{N}\frac{\sin(N\omega/2)}{\sin(\omega/2)}e^{-j(N-1)\omega/2} \tag{8.10}$$

$$\omega_1 = 2\pi f_1 T_s.$$

By applying the convolution theorem, we have

$$X_w[k] = \frac{1}{2N}\frac{\sin(N(\omega-\omega_1)/2)}{\sin((\omega-\omega_1)/2)}e^{-j(N-1)(\omega-\omega_1)/2}\bigg|_{\omega_k} + \frac{1}{2N}\frac{\sin(N(\omega+\omega_1)/2)}{\sin((\omega+\omega_1)/2)}e^{-j(N-1)(\omega+\omega_1)/2}\bigg|_{\omega_k}. \tag{8.11}$$

The spectrum of the windowing signal is then obtained by sampling Equation (8.11) at the desired frequencies (bins), that is, at

$$\omega_k = \frac{2\pi}{N}k.$$

Figure 8.2 depicts the two situations of (a) coherent sampling and (b) non-coherent sampling. In Figure 8.2a depicts a synchronous sampling case where the sinusoidal is 60 Hz sampled at 64 samples per cycle, the rectangular window length is equal to 1 cycle, the continuous line in the figure represents the DTFT of Equation (8.11) and the stems represent the bins. Note that the bin $k = 1$ has magnitude equal to the true sinusoid and the other bins have zero magnitude (the spectrum was multiplied by 2 to show directely the actual value of the cossine signal).

Figure 8.2b represents the non-coherent sampling example described in Example 8.1: a sinusoid with frequency of 55 Hz with sampling rate 64 × 60 Hz. Note that the DFT presents several values, all different from zero (compare with Figure 8.1), because of non-coherent sampling the DFT bins do not correspond to the zeroes of Equation (8.11). This situation brings great difficulties for power engineering, especially when the harmonic and interharmonic components need to be evaluated. Much care is needed when analyzing the spectrum content of a signal produced by FFT.

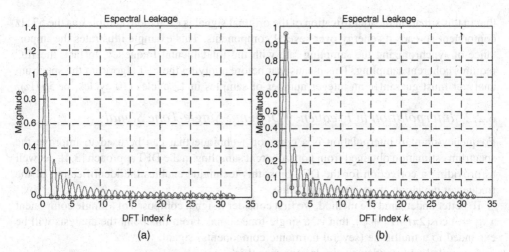

Figure 8.2 Leakage effect: (a) coherent sampling; (b) non-coherent sampling.

Example 8.2   Spectral Leakage

Let the discrete-time signal be given by $x[n] = A\cos(2\pi f_1 nT_s + \varphi)$, where $f_1 = 57$ Hz, $A = 1$ and $\varphi = 0$. The sampling frequency is $F_s = 5120$ Hz according to Equation (8.5), and a rectangular window of 12 cycles length is used according to the IEC specification. The spectrum for this case is plotted in Figure 8.3 using the MATLAB® program leakage_effect_iec.m (http://www.ufjf.br/pscope-eng/digital-signal-processing-to-smart-grids/). Note

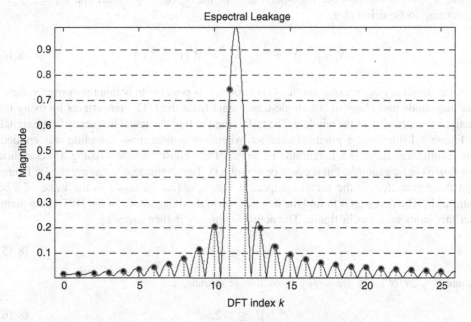

Figure 8.3 Leakage effect using IEC specification.

that in this situation the identification of the original signal is even more complex as the 57 Hz component spreads its energy over several components. This example illustrates the importance of synchronizing the sampling rate with the fundamental frequency. In fact, the IEC requires coherent sampling. This means that before applying the FFT algorithm the spectrum analyzer must guarantee an integer number of samples in 12 cycles (10 cycles for 50 Hz).

8.2.2 Interpolation in Frequency Domain: Single-Tone Signal

Frequency-domain interpolation is a common technique that can be used to improve the spectrum estimation obtained from non-coherent sampling in the DFT approach [3,4]. As well as providing a correction for the DFT bins, this technique enables the identification of the frequency deviation from a nominal value.

To better understand the method, we first consider a signal composed of a single component $x[n] = A \cos(2\pi f_a n T_s + \varphi)$, that is, a single-tone signal. From this point the analysis will be expanded to a multitone (several harmonic components) signal.

A sampled and windowed single-tone signal is

$$x_w[n] = A \cos(2\pi f_a n T_s + \varphi) w[n] \tag{8.12}$$

where A, φ and f_a are unknown parameters and $w[n]$ is any type of window (e.g. rectangular, Hamming or Hanning). By using the modulation theorem and representing only the positive spectrum (DTFT), we have:

$$X_w^e(e^{j\omega}) = \frac{A}{2} e^{j\varphi} W(e^{j(\omega - \omega_a)}) \tag{8.13}$$

where $X_w^e(e^{j\omega})$ represents the right-hand side of the $X_w(e^{j\omega})$ spectrum and $\omega_a = 2\pi f_a T_s$. According to Equation (8.9),

$$X_w^e[k] = X_w^e(e^{j\omega})\big|_{\omega_k} = \frac{A}{2} e^{j\varphi} W\left(e^{j(\omega_k - \omega_a)}\right). \tag{8.14}$$

If the frequency ω_a in Equation (8.14) is known, it is possible to obtain the correct value of the magnitude and phase of the single-tone signal from the DTF components by using the analytical expression of the window. The task is then to find the actual frequency of the signal.

Figure 8.4 illustrates a practical situation where an asynchronous sampling was applied. The continuous curve is a magnitude response of the chosen window (using a rectangular window as basic example, but it could be any other). The continuous stems are the DFT bins, and the dotted stem is the actual component. The question is, how can the value of δ be estimated? The first step is to find the index of the highest magnitude of the DFT component. Let this index be l_0, as in figure. The actual frequency is then given by

$$\omega_a = (l_0 + \delta)\theta \tag{8.15}$$

where $\theta = 2\pi/N$ is the frequency resolution in radians, and

$$|\delta| \le 1/2. \tag{8.16}$$

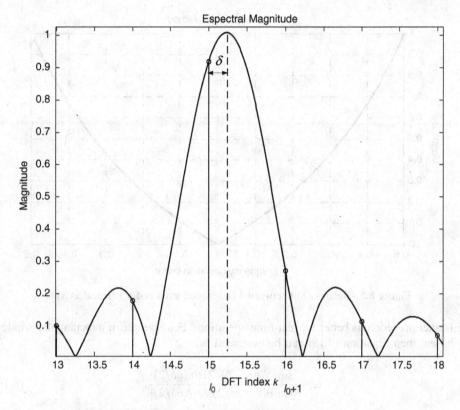

Figure 8.4 Interpolation method.

Note that Equation (8.16) indicates that the actual frequency can stay at the right (as in the figure) or at the left of the maximum DFT component.

To obtain δ we first need to find the ratio between the two highest DFT bins, defined:

$$\alpha = \frac{\left|X_w^e[l_0 + \varepsilon]\right|}{\left|X_w^e[l_0]\right|}. \tag{8.17}$$

The parameter ε can be $+1$ or -1 depending on whether the second-highest DFT bin is at the right or the left of the highest bin; hence α can be derived. By using Equation (8.14) we have the relationship:

$$\alpha = \frac{\left|W\left(e^{j(\varepsilon - \delta)\theta}\right)\right|}{\left|W\left(e^{-j\delta\theta}\right)\right|}. \tag{8.18}$$

Note that the only unknown in Equation (8.18) is the δ variable. The inversion of relationship (8.18) provides δ once $|W(e^{j\lambda})|$ is known for $|\lambda| \leq 1$. Finally, the actual frequency can be estimated using Equation (8.15).

The direct inversion of Equation (8.18) is generally a challenging task. Instead, a polynomial approximation can be adopted or a look-up table (LUT) can be constructed;

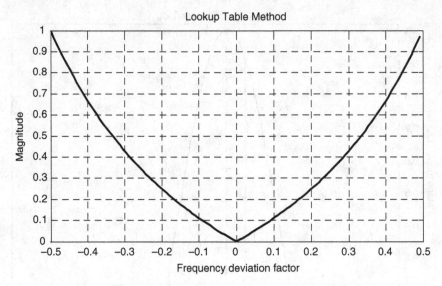

Figure 8.5 Relationship between δ (horizontal axis) and α (vertical axis).

the former procedure is better for real-time operations. For example, if a rectangular window is chosen, then Equation (8.18) can be evaluated as:

$$\alpha = \frac{|\sin(N\theta(\varepsilon - \delta)/2)\sin(\theta\delta/2)|}{|\sin(\theta(\varepsilon - \delta)/2)\sin(N\theta\delta/2)|}. \tag{8.19}$$

Figure 8.5 depicts the relationship between δ and α, obtained using the MATLAB® program freq_interp.m. Note that Figure 8.5 is symmetric for δ; as such, only the positive δ axis needs to be saved once it is know whether δ is positive or negative. When the factor α has been determined through Equation (8.17), the estimated value for δ can be obtained. Example 8.3 illustrates this method.

Example 8.3   Frequency Interpolation (frequency_interp.m)

Consider the signal as in Example 8.2, with $A = 1.5$ and $\varphi = 1.5$ rad. The actual frequency is still $f_1 = 57$ Hz. The first step is to build a normalized LUT, obtained from Equation (8.18) for L different values ($L = 100$) of δ in the range 0–0.5. A Gaussian noise of 0.1 of the standard deviation was added to the single-tone signal, shown in Figure 8.6a. The sampling frequency is $F_s = 5120$ Hz. Figure 8.6b shows the DFT bins of a windowed signal of length equal to 12 cycles of nominal frequency. Note that the maximum bin magnitude is 1.1468 (about 30% error in magnitude).

Table 8.1 lists the estimated results for three different frequencies. As can be observed, the estimation error is very small. Table 8.2 lists the results for amplitude and phase estimation. The actual amplitude and phase are equal to 1.5 and 1.5 rad, respectively. In the same table the amplitude and phase of the maximum bin are presented, showing a high deviation from the

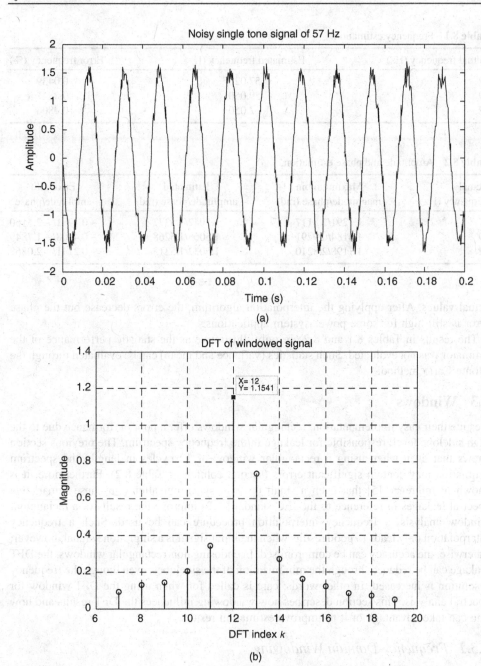

Figure 8.6 (a) Noisy single-tone signal of 57 Hz used in interpolation method; (b) DFT of the windowed signal.

Table 8.1 Frequency estimation.

Actual frequency (Hz)	Estimated frequency (Hz)	Error frequency (%)
57	57.025	−0.0439
59	59.050	−0.0847
62	62.050	−0.0806

Table 8.2 Amplitude and phase estimation.

Actual frequency (Hz)	Maximum bin magnitude/phase (rad)	Estimated amplitude/phase (rad)	Error (%) amplitude/phase
57	1.1291/2.8117	1.5042/1.5377	−0.2791/ −2.5160
59	1.4184/0.9291	1.5065/1.5268	−0.4308/ −1.7849
62	1.1198/2.8210	1.5032/1.5313	−0.2141/ −2.0863

actual values. After applying the interpolation algorithm, the errors decrease but the phase error is still high for some power system applications.

The results in Tables 8.1 and 8.2 are only illustrative as the statistic performance of the estimator was not evaluated. Such statistics (variance and mean) can be evaluated through the Monte Carlo method.

8.3 Windows

Despite their easy implementation, rectangular windows offer a poor performance due to the high sidelobe level, responsible for leakage in the frequency spectrum. The previous section shows that even when using a rectangular window of 12 cycles in length, the spectrum estimation may contain significant errors (second column of Table 8.2). Furthermore, it is shown in reference [5] that even a small error in synchronization can cause remarkable spectral leakages in reference to the IEC standards. To improve the results of a rectangular window analysis, a frequency interpolation procedure can be used. Such a frequency interpolation only leads to good results when the input signal is a single-tone signal, however; otherwise, the accuracy can be compromised. In choosing non-rectangular windows, the DFT leakage can be reduced through the minimization of the sidelobes, even though the frequency resolution is increased. In other words: care is called for when using the DFT window for spectral analysis. This section describes how windowing influences the DFT results and how one can take advantage of it to improve estimation results.

8.3.1 Frequency-Domain Windowing

Windowing is an integral part of any classical spectral estimation and is used to control the effect of sidelobes. The basic idea behind spectrum analysis through windowing is to pre-multiply the input signal by the window and then find the frequency spectrum of the new signal. However, if the window presents a suitable analytical function, we can perform the FFT of the unwindowed data (in fact, a rectangular windowed signal) and the frequency-domain windowing of that FFT result can be performed, as demonstrated in the following.

Table 8.3 Properties of some windows of length N.

Window	Discrete-time function	Frequency response
Rectangular	$w[n] = 1$	$W(e^{j\omega})$
Bartlett (triangle)	$w_B[n] = 1 - \|2n/(N-1) - 1\|$	$\frac{2}{N}W^2(e^{j\omega/2})$
Hanning (squared cosine)	$w_{Hn}[n] = 0.5 - 0.5\cos(2\pi n/N)$	$W_{Hn}(e^{j\omega}) = 0.5W(e^{j\omega})$ $- 0.25\left(W(e^{j(\omega-\theta)}) + W(e^{j(\omega+\theta)})\right)$
Hamming	$w_{Hm}[n] = 0.54 - 0.46\cos(2\pi n/N)$	$W_{Hm}(e^{j\omega}) = 0.54W(e^{j\omega})$ $- 0.23\left(W(e^{j(\omega-\theta)}) + W(e^{j(\omega+\theta)})\right)$

Consider an analytical expression of the form

$$w[n] = \alpha - \beta\cos(2\pi n/N), \qquad n = 0, 1, \ldots, N-1 \tag{8.20}$$

as used by the Hanning and Hamming window, as listed in Table 8.3 [6]. Note that the frequency responses of the previous windows are given as a function of the frequency response of the rectangular window. The demonstration of the Hanning and Hamming frequency response in Table 8.3 is as follows.

The product of the signal $x[n]$ by the window defined by Equation (8.20) is

$$x_w[n] = x[n](\alpha - \beta\cos(2\pi n/N)). \tag{8.21}$$

Taking the DTFT of Equation (8.21) results in

$$X_w[e^{j\omega}] = \sum_{n=0}^{N-1} x[n](\alpha - \beta\cos(2\pi n/N))e^{-j\omega n}. \tag{8.22}$$

Expanding the above equation yields

$$X_w(e^{j\omega}) = \alpha\sum_{n=0}^{N-1} x[n]e^{-j\omega kn} - \frac{\beta}{2}\sum_{n=0}^{N-1} x[n]e^{-jn(\omega-\theta)} - \frac{\beta}{2}\sum_{n=0}^{N-1} x[n]e^{-jn(\omega+\theta)} \tag{8.23}$$

or

$$X_w(e^{j\omega}) = \alpha X(e^{j\omega}) - \frac{\beta}{2}X\left(e^{j(\omega-\theta)}\right) - \frac{\beta}{2}X\left(e^{j(\omega+\theta)}\right) \tag{8.24}$$

where $\theta = 2\pi/N$.

The DFT of the windowing signal can be obtained by

$$X_w[k] = \alpha X[k] - \frac{\beta}{2}X[\langle k-1\rangle_N] - \frac{\beta}{2}X[\langle k+1\rangle_N] \tag{8.25}$$

where $X[\langle k\rangle_N]$ represents a circular shift.

Equation (8.25) shows that each bin of the windowing signal can be obtained as a linear combination of the DFT of the original signal (windowing in frequency domain). Now, if the input signal is a rectangular sequence,

$$x[n] = \begin{cases} 1, & n = 0, 1, \ldots N - 1 \\ 0, & \text{otherwise.} \end{cases}$$

According to Equation (8.10),

$$X(e^{j\omega}) = W(e^{j\omega})$$

where

$$W(e^{j\omega}) = \frac{1}{N}\frac{\sin(N\omega/2)}{\sin(\omega/2)}e^{-j(N-1)\omega/2} \tag{8.26}$$

and the frequency response of the Hanning or Hamming window can be obtained as a linear combination of the frequency response of the rectangular window through Equation (8.24) as:

$$W_w\left(e^{j\omega}\right) = \alpha W\left(e^{j\omega}\right) - \frac{\beta}{2}W\left(e^{j(\omega-\theta)}\right) - \frac{\beta}{2}W\left(e^{j(\omega+\theta)}\right). \tag{8.27}$$

Important information can be obtained from Equation (8.27) when $\omega = \theta$:

$$W_w\left(e^{j\theta}\right) = \alpha W\left(e^{j\theta}\right) - \frac{\beta}{2}W\left(e^{j(0)}\right) - \frac{\beta}{2}W\left(e^{j(2\theta)}\right). \tag{8.28}$$

Inserting these values into Equation (8.26) yields

$$W_w\left(e^{j\theta}\right) = -\frac{\beta}{2}. \tag{8.29}$$

In the same way,

$$W_w(e^{jk\theta}) = 0, \quad k > 1 \tag{8.30}$$

due to the fact that

$$W(e^{jk\theta}) = \begin{cases} 1, & k = 0 \\ 0, & k > 1. \end{cases} \tag{8.31}$$

Figure 8.7 depicts the magnitude response for the four windows described in Table 8.3. These plots were obtained using the MATLAB® code plot_window.m. As can be seen, the magnitude responses of the Hamming, Hanning and Bartlett windows have their first zero

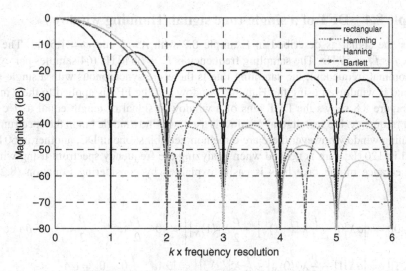

Figure 8.7   Magnitude response of four windows.

magnitude in the second bin. This observation is very relevant when investigating the harmonic content and applying it to one of the previous windows. It is worthwhile observing the gains of the windows at zero frequency that are normalized to be 1 or 0 dB. Its normalization factor is $1/\alpha$, as can be verified from Equation (8.27).

Another point to be highlighted is regarding the analytical expression of the windows. The windows used in filter design applications are slightly different from those used in spectral analysis. For example, in MATLAB® the sequence for the Hamming window is obtained by the command $w = \text{hamming}(N, sflag)$ where *sflag* can be symmetric (the default option) or periodic. The default option is used for filter design and the periodic option for spectrum analysis. Equation (8.20) is for a periodic sequence; the expression for symmetric sequencing is:

$$w[n] = \alpha - \beta \cos(2\pi n/(N-1)), \qquad n = 0, 1, \ldots, N-1. \qquad (8.32)$$

Table 8.4 lists some common windows with useful properties that can be used as design parameters. The most important information is the main lobe width and the sidelobe levels, related to frequency resolution and spectral leakage, respectively.

Table 8.4   Sidelobe and main lobe window properties.

Window	Highest sidelobe level (dB)	Main lobe width
Rectangular	−13.3	2θ
Triangle	−26.5	$\sim 4\theta$
Hanning	−31.5	4θ
Hamming	−43.0	4θ

Example 8.4 DFT of a single-tone signal (Hanning window)

Consider the signal as described in Example 8.2 with $A = 2$ and $\varphi = 1.5$ rad. The actual frequency is $f_1 = 57$ Hz. The sampling frequency is $F_s = 3840$ Hz (64 samples per cycle for 60 Hz nominal frequency); the sampling rate is therefore synchronous with a single period. The signal is windowed using the Hanning window and the FFT is applied to the windowed signal. Figure 8.8 shows the DFT bins of a windowed signal of length equal to 1 cycle of nominal frequency. If a rectangular window is used, we find a single bin in the spectrum. With the Hanning window however there are three non-zero bins: one in DC, another in 60 Hz and the third in 120 Hz; care is needed when analyzing the frequency spectrum using windows! The appearance of the fake bins is easily explained by considering Equation (8.25), for example,

$$|X_w[0]| = \left| \alpha X[0] - \frac{\beta}{2}X[\langle -1 \rangle_N] - \frac{\beta}{2}X[\langle 1 \rangle_N] \right| = \left| 0 - \frac{\beta}{2}Ae^{-j\varphi} - \frac{\beta}{2}Ae^{j\varphi} \right| = \beta A \cos(\varphi)$$

$$|X_w[1]| = \left| \alpha X[1] - \frac{\beta}{2}X[\langle 0 \rangle_N] - \frac{\beta}{2}X[\langle 2 \rangle_N] \right| = |\alpha Ae^{j\varphi} - 0 - 0| = \alpha A$$

$$|X_w[2]| = = \left| \alpha X[2] - \frac{\beta}{2}X[\langle 1 \rangle_N] - \frac{\beta}{2}X[\langle 3 \rangle_N] \right| = \left| 0 - \frac{\beta}{2}Ae^{j\varphi} - 0 \right| = \frac{\beta}{2}A.$$

To compute $|X_w[0]|$, $\left| X_w[\langle -1 \rangle_N] \right| = |X_w[N-1]| = |X_w[1]|$ was used. Furthermore, the spectrum in Figure 8.8 was normalized by $1/\alpha$.

8.4 Interpolation in Frequency Domain: Multitone Signal

The interpolation technique described in Section 8.2.2 can be used for a multitone signal. The utilization of a window improves its estimation due to the high attenuation of the

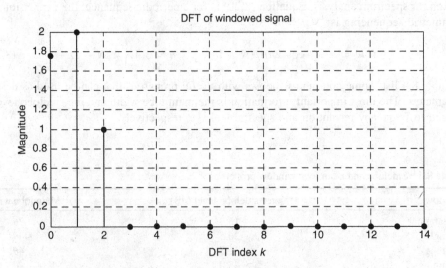

Figure 8.8 DFT of a single-tone signal using Hanning window.

sidelobe and consequently the reduction of leakage. An important point to take into account is with regard to the bins involved in the harmonic estimation. We need to apply a rectangular-like window in the frequency bins in order to work around the harmonics to be analyzed. After estimating the fundamental frequency, the definition of the next bins involved in the process becomes easy. The region under analysis needs to be validated for adequate energy in comparison to the fundamental component, to avoid the inference of a non-existent harmonic. The block diagram in Figure 8.9 illustrates the steps in the frequency interpolation estimation. We need to choose the sampling rate (number of points per cycle or N_{ppc}), the frequency resolution (M), the LUT and so on. Before being able to proceed with the estimation we need to define the bins around the harmonic of interest, starting from the fundamental component. If the energy of the bins is insignificant this is because there are no harmonics in that region; if this is the case, there is no need for further analysis. However, if the energy in the bins is significant, the window is applied to the signal and the FFT is performed.

Example 8.5   Multitone Signal

This example is used to estimate the parameters of a multitone signal given by

$$x[n] = A\cos(\omega_1 n + \varphi_1) + A\cos(3\omega_1 n + \varphi_3)/3 + A\cos(5\omega_1 n + \varphi_5)/5. \tag{8.33}$$

The MATLAB® program freq_interp_comparison.m (http://www.ufjf.br/pscope-eng/digital-signal-processing-to-smart-grids/) was used to generate the parameter estimation and the comparison between the rectangular window and Hanning window approach. The reader can make several modifications to the program in order to test other cases. Figure 8.10a shows the frequency estimation error when a rectangular window is used in the multitone signal described by Equation (8.33). The fundamental frequency is changed from 55 to 65 Hz. A noise of standard deviation 0.01 is added to the signal. The number of cycles used is $M = 4$ of the fundamental component, and the number of points per cycle is $N_{ppc} = 64$. Note that the error in the frequency estimation is higher than 1.5% depending on the actual frequency. Figure 8.10b depicts the same example but with a Hanning window; the estimation error is lower than 0.08% for all frequencies considered.

The amplitude and phase estimations are depicted in Figure 8.11a and b, respectively. Note that the error for the Hanning window is considerably lower than the error for the rectangular window. The error in phase estimation for the Hanning window is lower than 1.0%, while for a rectangular window this error can reach 20%.

Finally, to be consistent with the IEC standard 61000-4-7 [1], a window of 12 cycles is used. The plot for the amplitude estimation of the 5th harmonic is shown in Figure 8.12. The error overpass is 2.5% for a rectangular window, which represents the worst-case scenario among all components and frequency deviations. The frequency estimation error is lower than 0.12% in the worst case (for all harmonics) and the phase error overpass is 5% for the fundamental frequency and 15% for the other components. When a Hanning window approach is used, the accuracy is improved. In the worst-case scenario it is lower than 0.5% in magnitude estimation, as shown in Figure 8.12.

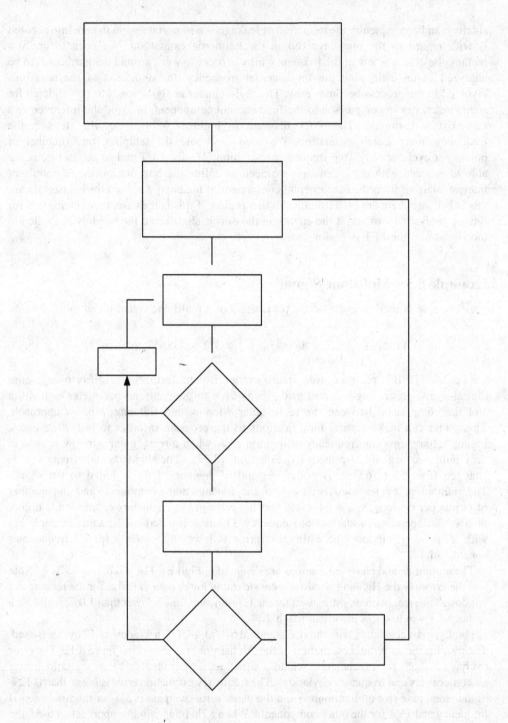

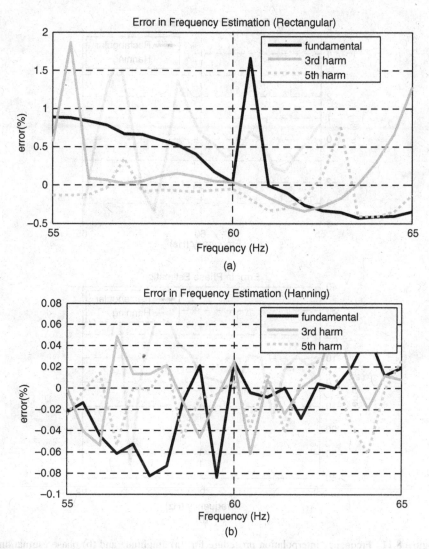

Figure 8.10 Error in frequency estimation: (a) rectangular window; (b) Hanning window.

8.5 Interharmonics

In addition to harmonics, interharmonics can be observed in an increasing number of loads. These loads include static frequency converters, cycloconverters, adjustable speed drivers (ASD) for induction or synchronous motors, arc furnaces and loads that do not pulsate synchronously with the fundamental frequency of the power system [7,8]. The IEC standard defines an interharmonic as 'any frequency which is not an integer multiple of the fundamental frequency'.

The presence of interharmonic components strongly increases the difficulties of modeling and measuring the distorted waveforms. An interharmonic has a direct influence on the

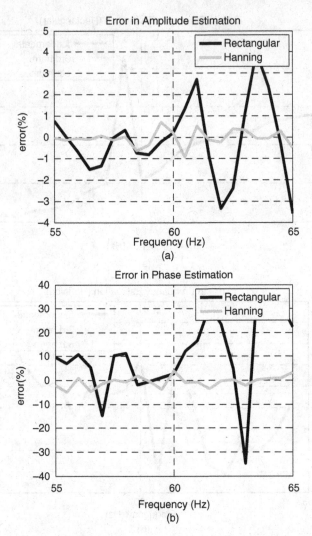

Figure 8.11 Frequency interpolation procedure for: (a) amplitude and (b) phase estimation.

accuracy of the standard digital signal algorithm, such as frequency estimators and harmonics spectrum analysis. As interharmonic is a non-integral order component. Traditional methods based on FFT need to use long windows to obtain adequate resolution for interharmonic detection and estimation. However, the non-stationary nature of waveforms can corrupt the spectrum analysis results and induce erroneous interpretation of the spectrum. In addition, frequency leakage due to non-coherent sampling introduces additional difficulties in analysis and measurements because we cannot be sure that the bins of the FFT exist due to leakage or the presence of real interharmonic components.

The resolution frequency defined by Equation (8.3) plays an important role in interharmonic detection. For example, if the IEC resolution of 5 Hz is used in its coherent spectral analysis, it is possible to detect bins at frequencies which are multiples of 5 Hz such as 50, 55,

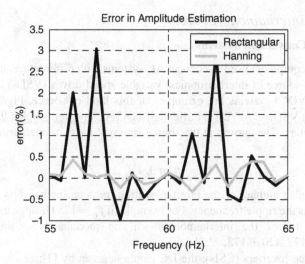

Figure 8.12 Frequency interpolation procedure for amplitude estimation: 5th harmonic (12 cycles).

60, 65 Hz and so on. If an interharmonic has one of these frequencies, it will be an authentic interharmonic and will be correctly detected.

There are however cases where a false interhamonic component is produced due to spectral leakage, for example, if a component of 67 Hz is present in the signal the spectral leakage will pollute the spectrum. Figure 8.13 shows the bins around the fundamental frequency (60 Hz; 12th bin). An interharmonic of 67 Hz frequency and 0.2 pu is added to the signal. The plot shows that the interharmonic spreads its energy to adjacent bins.

Before exploring some interhamonic detection and estimation techniques, we first discuss some of the possible sources of interharmonics in power systems. This information is important as it helps to understand how to choose the best signal processing tools for each scenario.

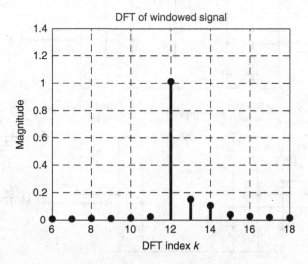

Figure 8.13 Spectrum with interharmonic of 67 Hz.

8.5.1 Typical Interhamonic Sources

8.5.1.1 Double Conversion Systems

Power electronic equipment used to connect AC systems with different frequencies through a DC link can be a source of interharmonics; variable speed drives (VSDs) and high-voltage direct-current (HVDC) systems are examples of this kind of source. Figure 8.14 shows a typical circuit of a VSD, an AC/DC rectifier and a DC/AC inverter coupled through a DC link (capacitor or reactor). The current at the source side will have an interharmonic component given by [7]:

$$f_i = (p_1 m \pm 1)f_1 \pm f_r \qquad (8.34)$$

where p_1 is the pulse number of the rectifier section, m is an integer, f_1 is the fundamental frequency and f_r is the ripple frequency. For example, if $f_r = 173$ Hz appears on the DC side, then for $p_1 = 6$ pulses the interhamonics will be modulated by this frequency as $60 \pm 173, 300 \pm 173, 420 \pm 173, \ldots$ Hz.

In current source inverters (CSIs), the DC ripple is given by [7]:

$$f_r = p_2 n f_0 \qquad (8.35)$$

where p_2 is the pulse number of the output section, n is an integer and f_0 is the output frequency.

In a voltage source inverter (VSI) with synchronous pulse width modulation (PWM) control strategy, the harmonic frequency generated by the inverter is [7]:

$$f_r(m_f, j, r) = |m_f j \pm r| f_0 \qquad (8.36)$$

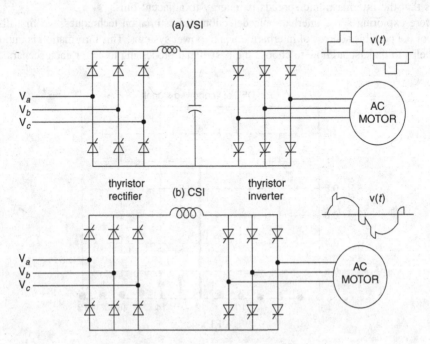

Figure 8.14 (a) VSI circuit and (b) CSI circuit.

where integers j and r depend on the modulation ratio and m_f is related to the switching strategy adopted. An example of VSI with PWM control strategy is given in [7] for $m_f = 9$ and $f_0 = 40\,\text{Hz}$.

8.5.1.2 Cycloconverters

The current spectral components introduced by cycloconverters are also defined by Equations (8.34) and (8.35).

8.5.1.3 Time-Varying Load

Another group of interharmonic sources are the time-varying loads included in regularly and irregularly fluctuating loads. Regular loads such as welder machines, laser printers and integral cycloconverters can be mathematically modeled. The resistance expression is given by [8]:

$$R(t) = 1 - r \sin \Omega_m t \qquad (8.37)$$

where $r < 1$ and Ω_m is the load varying frequency in rad/s. Consequently, the current drained from an ideal voltage source can be determined as:

$$i(t) = \frac{\sin \Omega_1 t}{1 - r \sin \Omega_m t} = \sin \Omega_1 t \left(1 + r \sin \Omega_m t + r^2 \sin^2 \Omega_m t + \ldots\right) \qquad (8.38)$$

After manipulation of Equation (8.38) it is possible to recognize that the current will contain components at frequencies of $\Omega_1 \pm \Omega_m, \Omega_1 \pm 2\Omega_m, \Omega_1 \pm 3\Omega_m, \ldots$ For example, if $r = 0.5$ and $\Omega_m = 16\pi\,\text{rad/s}$ (8 Hz of modulation frequency), there will be interhamonics at frequencies $36, 44, 52, 68, 76, 84 \ldots$ Hz; the most significant components are those near the fundamental frequency (60 Hz or 50 Hz).

Typical examples of irregular fluctuating loads are arc furnaces. The non-linear and time-varying behaviors of these loads are complex and difficult to model analytically. The interharmonics generated by arc furnaces are characterized by constantly changing chaotic frequencies, and there is no sense in trying to identify individual interharmonic components.

8.5.1.4 Wind Turbines

Wind turbines play an important role in the production of voltage interhamonics. The origin of these interharmonics are not electrical but of mechanical in origin and are caused by the wind variation and the tower shadow effect, resulting in power fluctuations. Rich interharmonic components around the fundamental frequencies are reported in [9]. No analytical model is available for predicting these interharmonic components.

8.5.2 The IEC Standard 61000-4-7

International standards are a general guide for harmonic and interharmonic measurements. They are applicable to instrumentation intended for measuring spectral components in the frequency range from DC up to 9 kHz and a frequency resolution of 5 Hz. The FFT of 12 (60 Hz) or 10 cycles (50 Hz) of the input signal is windowed by a rectangular window and

Table 8.5   IEC 61000-4-7 parameters.

Parameter	Comments
Spectral digital signal processing tools	FFT
Window	Rectangular window
Window length	12 cycles for 60 Hz and 10 cycles for 50 Hz
Frequency resolution	5 Hz
Frequency range	0–9 kHz
Harmonic measurement	Up to the 50th harmonic
Coherent sampling	Required
Synchronization technique	Not defined
Frequency sampling	Sufficiently high to allow the analysis of frequency components up to 9 kHz

coherent sampling is required. Table 8.5 lists the main characteristics defined by these standards.

The IEC has introduced the concept of groups and subgroups, attractive for compliance with monitoring and compatibility testing since compatibility levels can be fixed on the basis of the energy of the specified interharmonic group and subgroup instead of the specific bin measurements. Figure 8.15 illustrates the concept of harmonic group and subgroups. The 6th-order harmonic subgroup takes into account the bins 71, 72 and 73 or, generically, the hth-order harmonic subgroup takes into account the bins $k_h - 1, k_h, k_h + 1$ where k is the DFT index. Recall that the index for the hth harmonic is given mathematically by

$$k_h = hf_1/\Delta f \tag{8.39}$$

or simply

$$k_h = hM \tag{8.40}$$

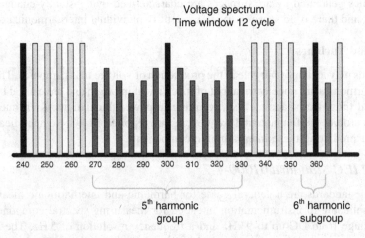

Voltage spectrum
Time window 12 cycle

5th harmonic group

6th harmonic subgroup

Figure 8.15   Definition of harmonic group and subgroup.

Table 8.6 IEC-61000-4-7 definitions of harmonic components.

Parameter	Notation	Definition
Harmonic frequency	f_h	$f_h = h f_1$
Harmonic order	h	$h = k_h / M$ (k_h is the index of the Fourier bin)
RMS	G_h	$G_h = C_{hM}$ (C_k is the RMS of the bin $k = hM$)
RMS harmonic group	$G_{g,h}$	See Equation (8.41) and Figure 8.14
RMS harmonic subgroup	$G_{sg,h}$	See Equation (8.42) and Figure 8.14
Total harmonic distortion	THD	$\text{THD} = \sqrt{\sum_{h=2}^{H} G_h^2 / G_1^2}$
Group total harmonic distortion	THDG	$\text{THDG} = \sqrt{\sum_{n=2}^{H} G_{g,n}^2 / G_{g,1}^2}$
Subgroup THD	THDS	$\text{THDS} = \sqrt{\sum_{n=2}^{H} G_{sg,n}^2 / G_{sg,1}^2}$

where M is the number of cycles in the window (10 or 12). The harmonic group uses six bins on the left and six on the right of the bin k_h (60 Hz example).

Table 8.6 summarizes the parameters regarding harmonic groups and subgroups. The concepts of RMS group and subgroup are defined according to Equations (8.41) and (8.42):

$$G_{g,h}^2 = \begin{cases} \dfrac{C_{k_h-6}^2}{2} + \displaystyle\sum_{i=-5}^{5} C_{k_h+i}^2 + \dfrac{C_{k_h+6}^2}{2}, & \text{for 60 Hz} \\[3mm] \dfrac{C_{k_h-5}^2}{2} + \displaystyle\sum_{i=-4}^{4} C_{k_h+i}^2 + \dfrac{C_{k_h+5}^2}{2}, & \text{for 50 Hz} \end{cases} \qquad (8.41)$$

and

$$G_{sg,h}^2 = \sum_{i=-1}^{1} C_{k_h+i}^2. \qquad (8.42)$$

The concept of the interharmonic group and subgroups is illustrated in Figure 8.16. While the interharmonic group of order h takes into account all bins between the harmonic h and $h+1$, the kth interharmonic subgroup includes the bins from $k_h + 2$ to $k_{h+1} - 2$.

Table 8.7 summarizes the parameters regarding interharmonic groups and subgroups. The concepts of RMS group and subgroup are defined according to Equations (8.43) and (8.44):

$$G_{ig,h}^2 = \begin{cases} \displaystyle\sum_{i=1}^{11} C_{k_h+i}^2, & \text{for 60 Hz} \\[3mm] \displaystyle\sum_{i=1}^{9} C_{k_h+i}^2, & \text{for 50 Hz} \end{cases} \qquad (8.43)$$

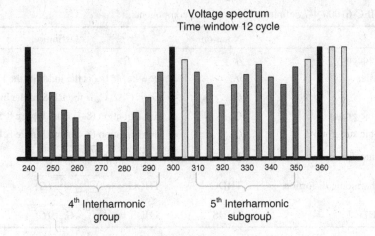

Figure 8.16 Definition of interharmonic group and subgroup.

Table 8.7 IEC 61000-4-7 definitions of interharmonic components.

Parameter	Notation	Definition
RMS interharmonic	G_k	$G_k = C_k,\ k \neq hM$
RMS interharmonic group	$G_{\text{ig},h}$	RMS value of all interharmonics between the harmonic h and $h+1$; see Equation (8.43) and Figure 8.15
RMS interharmonic subgroup	$G_{\text{isg},h}$	RMS value of all interharmonics between the harmonic h and $h+1$ excluding the interharmonic immediately adjacent to the both harmonic; see Equation (8.44) and Figure 8.15.
Interharmonic group frequency	$f_{\text{ig},h}$	$f_{\text{ig},h} = (f_h + f_{h+1})/2$
Interharmonic centered subgroup frequency	$f_{\text{isg},h}$	$f_{\text{isg},h} = (f_h + f_{h+1})/2$

and

$$G_{\text{isg},h}^2 = \begin{cases} \displaystyle\sum_{i=2}^{10} C_{k_h+i}^2, & \text{for 60 Hz} \\[2mm] \displaystyle\sum_{i=2}^{8} C_{k_h+i}^2, & \text{for 50 Hz}. \end{cases} \qquad (8.44)$$

8.6 Interharmonic Detection and Estimation Based on IEC Standard

The impacts of interharmonics are similar to those caused by harmonics. These include filter overload, overheating, ripple, voltage fluctuation, flicker, noise in audio amplifiers, additional torques in motors and generators and disturbances in zero-crossing detectors.

Solving the problems of interharmonics is a major challenge, since the interharmonics can take any values between the harmonic frequencies. Whatever technique is used to estimate the interharmonic, it must still have sufficient resolution in its frequency domain in addition to being able to follow the temporal variations of the interharmonic. This represents a difficult tradeoff.

As previously noted, the IEC standard defines the groups and subgroups for harmonics and interharmonics. The inclusion of adjacent lines in the spectrum harmonic group aims to provide a more accurate representation of harmonics in a case of voltage amplitude fluctuations in a power system. In the case of interharmonics, the group provides an overall value of interharmonic components between two discrete harmonics, including the effects of fluctuations of harmonic components. However, several authors have highlighted the draw-backs regarding the standard and proposed modifications in order to improve the method. The main concerns regarding the IEC standard are the poor sidelobe of the rectangular window, the non-correct identification of the interharmonic frequency (the frequency of the group or subgroup is used instead) and the fact that, in some cases, the inclusion of the adjacent bins at the groups and subgroups can lead to imprecise information [7,8].

In this context the Hanning window is frequently used instead of the rectangular window. Figure 8.17 shows the magnitude response of the Hanning window in comparison to the rectangular window, on a linear scale. These windows were discussed in Section 8.3.1. However, some important aspects need to be highlighted in the case of using a Hanning window with the IEC standard.

The Hanning window shown in Figure 8.18 is normalized to have unit magnitude at the zero frequency; positive and negative frequencies are also presented. As can be observed from the figure, the Hanning magnitude response is not zero at the second bin (5 Hz). When the harmonic subgroup is computed using Equation (8.42) lateral bins are taken into account, regardless of their existence. If we let the RMS of the kth harmonic be equal to C_h and assume

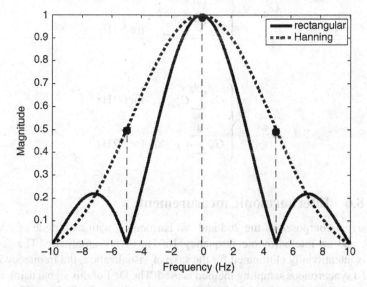

Figure 8.17 Comparison between Hanning and rectangular window.

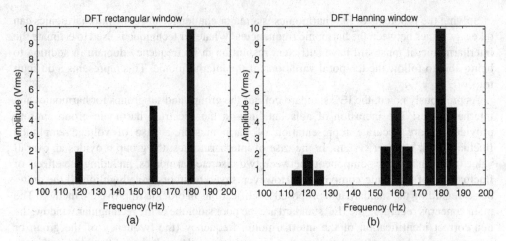

Figure 8.18 DFT of signal in Example 8.6: (a) rectangular window; (b) Hanning window.

that there are no interharmonics in the signal, then

$$G_{sg,h}^2 = \frac{C_h^2}{4} + C_h^2 + \frac{C_h^2}{4} = \frac{3}{2}C_h^2 = \sqrt{\frac{3}{2}}C_h. \tag{8.45}$$

To correct the value of a harmonic subgroup, all bins must be scaled by the factor

$$G_w = \sqrt{2/3}. \tag{8.46}$$

As such, the interharmonic group and subgroup are found using the expressions:

$$G_{ig,h}^2 = \begin{cases} G_w \displaystyle\sum_{i=1}^{11} C_{k_h+i}^2, & \text{for } 60\,\text{Hz} \\[4mm] G_w \displaystyle\sum_{i=1}^{9} C_{k_h+i}^2, & \text{for } 50\,\text{Hz} \end{cases} \tag{8.47}$$

and

$$G_{isg,h}^2 = \begin{cases} G_w \displaystyle\sum_{i=2}^{10} C_{k_h+i}^2, & \text{for } 60\,\text{Hz} \\[4mm] G_w \displaystyle\sum_{i=2}^{8} C_{k_h+i}^2, & \text{for } 50\,\text{Hz}. \end{cases} \tag{8.48}$$

Example 8.6   Interharmonic measurement

Consider a signal composed of the 2nd and 3rd harmonics, with amplitude of 2 V and 10 V respectively, and an interharmonic frequency 160 Hz of 5 V amplitude. The fundamental component is intentionally eliminated for the sake of visualization, the frequency is assumed constant and a synchronous sampling method is used. The DFT of this signal using rectangular and Hanning windows is depicted in Figure 8.18.

Table 8.8 Interharmonic group and subgroup.

Window	Interharmonic subgroup	Error (%)	Interharmonic group	Error (%)
Rectangular	5	0	5	0
Hanning	5.0031	0.0623	6.5135	30.271

Table 8.8 lists the interharmonic group and subgroup of the signal. As shown, the rectangular window presents a better result due to the signal's synchronous sampling. In the Hanning window the lateral bins cause a significant error in the interharmonic group.

Example 8.7   Interharmonic measurement

The resulting DFT from changing only the frequency of the interharmonic component to 167 Hz is depicted in Figure 8.19. Table 8.9 lists the values of the interharmonic group and subgroup. In this case it is possible to note that, because the signal period is not an integer multiple of the fundamental frequency, the effects of spectral leakage due to windowing are apparent by the appearance of various frequency components that do not exist in the original signal. Note that errors occur for both windows. The interharmonic subgroup error for the Hanning is lower than for the rectangular window. The opposite occurs for the interharmonic

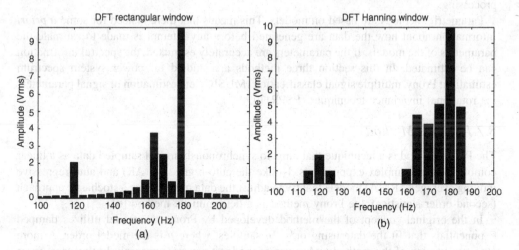

Figure 8.19 DFT of signal in Example 8.7: (a) rectangular window; (b) Hanning window.

Table 8.9 Interharmonic group and subgroup.

Window	Interharmonic subgroup	Error (%)	Interharmonic group	Error (%)
Rectangular	4.7556	4.8888	4.8543	2.9135
Hanning	4.9771	0.4586	6.4189	28.3778

group error, because the interharmonic group error takes into account the bins near the harmonic bin.

In general, it is difficult to determine which window has the best performance. Everything depends on the condition of the interharmonic. What can be observed is that the DFT is very sensitive to temporal variations in amplitude and frequency domains, and that the subgroup has an error smaller than the that presented by the group in case of the Hanning window.

8.7 Parametric Methods for Spectral Estimation

Spectrum estimation using the non-parametric method was partially described in the previous section; these important non-parametric methods are well established in signal processing research for power spectrum density (PSD) estimation. In power system applications however, non-parametric methods are mostly used to obtain an estimation of frequency, magnitude and phase of the spectrum content and the PSD approach hides the phase information. Nonetheless, the non-parametric methods for PSD estimation can be used when the SNR is low and the phase information is not of interest. For more information about these, the reader is invited to read the specialized literature such as [6,10,11].

Non-parametric methods are relatively simple, well understood and easily implemented via the FFT; however, they need long windows to achieve good resolution and they experience spectral leakage due to the use of windows. The major limitation of their use in power quality determination is the finite record length of the signal under analysis for the quality of the spectrum estimation; as such, non-parametric methods are limited to applications in real-time processing.

Parametric methods are based on models. This means that there is need for some *a priori* information about how the data are generated before any attempt is made to estimate the parameters of the models. If the parameters are accurately estimated, the spectral information can be estimated. In this section three methods are studied for power system spectrum estimation: Prony, multiple signal classification (MUSIC) and estimation of signal parameters via rotational invariance techniques (ESPRIT).

8.7.1 Prony Method

The Prony method is a technique that aims to synchronously model sampled data as a linear combination of complex exponentials. Unlike the auto-regressive (AR) and auto-regressive moving average (ARMA) models that try to adjust the data model using a stochastic approach (second-order statistics), the Prony method is a deterministic method [6].

In the original concept of the method developed by Prony, the model utilizes damped exponentials that fit the data using only $2p$ samples, where p is the model order. A more common version of the method makes use of the least-squares approach. In this case, more samples are used than the model order itself to better fit the model, that is, an amount greater than $2p$ samples. The methodology can be briefly described in three steps:

Step 1: Determine the coefficients of a linear predictive model that models the samples.
Step 2: Determine the roots of a characteristic polynomial associated with the linear predictive equation. Through this process, the damping factors and the frequency of each exponential term are found.
Step 3: Estimate the amplitude and phase of each exponential term.

These steps are explained in the following sections.

8.7.1.1 Original Prony Method

Assume N samples of the original signal $x[n], n = 1, \ldots, N$ where N is the number of samples, and assume that the model that generates these samples is represented as a linear combination of damping exponentials, that is,

$$\hat{x}[n] = \sum_{k=1}^{p} A_k e^{(\alpha_k + j\Omega_k)(n-1)T_s + j\theta_k} \qquad n = 1, \ldots, N \qquad (8.49)$$

where T_s is the sampling period, A_k the amplitude of the kth complex exponential, α_k the time constant in seconds, Ω_k the angular frequency in rad/s, θ_k the phase and p the model order.
  Equation (8.49) can be rewritten as

$$\hat{x}[n] = \sum_{k=1}^{p} h_k z_k^{n-1} \qquad (8.50)$$

where the complex constants are defined:

$$h_k = A_k e^{j\theta_k} \qquad (8.51)$$

$$z_k = e^{(\alpha_k + j\Omega_k)T_s}. \qquad (8.52)$$

The basic problem is minimizing the square error:

$$\min_{p, h_k, z_k} \rho = \sum_{n=1}^{p} |\varepsilon[n]|^2$$

where

$$\varepsilon[n] = x[n] - \hat{x}[n] = x[n] - \sum_{k=1}^{p} h_k z_k^{n-1}.$$

The optimum solution is difficult to obtain because the problem is complex and non-linear. The Prony contribution was to propose a suboptimum solution. If as many samples are used as there are exponential terms, that is, $N = p$, then the exponential fitting given by Equation (8.50) can be written in matrix form as

$$
\begin{bmatrix}
z_1^0 & z_2^0 & \cdots & z_p^0 \\
z_1^1 & z_2^1 & \cdots & z_p^1 \\
\vdots & \vdots & \ddots & \vdots \\
z_1^{p-1} & z_2^{p-1} & \cdots & z_p^{p-1}
\end{bmatrix}
\cdot
\begin{bmatrix}
h_1 \\
h_2 \\
\vdots \\
h_p
\end{bmatrix}
=
\begin{bmatrix}
x[1] \\
x[2] \\
\vdots \\
x[p]
\end{bmatrix}
\qquad (8.53)
$$

or in matrix notation

$$\mathbf{Z.h} = \mathbf{x}. \tag{8.54}$$

The $\mathbf{Z}$ matrix is the classic Vandermonde matrix and is consequently invertible. However, Equation (8.53) has $2p$ unknown parameters and only p lines, so it is not possible to find all parameters from this equation. If the $\mathbf{Z}$ matrix is assumed known, then the coefficients $\mathbf{h}$ can be easily determined as:

$$\mathbf{h} = \mathbf{Z}^{-1}\mathbf{x}. \tag{8.55}$$

The problem is then how to find the z_k parameters. The key is to recognize that Equation (8.50) is the solution of the homogeneous difference equation given by

$$\sum_{i=0}^{p} a_i x[n - i] = 0. \tag{8.56}$$

The solution is obtained using the fact that $x[n] = cz^n$ is the solution of Equation (8.56). Substituting this generic solution into Equation (8.56) yields

$$\sum_{i=0}^{p} a_i cz^{n-i} = 0 \tag{8.57}$$

$$cz^n \left(a_0 + a_1 z^{-1} + \cdots + a_p z^{-p}\right) = 0. \tag{8.58}$$

The characteristic polynomial is

$$Q(z) = \left(a_0 + a_1 z^{-1} + \cdots + a_p z^{-p}\right) = \prod_{i=0}^{p} \left(z^{-1} - z_i\right) = 0. \tag{8.59}$$

In power system applications generaly the characteristic polynomial has only single roots. The characteristic polynomial can therefore be rewritten

$$Q(z) = \left(a_0 z^p + a_1 z^{p-1} + \cdots + a_p\right) = \sum_{m=0}^{p} a[m] z^{p-m} \tag{8.60}$$

where $a[0] = 1$.

Changing the indexes in Equation (8.50) from n to $n - m$ and multiplying both sides by $a[m]$ yields

$$a[m]x[n - m] = a[m] \sum_{k=1}^{p} h_k z^{n-m-1}. \tag{8.61}$$

By calculating the summation of the products, $a[1]x[n - 1], \ldots, a[m - 1]x[n - m + 1]$ we arrive at

$$\sum_{m=0}^{p} a[m]x[n - m] = \sum_{k=0}^{p} h_k \sum_{m=0}^{p} a[m] z_k^{n-m-1} \tag{8.62}$$

which is valid for $p+1 \leq n \leq 2p$. Making the substitution $z_k^{n-m-1} = z_k^{n-m} z_k^{p-m-1}$ yields

$$\sum_{m=0}^{p} a[m]x[n-m] = \sum_{k=0}^{p} h_k z_k^{n-p} \sum_{m=0}^{p} a[m]z_k^{p-m-1}. \tag{8.63}$$

By comparing the last summation of the above equation with Equation (8.60), assuming z_k is the root of the characteristic polynomial,

$$\sum_{m=0}^{p} a[m]x[n-m] = 0$$

or

$$\sum_{m=1}^{p} a[m]x[n-m] = -x[n] \quad \text{for} \quad p+1 \leq n \leq 2p. \tag{8.64}$$

The matrix

$$\begin{bmatrix} x[p] & x[p-1] & \cdots & x[1] \\ x[p+1] & x[p] & \cdots & x[2] \\ \vdots & \vdots & \ddots & \vdots \\ x[2p-1] & x[2p-2] & \cdots & x[p] \end{bmatrix} \cdot \begin{bmatrix} a[1] \\ a[2] \\ \vdots \\ a[p] \end{bmatrix} = - \begin{bmatrix} x[p+1] \\ x[p+2] \\ \vdots \\ x[2p] \end{bmatrix} \tag{8.65}$$

can be formed from the data or, in matrix notation

$$\mathbf{X.a} = \mathbf{x} \tag{8.66}$$

where each term in the previous equation can be indentified from Equation (8.65). The matrix $\mathbf{X}$ is a Toeplitz matrix, a well-known format matrix in linear algebra that has efficient invertible algorithms.

Note that the samples of $x[n]$ used to obtain the vector $\mathbf{a}$ are from $x[1]$ to $x[2p]$. This vector contains the coefficients of the characteristic equation. Once they are found, the roots of the characteristic equation $Q(z)$ can be found. The Prony method can be summarized as follows.

- Use the samples from 1 to $2p$ to solve Equation (8.66) and the coefficients of characteristics equation.
- Determine the roots of the characteristic polynomial (8.60). Frequency f_k and damping factor α_k are consequently found by

$$\alpha_k = \frac{\ln|z_k|}{T_s} \tag{8.67}$$

$$f_k = \frac{1}{2\pi T_s} \tan^{-1}\left(\frac{\text{Im}(z_k)}{\text{Re}(z_k)}\right) \tag{8.68}$$

- Equation (8.55) is then solved to obtain the magnitude and phase of the damping factors from

$$A_k = |h_k| \tag{8.69}$$

$$\theta_k = \tan^{-1}\left(\frac{\mathrm{Im}(h_k)}{\mathrm{Re}(h_k)}\right). \tag{8.70}$$

8.7.1.2 Least-Squares Prony Method

For practical situations, the number of data points is more than twice the number of parameters in the model, that is, $N > 2p$. In this case Equation (8.64) can be modified as

$$x[n] + \sum_{m=1}^{p} a[m]x[n-m] = \varepsilon[n] \quad \text{for} \quad p+1 \le n \le N. \tag{8.71}$$

The above equation is recognized as a linear predictor filter from adaptive filter theory; the term $\varepsilon[n]$ is the prediction error and the summation is the forward predictor filter. Figure 8.20 depicts the forward prediction block diagram.

The forward linear predictor filter is a FIR filter; its coefficients need to be optimized to minimize the cost function. The cost function can be written:

$$J(\mathbf{a}) = \sum_{n=p+1}^{N} |\varepsilon[n]|^2 \tag{8.72}$$

which is recognized as the least-squares cost function discussed in Chapter 7. As the model is linear, the solution is given by

$$\hat{\mathbf{a}} = (\mathbf{H}'\mathbf{H})^{-1}\mathbf{H}'\mathbf{x} \tag{8.73}$$

or

$$\begin{bmatrix} x[p] & x[p-1] & \cdots & x[1] \\ x[p+1] & x[p] & \cdots & x[2] \\ \vdots & \vdots & \ddots & \vdots \\ x[N-1] & x[N-2] & \cdots & x[N-p] \end{bmatrix} \cdot \begin{bmatrix} a[1] \\ a[2] \\ \vdots \\ a[p] \end{bmatrix} = - \begin{bmatrix} x[p+1] \\ x[p+2] \\ \vdots \\ x[N] \end{bmatrix}. \tag{8.74}$$

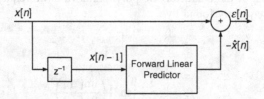

Figure 8.20 Block diagram of the forward linear predictor filter.

The matrix $\mathbf{H}$ is identified as the first matrix in the previous equation, the vector $\mathbf{x}$ is the right-hand side vector and the estimated parameters are the coefficients of the characteristic equation. After finding these coefficients, the zeros of the characteristic equations can be obtained. Equation (8.53) can then be written

$$
\begin{bmatrix}
z_1^0 & z_2^0 & \cdots & z_p^0 \\
z_1^1 & z_2^1 & \cdots & z_p^1 \\
\vdots & \vdots & \ddots & \vdots \\
z_1^{N-1} & z_2^{N-1} & \cdots & z_p^{N-1}
\end{bmatrix}
\cdot
\begin{bmatrix}
h_1 \\
h_2 \\
\vdots \\
h_p
\end{bmatrix}
=
\begin{bmatrix}
x[1] \\
x[2] \\
\vdots \\
x[N]
\end{bmatrix}. \tag{8.75}
$$

Equation (8.75) is no longer square and the least-squares solution must be used. The vector $\mathbf{h}$ is obtained as

$$
\mathbf{h} = (\mathbf{Z}^t \mathbf{Z})^{-1} \mathbf{Z}^t \mathbf{x}. \tag{8.76}
$$

8.7.1.3 Modified Least-squares Prony Method

The Prony method can be applied to signals where some frequency components are known *a priori*, as in power system applications. This assumption reduces the computational complexity when compared to the conventional method. The procedure is developed in reference [6] and is summarized below.

Assume that q exponential components z_k, are known, that is, $z_1, z_2, \ldots z_q$. Knowledge of a sinusoidal component corresponds to knowledge of two exponential components. The characteristic polynomial associated with q known components is:

$$
\prod_{m=1}^{q} (z - z_k) = \sum_{k=0}^{q} c[m]z^k \tag{8.77}
$$

where $c[q] = 1$. The characteristic polynomial relating all exponentials can be written:

$$
Q(z) = \sum_{m=0}^{p} a[m]z^m = \left(\sum_{k=0}^{q} c[m]z^k \right) \left(\sum_{i=0}^{p-q} \alpha[m]z^i \right) \tag{8.78}
$$

where $\alpha[p - q] = 1$. The coefficients $a[m]$ are then given by

$$
a[m] = \sum_{k=0}^{q} c[m]\alpha[m - k] \tag{8.79}
$$

where $\alpha[i] = 0$ for $i > p - q$ or $i < 0$. Applying Equation (8.79) to Equation (8.63) yields

$$
\sum_{m=1}^{p} a[m]x[n - m] = \sum_{m=1}^{p} \left(\sum_{k=0}^{q} c[m]\alpha[m - k] \right) x[n - m] = 0 \tag{8.80}
$$

which is defined for $p + 1 \leq n \leq 2p$. Equation (8.80) can be rewritten

$$\sum_{m=0}^{p-q} \alpha[m]y[n-m] = 0 \tag{8.81}$$

where

$$y[n] = \sum_{k=0}^{q} c[k]x[n-k]. \tag{8.82}$$

Note that Equation (8.82) represents a convolution process, that is, a filtering process. The sequence $y[n]$ is therefore obtained by filtering the input signal $x[n]$ with a filter that has an impulse response of $c[n]$. Note that once $y[n]$ is found, the values of $\alpha[n]$ can be determined by using Equation (8.81).

The procedure can be summarized as follows.

- The data window is filtered through Equation (8.82). The filter coefficients are composed of the known poles obtained from Equation (8.77).
- By using the filtered data $y[n]$, we can find $\alpha[n]$ using Equation (8.81). The characteristic polynomial with the $p - q$ unknown poles is then established.
- After finding the complete coefficients of the polynomial characteristics, Equation (8.76) can be applied to estimate the amplitude and phase of all components.

The Prony method can be used to estimate harmonic and interharmonic components as well as damping components. The accuracy of the estimation depends on the level of signal distortion, the observation window and the number of samples used in the estimation process, as well as the order of the model. Some of these topics are discussed in reference [12]. The application of the proposed method makes it possible to estimate the components whose frequencies differ insignificantly. In most cases, superior performance is achieved when compared with a Fourier algorithm. Its limitations are sensitivity to noise and computational cost. The next section includes examples to evaluate the Prony performance and its limitations.

Example 8.8

Figure 8.21 shows the performance of a synthetic signal containing the second harmonic, third harmonic and an interhamonic of frequency 147 Hz. The amplitudes of each component are 10, 20 and 5 respectively. The fundamental component has an amplitude of 100. The MATLAB® program test_prony.m was used in this example. It calls the function prony_-least_square.m which returns the parameters h (Equation (8.76)) and z (the zeros of characteristic equation). Figure 8.21 shows the frequency estimated for this example; the fundamental component was omitted for the sake of clarity. The order of the model used is $p = 8$, the number of samples per cycle is 10, the total number of samples is $N = 40$ and a noise of 0.05 of standard deviation was added to the signal. Figure 8.22 shows the reconstructed and the original signal.

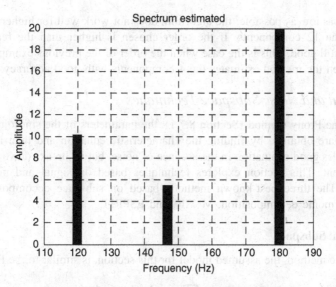

Figure 8.21   Frequency estimated in Example 8.8.

8.7.1.4   Important Considerations in Prony Estimation

The prony_least_square.m function has several limitations when noise is present. Firstly, a low sampling rate such as 10 samples per cycle has to be used as a higher number of samples per cycle leads to inaccurate estimations. If the noise level is increased the accuracy of the estimation is poor; SNR must generally be higher than 45 dB for a good estimation. The model

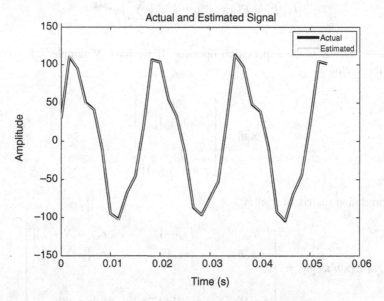

Figure 8.22   Actual and estimated signal of Example 8.8.

order must be as low as possible; the algorithm does not work well for higher-order models with more than 12 components. If the order chosen is higher than the real model, the estimation is still good. This is the case when the order of the previous example is set to 10 instead 8, when the method estimates the 4 components with good accuracy.

8.7.2 Signal and Noise Subspace Techniques

When using the Prony method (Section 8.7.1), the parameters of the sinusoidal signals (or exponentials) are obtained by finding the characteristic equation and then its roots. This method provides good estimation in the absence of noise, but its breaks down quickly when noise is present. This section explores techniques based on signal and noise subspace components. The three best-known methods based on subspace decomposition are the Pisarenko Harmonic decomposition, MUSIC and ESPRIT.

8.7.2.1 Signal Subspace

The parametric method, the assumed model for this section, is similar to the Prony method:

$$x[n] = \sum_{k=1}^{p} A_k e^{j(2\pi f_k T_s n + \phi_k)}. \tag{8.83}$$

Using the digital frequency $\omega = 2\pi f T_s$ (rad) we have

$$x[n] = \sum_{k=1}^{p} A_k e^{j(\omega_k n + \phi_k)}. \tag{8.84}$$

The autocorrelation function for $x[n]$ is

$$r_{xx}[k] = E(x[n]x^*[n-k]) = \sum_{i=1}^{p} p_i e^{j\omega_i k} \tag{8.85}$$

where $p_i = A_i^2$ and E is the expectation operator. If we have N samples of $x[n]$ and then assemble the vector

$$\mathbf{x}[n] = \begin{bmatrix} x[n] \\ x[n+1] \\ \vdots \\ x[n+N-1] \end{bmatrix} \tag{8.86}$$

the autocorrelation matrix of $x[n]$ is

$$\mathbf{R}_{xx} = E(\mathbf{x}[n]\mathbf{x}^H[n]) = \begin{bmatrix} r_{xx}[0] & r_{xx}[-1] & \cdots & r_{xx}[-(N-1)] \\ r_{xx}[1] & r_{xx}[0] & \cdots & r_{xx}[-(N-2)] \\ \vdots & \vdots & \ddots & \vdots \\ r_{xx}[N-1] & r_{xx}[N-2] & \cdots & r_{xx}[0] \end{bmatrix}. \tag{8.87}$$

where superscript H represents the transpose conjugate. Using Equation (8.85), the auto-correlation matrix can be written as

$$\mathbf{R}_{xx} = \sum_{k=1}^{p} p_k \mathbf{s}_k \mathbf{s}_k^H \tag{8.88}$$

where

$$\mathbf{s}_i = \begin{bmatrix} 1 \\ e^{j\omega_i} \\ e^{j2\omega_i} \\ \vdots \\ e^{j(N-1)\omega_i} \end{bmatrix}. \tag{8.89}$$

If we define

$$\mathbf{S} = \begin{bmatrix} \mathbf{s}_1 & \mathbf{s}_2 & \cdots & \mathbf{s}_p \end{bmatrix} \tag{8.90}$$

and

$$\mathbf{P} = diag\begin{pmatrix} p_1 & p_2 & \cdots & p_p \end{pmatrix}, \tag{8.91}$$

then Equation (8.88) can be rewritten in matrix form as

$$\mathbf{R}_{xx} = \mathbf{SPS}^H. \tag{8.92}$$

The vector space $\mathbb{S} = span\{\mathbf{s}_1 \quad \mathbf{s}_2 \quad \cdots \quad \mathbf{s}_p\}$ is said to be the signal subspace of the signal $x[n]$. The eigenvalues of the autocorrelation matrix(in descending order) are $\lambda_1, \lambda_2, \ldots \lambda_p, \lambda_{p+1} \cdots \lambda_N$. However, $N > p$ implies that rank$(\mathbf{R}_{xx}) = p$ and the eigenvalues $\lambda_{p+1} = \lambda_{p+2} = \cdots \lambda_N = 0$, meaning that the subspace $\mathbb{S}$ can be spanned by using only the first p eigenvectors $\mathbf{u}_1, \mathbf{u}_2, \ldots, \mathbf{u}_p$. These eigenvectors are known as the *principal eigenvectors* of $\mathbf{R}_{xx}$.

The autocorrelation in this case can be obtained by the expression

$$\mathbf{R}_{xx} = \sum_{k=1}^{p} \lambda_k \mathbf{u}_k \mathbf{u}_k^H. \tag{8.93}$$

8.7.2.2 Noise Subspace

If signal $x[n]$ is corrupted by a stationary zero-mean white-noise of variance σ_w^2, then we can write

$$y[n] = x[n] + w[n] \tag{8.94}$$

and

$$r_{yy}[k] = r_{xx}[k] + \sigma_w^2 \delta[k] = \sum_{i=1}^{p} p_i e^{j\omega_i k} + \sigma_w^2 \delta[k]. \tag{8.95}$$

If we have N samples of $y[n]$ and assemble the vector as for Equation (8.86), we obtain

$$\mathbf{y}[n] = [y[n] \quad y[n+1] \quad \cdots \quad y[n+N]]^t \tag{8.96}$$

and

$$\mathbf{w}[n] = [w[n] \quad w[n+1] \quad \cdots \quad w[n+N]]^t, \tag{8.97}$$

yielding

$$\mathbf{R}_{yy} = E(\mathbf{y}[n]\mathbf{y}^H[n]) = \mathbf{R}_{xx} + \sigma_w^2 \mathbf{I}. \tag{8.98}$$

Now the $\text{rank}(\mathbf{R}_{yy}) = N$ because of the last term in Equation (8.98). Let the eigenvalue of $\mathbf{R}_{yy}$ be $\mu_1 \geq \mu_2 \geq \cdots \geq \mu_N$. The eigenfunction is

$$\mathbf{R}_{yy}\mathbf{u}_i = \mu_i \mathbf{u}_i. \tag{8.99}$$

Substituting Equation (8.98) into Equation (8.99) we obtain

$$\mathbf{R}_{xx}\mathbf{u}_i + \sigma_w^2 \mathbf{u}_i = \mu_i \mathbf{u}_i. \tag{8.100}$$

For the first p eigenvalues of $\mathbf{R}_{yy}$ we have the relationship

$$\mu_i = \lambda_i + \sigma_w^2, \quad i = 1, 2, \ldots, p \tag{8.101}$$

and the corresponding eigenvectors are equals. Furthermore, the eigenvalues $\mu_{p+1} = \mu_{p+2} = \mu_N = \sigma_w^2$. Finally the autocorrelation matrix can be written as

$$\mathbf{R}_{yy} = \sum_{k=1}^{p} (\lambda_k + \sigma_w^2)\mathbf{u}_k \mathbf{u}_k^H + \sum_{k=p+1}^{N} \sigma_w^2 \mathbf{u}_k \mathbf{u}_k^H. \tag{8.102}$$

The noise subspace $\mathcal{N}$ is spanned by the eigenvectors $\mathbf{u}_i, \quad i = p+1, \ldots, N$:

$$\mathcal{N} = \text{span}\{\mathbf{u}_{p+1} \quad \mathbf{u}_{p+2} \quad \cdots \mathbf{u}_N\}. \tag{8.103}$$

8.7.2.3 Pisarenko Harmonic Decomposition

The Pisarenko harmonic decomposition (PHD) is one of the first procedures for spectral estimation based on eigenanalysis. The PHD method is based on the observation that the signal subspace is orthogonal to the noise subspace and that the noise subspace is spanned only by one eigenvector $\mathbf{u}_N$. This means that the rank of $\mathbf{R}_{yy}$ is assumed to be $N = p + 1$. We then have

$$\mathbf{s}_i^H \mathbf{u}_N = 0 \quad \text{for} \quad i = 1, 2, \dots, p. \tag{8.104}$$

If the eigenvalue associated with the eigenvector $\mathbf{u}_N$ corresponds to the minimum eigenvalue of $\mathbf{R}_{yy}$, then the first step in the algorithm is to find the eigenvalue μ_N and the corresponding eigenvector $\mathbf{u}_N = \begin{bmatrix} u_{N,0} & u_{N,1} & \cdots & u_{N,N-1} \end{bmatrix}^t$. Equation (8.104) can then be rewritten as

$$\sum_{k=0}^{N-1} u_{N,k} e^{-j\omega_i k} = 0 \tag{8.105}$$

which is a polynomial in $e^{-j\omega_i}$. The roots of this polynomial lie within the unit circle and consequently are the frequencies of the sinusoid signal. Once the frequencies have been determined, the sinusoidal power can be found from Equation (8.95) for $k = 1, 2, \dots, p$, or

$$\begin{bmatrix} e^{j\omega_1} & e^{j\omega_2} & \cdots & e^{j\omega_p} \\ e^{j2\omega_1} & e^{j2\omega_2} & \cdots & e^{j2\omega_p} \\ \vdots & \vdots & \ddots & \vdots \\ e^{jp\omega_1} & e^{jp\omega_2} & \cdots & e^{jp\omega_p} \end{bmatrix} \begin{bmatrix} p_1 \\ p_2 \\ \vdots \\ p_p \end{bmatrix} = \begin{bmatrix} r_{yy}[1] \\ r_{yy}[2] \\ \vdots \\ r_{yy}[p] \end{bmatrix}. \tag{8.106}$$

Of course, in practice the correlation matrix $\mathbf{R}_{yy}$ must be estimated from the received samples.

When the number of sinusoids is unknown, the determination of the model order p may be difficult, especially if the SNR is low. In theory if $N > p + 1$ there is a multiplicity $(N - p)$ of the minimum eigenvalue. Checking for the repeating eigenvalue could be one method of order selection. In practice however, the $(N - p)$ eigenvalues will probably be different and this criterion cannot be used. An alternative procedure computes all eigenvalues and determines p by grouping the $(N - p)$ eigenvalues, where small eigenvalues theoretically correspond to noise. The average of these $(N - p)$ eigenvalues is used to obtain the corresponding eigenvector.

8.7.2.4 MUSIC Algorithm

The multiple signal classification (MUSIC) algorithm is also a noise subspace frequency estimator. Let the spectral estimation be:

$$P(\omega) = \sum_{k=p+1}^{N} \left| \mathbf{s}^H(\omega) \mathbf{u}_k \right|^2 \tag{8.107}$$

where $\mathbf{u}_k$ $(k = p + 1, \dots, N)$ are the eigenvectors in the noise subspace and $\mathbf{s}(\omega)$ is the complex sinusoidal vector given by

$$\mathbf{s}(\omega) = \begin{bmatrix} 1 & e^{j\omega} & e^{j2\omega} & \cdots & e^{j(N-1)\omega} \end{bmatrix}^t. \tag{8.108}$$

When $\omega = \omega_i$, where ω_i is one of the p frequencies of the signal, then the vector $\mathbf{u}_k$ and $\mathbf{s}(\omega_i)$ will be orthogonal and consequently $P(\omega_i) = 0$.

The reciprocal of $P(\omega)$ therefore presents peaks at the signal frequencies and provides a method for estimating such frequencies. The reciprocal of $P(\omega)$ is represented by a MUSIC spectrum of ω (or pseudospectrum):

$$P_{\text{MUSIC}}(\omega) = \frac{1}{\sum\limits_{k=p+1}^{N} |\mathbf{s}^H(\omega)\mathbf{u}_k|^2}. \tag{8.109}$$

By locating the peaks, the frequencies can be identified. Knowing the frequencies, the amplitude of each component can be obtained from Equation (8.106). The key point in the MUSIC method is how to separate the signal and noise subspaces. However, examination of the relative magnitude of the eigenvalues of the autocorrelation matrix $\mathbf{R}_{yy}$ and grouping these eigenvalues to construct the signal and noise subspace does not work well in practice. One criterion for estimating the model order is described in reference [6].

Comparing the Pisarenko method with the MUSIC algorithm it can be noted that the Pisarenko method selects $N = p + 1$ and projects the signal vectors onto a single noise eigenvector. The Pisarenko method assumes precise knowledge of the order of the model. On the other hand, the MUSIC method assumes $N > p + 1$ and, after performing the eigenanalysis, subdivides the eigenvalues into signal and noise subspace. The signal vectors are then projected onto the $N - p$ noise subspace.

MATLAB® has two functions that estimate frequencies via the MUSIC eigenvector method: (a) pmusic and (b) rootmusic. Table 8.10 describes each function.

8.7.2.5 ESPRIT Algorithm

ESPRIT is another method for estimating the frequencies of a sum of sinusoids by the use of eigenanalysis. The methodology exploits the rotational invariance propriety of signal

Table 8.10 MATLAB® functions to estimate frequencies viaMUSIC algorithm.

Function	Comments
$[S,w] = pmusic(x, p)$	*pmusic* estimates frequency via the MUSIC eigenvector method. It returns the pseudospectrum S of a discrete-time signal x; w is the vector of a normalized angular frequency (rad/sample) at which the pseudospectrum is evaluated; p is the number of complex sinusoidal in x. If p is a two-element vector, $P(2)$ is used as a cutoff for signal and noise subspace separation. All eigenvalues greater than $P(2)$ times the smallest eigenvalue are designated signal eigenvalues. In this case, the signal subspace dimension is at most $P(1)$. Other forms of *pmusic* can be found through MATLAB® help.
$[w,P] = rootmusic(x, p)$	The *rootmusic* function returns the estimation of frequencies w in rad/sample and the square of the amplitudes P; x and p are as above.

subspaces spanned by two delayed vectors. As for MUSIC, the method assumes that there are p sinusoids in white noise and that $\mathbf{y}[n] = \mathbf{x}[n] + \mathbf{w}[n]$ according to Equations (8.94)–(8.97). Additionally, a new vector of delayed samples is introduced:

$$\mathbf{z}[n] = [y[n+1] \quad y[n+2] \quad \cdots \quad y[n+N]]^t. \tag{8.110}$$

Recall that $x[n]$ is defined as

$$x[n] = \sum_{k=1}^{p} A_k e^{j(\omega_k n + \phi_k)} = \sum_{k=1}^{p} a_k e^{j\omega_k n}, \tag{8.111}$$

where

$$a_k = A_k e^{j\phi_k}.$$

With these definitions we can express the vector $\mathbf{y}[n]$ and $\mathbf{z}[n]$ as

$$\mathbf{y}[n] = \mathbf{Sa} + \mathbf{w}[n] \tag{8.112}$$

$$\mathbf{z}[n] = \mathbf{S\Phi a} + \mathbf{w}[n] \tag{8.113}$$

where $\mathbf{a} = [a_1 \quad a_2 \quad \cdots \quad a_p]^t$, $\mathbf{S}$ is defined by Equation (8.90) and $\mathbf{\Phi}$ is a diagonal matrix that represents the phase shift between successive samples, given by

$$\mathbf{\Phi} = \mathrm{diag}\left[e^{j\omega_1} \quad e^{j\omega_2} \quad \cdots e^{j\omega_p}\right] \tag{8.114}$$

where $\mathbf{\Phi}$ is a rotational operator.

As before, we can write

$$\mathbf{R}_{yy} = E(\mathbf{y}[n]\mathbf{y}^H[n]) = \mathbf{SPS}^H + \sigma_w^2 \mathbf{I},$$

$$\mathbf{R}_{yz} = E(\mathbf{y}[n]\mathbf{z}^H[n]) = \mathbf{SP\Phi}^H \mathbf{S}^H + \mathbf{R}_w$$

where

$$\mathbf{R}_w = E\left(\mathbf{w}[n]\mathbf{w}^H[n+1]\right) = \sigma_w^2 \begin{bmatrix} 0 & 0 & 0 & \cdots & 0 & 0 \\ 1 & 0 & 0 & \cdots & 0 & 0 \\ 0 & 1 & 0 & \cdots & 0 & 0 \\ \vdots & & & \ddots & & \\ 0 & 0 & 0 & \cdots & 1 & 0 \end{bmatrix}$$

and $\mathbf{S}$ and $\mathbf{P}$ are defined by Equations (8.90) and (8.91), respectively.

If $N > p$, then the matrix

$$\mathbf{R}_{xx} = \mathbf{R}_{yy} - \sigma_w^2 \mathbf{I} = \mathbf{SPS}^H \tag{8.115}$$

has rank p. Let

$$\mathbf{C}_{yz} = \mathbf{R}_{yz} - \mathbf{R}_w = \mathbf{SP\Phi}^H \mathbf{S}^H, \tag{8.116}$$

then

$$\mathbf{R}_{xx}\mathbf{u} = \lambda \mathbf{C}_{yz}\mathbf{u} \tag{8.117}$$

or

$$\mathbf{SP}(\mathbf{I} - \lambda \mathbf{\Phi}^H)\mathbf{S}^H \mathbf{u} = 0. \tag{8.118}$$

This is known as a generalized eigenanalysis, where λ is the generalized eigenvalue and $\mathbf{u}$ is the corresponding generalized eigenvector. The MATLAB$^®$ function named $eig(A)$ is used to calculate the conventional eigenvalue and eigenvector of matrix A. When this is used with two arguments (e.g. as eig(A,B)) then it computes the generalized values. This can be observed from the function help in the MATLAB$^®$ command line (help eig): '$E = eig(A,B)$ is a vector containing the generalized eigenvalues of square matrices A and B . . . $[V,D] = eig(A,B)$ produces a diagonal matrix D of generalized eigenvalues and a full matrix V whose columns are the corresponding eigenvectors so that $A*V = B*V*D$.'

Note that the solution of Equation (8.117) is obtained using $[V,D] = eig(R_{xx}, C_{yz})$. Since $\mathbf{\Phi}$ is diagonal, it is clear from Equation (8.118) that $\lambda = e^{j\omega_i}$, that is, the p generalized eigenvalues λ_i $(i = 1, 2, \ldots, p)$ that lies on the unit circle corresponds to the elements of the rotational operator $\mathbf{\Phi}$; consequently, the frequencies of the signal are ω_i. The $N-p$ remaining generalized eigenvalues lie near the origin.

The algorithm used to determine the frequencies via ESPRIT can be summarized as follows [6,13,14].

1. From data, estimate the autocorrelation matrix $\mathbf{R}_{yy}$ and $\mathbf{R}_{zy}$
2. Compute the eigenvalue of $\mathbf{R}_{yy}$, for example by using $E = eig(\mathbf{R}_{yy})$. The minimum eigenvalue of E is an estimate of σ_w^2
3. Compute the matrix $\mathbf{R}_{xx}$ and $\mathbf{C}_{yz}$.
4. Compute the generalized eigenvalues, for example by using $[V,D] = eig(\mathbf{R}_{xx}, \mathbf{C}_{yz})$. V is a matrix where each column is a generalized eigenvector $\mathbf{v}_i$ and D is a diagonal matrix with the generalized eigenvalues.
5. The generalized eigenvalues on or near the unit circle determine the frequencies of the signal. The $N-p$ remaining eigenvalues will lie at or near the origin.

The signal power can be obtained from the generalized eigenvector $\mathbf{v}_i$ [6].

$$P_i = \frac{\mathbf{v}_i^H \mathbf{C}_{yy}\mathbf{v}}{\left|\mathbf{v}_i^H \mathbf{s}_i\right|^2}, \quad \text{for} \quad i = 1, 2, \ldots, p. \tag{8.119}$$

Example 8.9

A synthetic signal consisting only of odd harmonics is analyzed. The frequencies, magnitude and amplitude of the synthetic signal are listed in Table 8.11. Four cycles of the fundamental component (60 Hz) are generated using the sampling rate of 32 samples per cycle. The

Table 8.11 Synthetic example.

Harmonic number	3	5	7	9	11	13
Magnitude (%)	1.5	4.0	4.0	0.8	2.5	2.0
Phase (%)	0	30	60	90	120	180

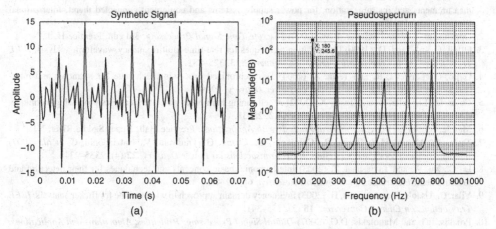

Figure 8.23   (a) Synthetic signal and its (b) pseudospectrum obtained from the MUSIC algorithm using MATLAB®.

Table 8.12   Estimated values from MUSIC algorithm.

Harmonic number	Frequency (Hz)
3	180.0991
5	299.4015
7	420.0776
9	537.4448
11	660.3513
13	779.0109

fundamental component is not included in the model signal, for the sake of clarity. Gaussian noise of zero mean and standard deviation of 0.5% is included. Figure 8.23a shows the synthetic signal and Figure 8.23b shows the equivalent pseudospectrum obtained through the MATLAB® function $[S,w] = $ pmusic$(x,12)$. The model order used was 12. The peaks in the pseudospectrum indicate the frequencies of the harmonics. Table 8.12 lists the estimated frequency values converted to Hz.

8.8   Conclusions

Spectral estimation becomes more significant as the waveforms of voltages and currents become more complex and time varying in the new electric smart-grid context. This chapter helps engineers and investigators to understand the details of the spectral estimation process in the presence of harmonics and interharmonics and applies the basic concepts to practical power system situations.

References

1. IEC 61000-4-7 (2002) Testing and measurement techniques. General guide on harmonics and interharmonics measurements and instrumentation, for power supply systems and equipment connected theret. International Electrotechnical Commission.
2. Oppenheim, A.V. and Schafer, R.W. (2009) *Discrete-Time Signal Processing*, 3rd edn, Prentice Hall.
3. Offeli, C. and Petri, D. (1990) Interpolation techniques for real-time multifrequency waveform analysis. *IEEE Transactions on Instrumentation and Measuremement*, **3**, 325–331.
4. Gallo, D., Langella, R. and Testa, A. (2004) Desynchronized processing technique for harmonic and interharmonic analysis. *IEEE Transactions on Power Delivery*, **39**, 993–1001.
5. Gallo, D., Langella, R. and Testa, A. (2002) A self tuning harmonics and interhamonics processing technique. *European Transactions on Electrical Power*, **12**, 25–31.
6. Marple, S.L. (1987) *Digital Spectral Analysis with Applications*, Prentice Hall, Upper Saddle River, NJ.
7. Testa, A., Akram, M.F., Burch, R., Carpinelli, G., Chang, G., Dinavahi, V., Hatziadoniu, C. *et al.* (2007) Interharmonics: Theory and modeling. *IEEE Transactions on Power Delivery*, **22** (4), 2335–2348.
8. Li, C., Xu, W. and Tayjasanat, T. (2003) Interhamonic: basic concepts and techniques for their detection and measurement. *Electronic Power Systems Research*, **66**, 39–48.
9. Vilar, C., Usaola, J. and Amaris, H. (2003) Frequency domain approach to wind turbine for flicker analysis. *IEEE Transactions on Energy Conversion*, **18** (2), 335–341.
10. Proakis, J.G. and Manolakis, D.G. (2007) *Digital Signal Processing: Principles, Algorithms and Applications*, Prentice Hall.
11. Bracale, A., Carpinelli, G., Leonowicz, Z., Lobos, T. and Rezmer, J. (2008) Measurement of IEC groups and subgroups using advanced spectrum estimation methods. *IEEE Transactions on Instrumentation and Measurement*, **57** (4), 672–681.
12. Bracale, A. and Carpinelli, G. (2007) Adaptive Prony method for waveform distortion detection in power system. *International Journal of Electrical Power and Energy Systems*, **29** (5), 371–379.
13. Stoica, P. and Moses, R. (2005) *Spectal Analysis of Signals*, Pearson.
14. Moon, T.K. and Stirling, W.C. (1999) *Mathematical Methods and Algorithms for Signal Processing*, Prentice Hall.

9

Time-Frequency Signal Decomposition

9.1 Introduction

A signal can be decomposed in different methods and domains, and alternative representations can be used for different applications. Signals obtained from electrical power systems are time and frequency dependent. In some applications such as power quality and protection, frequency domain analysis is used due to the fact that features and information can easily be obtained and characterized for possible transient conditions associated with the presence of high-frequency harmonics or other disturbances. As the electric smart grid of the future becomes more complex in terms of the variability of loads and generation, the response to market incentives and the increased utilization of power electronics for energy processing, electrical signals will require a broader set of tools and methods for signal processing.

The basic bridge between time and frequency domains is the Fourier transform (FT). The FT decomposes the original signal into a sum of the weighted sinusoidal functions. These sinusoidal functions are called bases of the decomposition and have infinite support, which means that each function of the base has infinite duration in the time domain. Because of this infinite support, the FT is not the best tool to analyze power system signals; power system signals are non-stationary signals but the Fourier transform assumes that the signals under analysis are stationary.

Figure 9.1 depicts a non-stationary continuous-time signal to be analyzed using FT in a computerized approach. This figure represents a signal that, after 0.25 s, is corrupted with a third harmonic of 0.3 pu amplitude and an interharmonic of 150 Hz and 0.1 pu amplitude. Figure 9.1b illustrates the FT applied to the windowing signal. Note that the magnitudes of the harmonic and interharmonic are different from its original signal. Furthermore, the FT does not carry any time information. As such, it cannot identify from its spectrum the instant at which the new frequency component appears.

In order to overcome this limitation, alternative methods have been proposed such as the short-time Fourier transform (STFT), wavelets and filters banks [1–4]. These techniques are commonly known as joint time-frequency analysis. A time-frequency decomposition is one of

Power Systems Signal Processing for Smart Grids, First Edition. Paulo Fernando Ribeiro, Carlos Augusto Duque, Paulo Márcio da Silveira and Augusto Santiago Cerqueira.
© 2014 John Wiley & Sons, Ltd. Published 2014 by John Wiley & Sons, Ltd.
Companion Website: http://www.wiley.com/go/signal_processing/

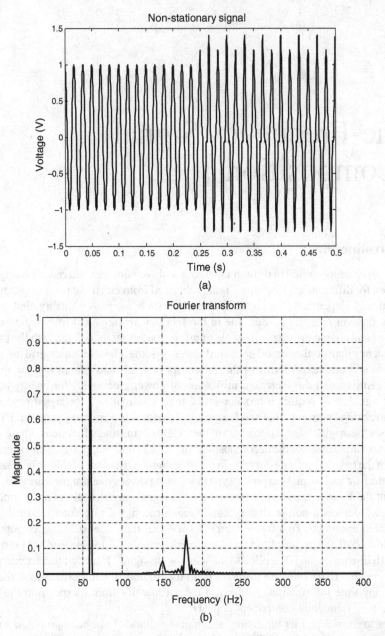

Figure 9.1 FT applied to non-stationary signal: (a) non-stationary signal; (b) spectrum of the signal.

the results of time-frequency analysis where there is access to individual components of the signal. Figure 9.2 depicts the decomposition of the previous signal using the STFT decomposition approach. The fundamental component can be seen at the upper section, the interharmonic component in the center and the third harmonic at the lower section. Note that both time and frequency information are available now, so that the instant that each

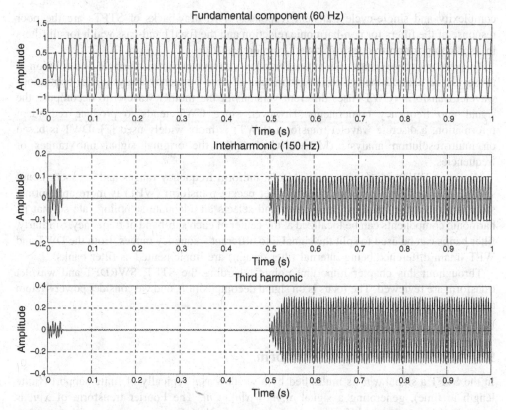

Figure 9.2 Time-frequency decomposition of signal in Figure 9.1a, by using short-time Fourier transform. From the top to bottom: fundamental component; interharmonic; and third harmonic.

component changes can be identified. Note that the amplitude of each decomposition is equal to its original value.

The decomposition based on STFT as presented in Figure 9.2 is not commonly used in STFT theory and its results are shown in complex numbers. Concerns about time-varying harmonics have increased in the last decade, and tools for the visualization of the individual components are useful for both power system diagnosis and understanding phenomena involved with the generation of time-varying harmonics.

There are two ways to analyze time-varying harmonics in power systems. The first is based on estimation techniques, which are concerned with extracting useful information from signals such as amplitude, phase and frequency. Alternatively, we can use the concept of decomposition as illustrated in Figure 9.2. Decomposition techniques concern how the original signal can be separated into its individual components including harmonics, interharmonics and subharmonics, and provides a new approach for harmonic analysis. Since the signal decomposition is carried out in the time domain, it allows for a better observation of the time-varying nature of the signals.

For a brief review of signal decomposition techniques, refer to the sliding-window recursive DFT (SWRDFT) [5,6]. The SWRDFT is a special case of STFT that enables the decomposition of a signal in its time-varying harmonic components by way of low computational complexity and single-cycle transient responses. Two drawbacks of STFTs are the poor response of the filters for non-harmonic rejection and the fixed band-pass width for all filters in its decomposition tree, as explained in reference [6].

To overcome these issues with a 'fixed window', a wavelet transform could be a solution as it uses the so-called time-scale analysis instead of time-frequency analysis. The continuous wavelet transform (CWT) uses different versions of the mother wavelet to decompose the signal into frequency components. However, since CWT tends to produce redundant information, a discrete wavelet transform (DWT) is more widely used [7]. DWT is based on multi-resolution analysis (MRA), which splits the original signal into ranges of frequencies.

However, DWT splits the signal into non-uniform frequency ranges and, as such, is not suitable for harmonic analyses. The wavelet packet transform (WPT) is more appropriate since it divides the frequency uniformly and selects an adequate sampling rate so that its harmonic components can be localized at the center of each sub-band of frequency. Similarly, filter banks can be used to split the signal into different frequency ranges. Both the DWT and WPT (main difference being internal filter design) are implemented as filter banks.

Throughout this chapter important points regarding the STFT, SWRDFT and wavelet transform are reviewed. The focus is on signal decomposition, and we consider power system applications whenever relevant.

9.2   Short-Time Fourier Transform

In the STFT a signal $x[n]$ is multiplied by a window $v[n]$ typically of finite support (finite length in time), generating a signal $x_0[n] = x[n] \cdot v[n]$. The Fourier transform of $x[n]$ is computed, the window is shifted in uniform amounts and the Fourier transform of the new data $x_i[n] = x[n] \cdot v[n - iK]$ is computed, where K is a constant shift imposed on the window. The duration of the window governs the time localization of the analysis, whereas the bandwidth of the window governs the frequency resolution. Figure 9.3 illustrates the basic concept of STFT. The figure shows a $x[n]$ signal being multiplied by the time-shifted version of the window $v[n]$ (continuous curves). Note that the shifted windows can overlap, depending on the value of K and the length of the window L. The FT of each windowing signal $x_i[n]$ is then performed. If m is the center of the window the time-frequency representation is given by $X_{\text{STFT}}(e^{j\omega}, m)$, where ω is a digital frequency in the range of $-\pi \leq \omega < \pi$ and m is typically an integer multiple of K. Mathematically, the STFT is defined

$$X_{\text{STFT}}(e^{j\omega}, m) = \sum_{k=-\infty}^{\infty} x[k]v[k - m]e^{-j\omega k}. \tag{9.1}$$

9.2.1   Filter Banks Interpretation

The filter banks interpretation is a convenient approach to study STFT since it gives a practical structure for implementation, it helps the generalization of the STFT and it leads to a better understanding of other decomposition methods such as wavelet transforms [1]. Equation (9.1) can be rewritten for a specific frequency ω_0, for which a graphical

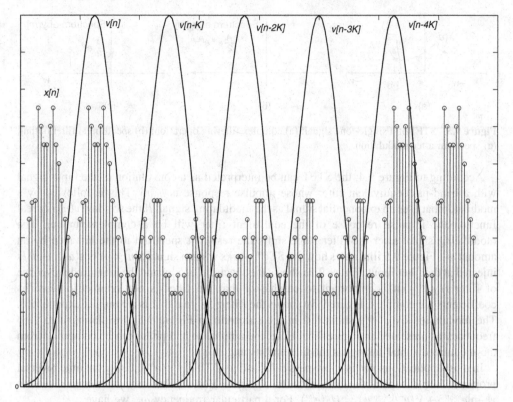

Figure 9.3 Short-time Fourier transform: illustration.

representation is provided in Figure 9.4:

$$X_{\text{STFT}}(e^{j\omega_0}, m) = e^{-j\omega_0 m} \sum_{k=-\infty}^{\infty} x[k] v[k-m] e^{-j\omega_0(k-m)} \qquad (9.2)$$

$$y_a[m] = \sum_{k=-\infty}^{\infty} x[k] h_0[m-k] \qquad (9.3)$$

$$h_0[m] = v[-m] e^{j\omega_0 m}. \qquad (9.4)$$

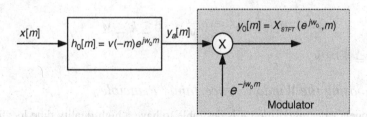

Figure 9.4 STFT filter interpretation.

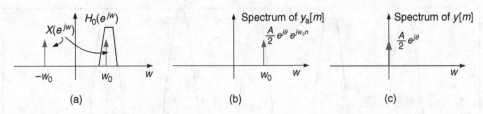

Figure 9.5 STFT of a single-tone signal: (a) complex filtering operation; (b) spectrum at filter output; (c) spectrum after modulation.

According to Figure 9.4, the STFT can be interpreted as a convolution of the input signal with a band-pass equivalent filter whose impulse response is $h_0[n]$. This is followed by a modulator that uses an exponential signal as the modulating signal. If the window $v(n)$ is a real function the impulse response of the equivalent filter will be a complex number. The modulating signal after the filter stage shifts the resultant spectrum to the left side by an amount ω_0. Figure 9.5 illustrates how the STFT works when a signal $x[n] = A\cos(\omega_0 n + \theta)$ is injected at the filter input. Figure 9.5a shows the spectrum of $x[n]$ and the magnitude response of filter $H_0(e^{j\omega})$. It is important to emphasize that the filter is not symmetric since its coefficients are complex. Figure 9.5b shows the filtered signal that is a complex exponential. The modulation by $e^{-jn\omega_0}$ shifts left in the spectrum in Figure 9.5b translating it to zero frequency as presented in Figure 9.5c. Here, the amplitude and phase can be computed from the modulated signal taking its module and its angle.

In most applications $v[n]$ is a low-pass filter and its DTFT is $V(e^{j\omega})$. Using the time reversal property (see Chapter 4) this leads to $v[-n] \leftrightarrow V(e^{-j\omega})$ and the modulated version $v[-n]e^{j\omega_0 n} \leftrightarrow V\left(e^{-j(\omega-\omega_0)}\right) = H_0(e^{j\omega})$. For a particular frequency ω_k, we have

$$v[-n]e^{j\omega_k n} \leftrightarrow V\left(e^{-j(\omega-\omega_k)}\right) = H_k(e^{j\omega}). \qquad (9.5)$$

For every frequency ω_k the STFT performs the filtering operation as illustrated in Figure 9.4 and is used to produce the output sequence $X_{STFT}(e^{j\omega_k}, n)$. The STFT can therefore be used as a filter bank, where each filter in the bank has the same shape (the same bandwidth) and is obtained by the modulation of the mother window as expressed in Equation (9.5). Figure 9.6a shows a filter bank used to decompose the signal in M components, each centered at ω_k, $k = 0, 1, \ldots, M - 1$. Figure 9.6b presents the magnitude response of the filter bank.

When the frequencies ω_k are uniformly spaced over the range $-\pi$ to π, the filter bank is referred to as a *uniform filter bank* [1]. In this case the filters are obtained in the z domain as

$$H_k(z) = V(z^{-1}W_M^{-k}), \quad 0 \le k \le M - 1 \qquad (9.6)$$

where $W_M = e^{-j2\pi/M}$.

9.2.2 Choosing the Window: Uncertainty Principle

In time-frequency decomposition it is desirable to have a high-quality time localization and frequency resolution. The window choice can control these two parameters. Unfortunately it

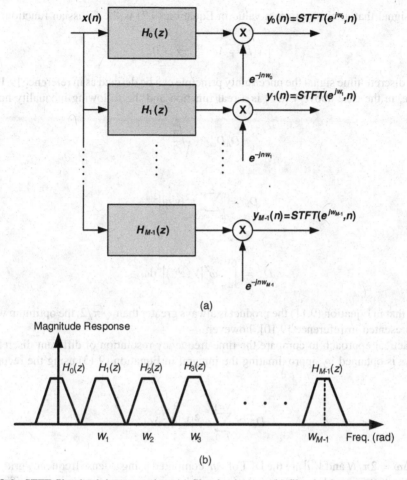

(a)

(b)

Figure 9.6 STFT filter bank interpretation: (a) filter banks tree; (b) filter banks magnitude responses.

is not possible to simultaneously have a good time localization (Δt) and frequency resolution ($\Delta\Omega$). The *uncertainty principle* relates Δt and $\Delta\Omega$ and is limited by [3,8]

$$\Delta t\, \Delta\Omega \geq 1/2. \tag{9.7}$$

The above relationship is valid for a continuous-time signal with unitary energy $\|v(t)\| = 1$. For a continuous-time signal, time and frequency resolution are defined [3,8]:

$$\Delta t^2 = \int_{-\infty}^{\infty} t^2 v^2(t)\mathrm{d}t \tag{9.8}$$

and

$$\Delta\Omega^2 = \frac{1}{2\pi}\int_{-\infty}^{\infty} \Omega^2 \left|V^2(j\Omega)\right|\mathrm{d}\Omega. \tag{9.9}$$

The signal that attains optimum value in Equation (9.7) is the Gaussian function:

$$v(t) = Ae^{-\alpha t^2}, \ \alpha > 0. \tag{9.10}$$

For a discrete-time signal the uncertainty principle can be defined as in reference [9,10]. For example, in the case above $V(e^{j\omega})$ is a real function and the following inequality holds:

$$D_n D_\omega > \sqrt{\frac{\pi}{2}} \tag{9.11}$$

where

$$D_n^2 = \sum_{n=-\infty}^{\infty} n^2 |v[n]|^2 \tag{9.12}$$

and

$$D_\omega^2 = \int_{-\pi}^{\pi} \omega^2 |V(e^{j\omega})|^2 d\omega. \tag{9.13}$$

Note that in Equation (9.11) the product is always greater than $\sqrt{\pi/2}$; the optimum window is not presented in reference [9,10], however.

A practical approach to compare the time-frequency resolution of different discrete-time windows is obtained by approximating the integral in Equation (9.13) using the rectangular rule:

$$D_\omega^2 \cong \sum_{i=0}^{N-1} \omega_i^2 |V[i]|^2 \Delta\omega \tag{9.14}$$

where $\Delta\omega = 2\pi/N$ and $V[i]$ are the DFT of $v[n]$ computed using a dense frequency grid, that is, N must be large enough in order to obtain a good approximation for the integral. The dense grid of frequency can be obtained using a zero pad approach to $v[n]$, and is included in the FFT MATLAB® command

$$V[i] = fft(v, N). \tag{9.15}$$

A simple example is for a rectangular window of length $2L+1$ where

$$v[n] = \begin{cases} 1, & -L \le n < L \\ 0, & \text{otherwise.} \end{cases} \tag{9.16}$$

The DTFT of Equation (9.16) is

$$V(e^{j\omega}) = \frac{\sin[(2L+1)\omega/2]}{\sin(\omega/2)}. \tag{9.17}$$

Table 9.1   Some windows and their time-frequency resolutions.

Window $(2L+1)$	Main lobe width $\Delta\omega$	Time-frequency resolution $(L=50)$	Time-frequency resolution $(L=100)$
Rectangular	$4\pi/(2L+1)$	3.22	3.21
Barlett	$4\pi/(L+1)$	1.78	1.77
Hann	$8\pi/(2L+1)$	1.58	1.57
Blackman	$12\pi/(2L+1)$	1.32	1.32

Table 9.1 shows some common windows and their time-frequency resolutions for $N=1000$ in Equation (9.11), the L value is specified in Table 9.1. Note that the rectangular window represents the worst time-frequency resolution despite the fact it has the smallest main lobe width. Conversely, the Blackman window offers the best time-frequency resolution and the highest main lobe width. As such, Table 9.1 shows that a time-frequency resolution depends on the chosen window and not on the L value. The small difference between the two last columns in Table 9.1 is due to a numerical error that arose during the evaluation of Equation (9.14).

Although the rectangular window represents the poorest time-frequency resolution in practice, it is frequently used due to its simplicity.

9.2.3   The Time-Frequency Grid

Figure 9.5 shows an example in which the signal is a single-tone component at frequency ω_0. After modulation, that component is moved to the zero frequency as depicted in Figure 9.5a. However, if the signal is not single-tone but spreads over the spectrum, only the section confined inside the corresponding band-pass filter is translated around the zero frequency as depicted in Figure 9.7. Figure 9.7a represents the portion of the signal filtered by the kth band-pass filter. The resultant signal is translated to a zero frequency after the modulation, as shown in Figure 9.7b. The sampling frequency of the system must be

$$F_s > MB_w \tag{9.18}$$

because it has M bands in the tree (see Figure 9.6) and B_w is the bandwidth of the band-pass filter. After the input signal passes through the corresponding filter, only a small portion of the spectrum is represented as can be seen in Figure 9.7b. Note that the sampling frequency in the corresponding branch can be reduced by a factor of M_k without any loss of information. The sampling rate reduction is performed through the down-sampler device inserted after the modulator. However, the down-sampler can however be moved prior to the modulator if the structure is modified appropriately, as shown in Figure 9.8.

Applying the decimation in Equation (9.3) according to Section 6.4.2,

$$y_{ak}[mM_k] = \sum_{k=-\infty}^{\infty} x[k]h_k[mM_k - k] \tag{9.19}$$

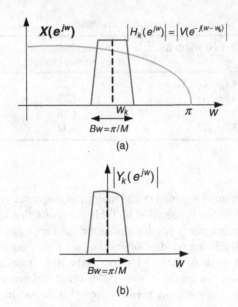

Figure 9.7 kth band base of the input signal: (a) kth band-pass filter applied to input signal; (b) resultant signal after modulation.

meaning that the window in the time domain has shifted in M_k samples at time. All filters in the filter bank are now a shifted version of the same window. Consequently, all filters have the same bandwidth B_w and the time shift for each branch is M, where M represents the time resolution. Furthermore, the distance between the center frequency of each band-pass filter represents its frequency resolution. In the present case this distance is equivalent to B_w.

Figure 9.9 shows a two-dimensional time-frequency grid corresponding to its short-time Fourier transform. The intersection of the lines represent the location of $X_{STFT}(e^{j\omega_k}, n)$.

9.3 Sliding Window DFT

Table 9.1 shows that the rectangular window is the poorest window with regards to time-frequency resolution; it is commonly used in practice due to its simplicity, however. The rectangular window used here is

$$v[n] = \begin{cases} 1, & -(N-1) \leq n \leq 0 \\ 0, & \text{otherwise.} \end{cases} \tag{9.20}$$

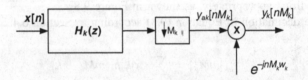

Figure 9.8 Down-sampling STFT structure for branch k.

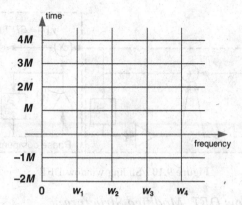

Figure 9.9 The time-frequency grid for SFTF.

Note that a shifted version of Equation (9.16) has been used. The z transform of Equation (9.20) is given by

$$V(z) = \frac{z^{(N-1)} - z^{-1}}{1 - z^{-1}}.$$ (9.21)

Representing Equation (9.5) in the z domain yields

$$v[-n]e^{j\omega_k n} \leftrightarrow V(z^{-1}e^{j\omega_k}) = H_k(z).$$ (9.22)

After the algebraic manipulation of Equation (9.21), taking into account Equation (9.22),

$$H_k(z) = \frac{1 - \alpha_k^N z^{-N}}{1 - \alpha_k z^{-1}}$$ (9.23)

where

$$\alpha_k = e^{j\omega_k}.$$ (9.24)

If $\omega_k = 2\pi k/N$ (the DFT bins) then

$$\alpha_k = e^{j\frac{2\pi}{N}k} = W_N^{-k}.$$ (9.25)

In this case the implementation of the pseudo-recursive block is shown in Figure 9.10. The name 'pseudo-recursive' is used here because the pole in $z = \alpha_k$ of Equation (9.23) is canceled by the zero at the same location [11]. The reader can compare this structure with that depicted in Figure 4.2: they are the same. To arrive at the same form the initial memory must be set up with unity instead of zero and the delay W_N^{-k} moved back. After a block manipulation, Figure 9.10 becomes identical to Figure 4.2.

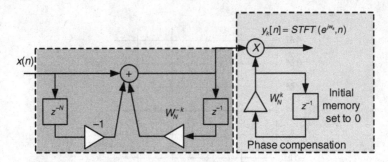

Figure 9.10 Sliding window DFT.

9.3.1 Sliding Window DFT: Modified Structure

Figure 9.10 is composed of two main blocks. The darker block is a digital filter with complex coefficients and the lighter one is a modulator block (see also Figure 9.4). The composed structure is a time-variant structure; as such, the cascade order cannot be changed. However, it is not difficult to verify that Figure 9.11 is an equivalent structure, where $\theta = 2\pi/N$ and $H_0(z)$ is obtained from Equation (9.23).

By taking the real and imaginary portion of the STFT from Figure 9.11, we obtain

$$Y_c^k[n] = Y_c^k[n-1] + (x[n] - x[n-N])\cos(\theta k n) \tag{9.26}$$

$$Y_s^k[n] = Y_s^k[n-1] - (x[n] - x[n-N])\sin(\theta k n) \tag{9.27}$$

where Y_c^k and Y_s^k are the real and imaginary components, respectively, of the kth STFT, also referred to as the quadrature terms of the sliding window DFT. Furthermore, Figure 9.12 is the equivalent time-varying filter structure obtained from Equations (9.26) and (9.27) used to compute the quadrature terms.

9.3.2 Power System Application

For protections purposes, normally a sliding-window DFT is used to extract and compute the amplitude and phase of the fundamental component [12], but not the waveform. The main objective of harmonic decomposition is however to obtain the fundamental waveform, as well as the waveform of each individual harmonic. This task can be performed by considering and using the quadrature expressions Equations (9.26) and (9.27). The implementation of this approach can be accomplished in two ways. First, the sine and cosine coefficients previously calculated and stored can be used. In this case, the algorithm must perform an internal product

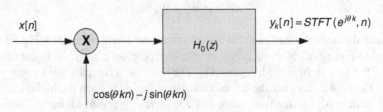

Figure 9.11 Sliding window DFT.

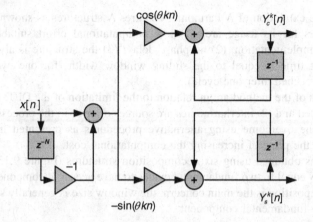

Figure 9.12   Modified sliding window DFT: quadrature computation.

using a vector of coefficients in each observable window. Secondly, a digital sine-cosine generator can be used; this method been adopted for the decomposition and analysis of some signals from power system events, as will be demonstrated.

After some minor modifications, the digital sine-cosine generator described in [11] can be implemented according to the matrix equation:

$$\begin{bmatrix} s_1[n] \\ s_2[n] \end{bmatrix} = \begin{bmatrix} \cos(\theta h) & \sin(\theta h) \\ -\sin(\theta h) & \cos(\theta h) \end{bmatrix} \cdot \begin{bmatrix} s_1[n-1] \\ s_2[n-1] \end{bmatrix} \tag{9.28}$$

where $s_1[n]$ is a sine function and $s_2[n]$ is a cosine function.

In adopting this sine-cosine generator the decomposition and the reconstruction tasks can run parallel to each other, as shown in Figure 9.13. These structures are referred to as the sliding-window recursive DFT (SWRDFT) [6].

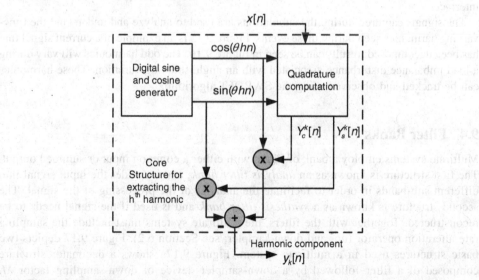

Figure 9.13 Sliding window recursive DFT (SWRDFT): decomposition core structure.

Of course, the extraction of N harmonics requires N structures as shown in Figure 9.13. Some advantages for its usage are: (1) low computational effort suitable for real-time decomposition implementation; (2) no phase delay; (3) the structure is always stable; and (4) the transient time is equal to the sliding window width (for one cycle window the convergence is reached after one cycle).

Disadvantages of the method are in relation to the limitation of the DFT: (1) synchronous sampling is needed and (2) interharmonics are sources of error in the process. Some of these drawbacks can be overcome using alternative procedures as presented in Section 9.4.6, unfortunately at the price of increasing the computational cost.

Figure 9.2 was obtained using six decomposition structures (Figure 9.13) and using the window length N equal to two fundamental cycles to have access at component 30 Hz. When harmonic decomposition is the main concern, the window size is generally set to be equal to one cycle of the fundamental component.

9.3.2.1 SWRDFT in Simulated Signal

Simulated and synthetic signals are used to check the SWRDFT algorithms. While synthetic signals can be generated directly, as for example in MATLAB®, a simulated signal can be obtained from different electromagnetic transient programs such as ATP, SimPower-MATLAB® and PSCAD™. Depending on the precision of the models, these signals represent the real world with great fidelity. For this reason, it is very useful to utilize them to test any class of algorithm to be implemented in intelligent electronic devices (IEDs).

Figure 9.14 shows a portion of a system that has been modeled in PSCAD. It contains a source and two sections of a transmission line feeding transformers, linear and non-linear loads, as a six-pulses bridge. Some disturbances are generated in this system such as a load imbalance, load rejection and failures in the converter pulse system through the control interface.

The signals captured during the simulations are used to analyze and understand the time-varying harmonics seen during these events. Figure 9.15 is an example of a current signal that has been decomposed; results can be seen in Figure 9.16. The odd harmonics will vary during a load imbalance disturbance associated with an angle shooting variation. These harmonics can be tracked and observed using the SWRDFT algorithm.

9.4 Filter Banks

Multirate systems employ a bank of filters with either a common input or summed output. The first structure is known as an *analysis filter bank*, and it divides the input signal into different sub-bands in order to facilitate the analysis or the processing of the signal. The second structure is known as a *synthesis filter bank* and is used if the signal needs to be reconstructed. Together with the filters, the multirate systems must include the sampling rate alteration operator (up- and down-sampler; see Section 6.2). Figure 9.17 depicts two basic structures used in a multirate system. Figure 9.17a shows a decimator structure composed of a filter followed by a down-sampler device of down-sampling factor M.

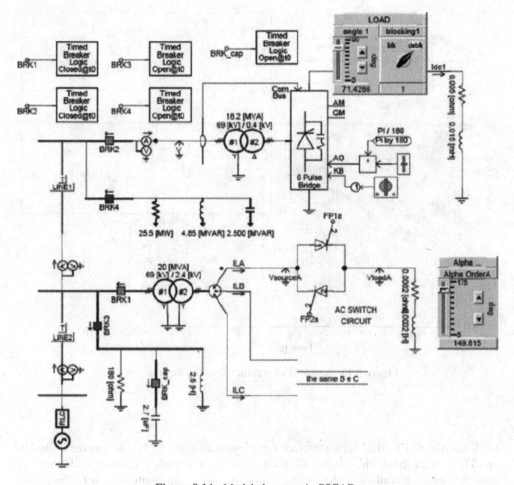

Figure 9.14 Modeled system in PSCAD.

Figure 9.17b shows the interpolator structure composed of an up-sampler (with an up-sampling factor of L) followed by a filter. The decimator structure is responsible for reducing the sampling rate by M and the interpolator structure for increasing it by L. On the right side of Figure 9.17 a simplified representation of the decimator and interpolator are shown where $M = L = 2$.

The division of the input signal into its odd harmonic components by a straightforward method of building an analysis filter bank, is depicted in Figure 9.18a. In this structure $H_k(z)$ is the transfer function in the z domain for the kth band-pass filter centered at the kth harmonic. Note that Figure 9.18a is not a multirate system because the structure does not include sampling rate alternation. This means that there is only one sampling rate in the whole system. The reader may ask what the difference is between the analysis filter

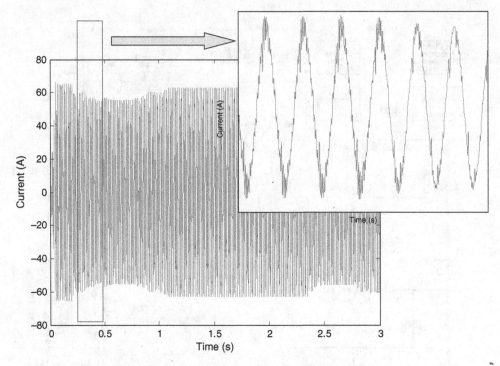

Figure 9.15   Variation of current caused by variable load.

bank and the STFT filter bank structure represented in Figure 9.6. The answer is that in an STFT filter bank the output of each filter is a complex number that contains information (magnitude and phase) about its harmonic components instead of the signal itself, and the filters have complex coefficients. In a filter bank structure, such as that depicted in Figure 9.18a, the output corresponds to the decomposed component in its time domain.

The practical problem concerning the structure shown in Figure 9.18a is the difficulty in designing each individual band-pass filter. This issue becomes even more challenging when a high sampling rate must be used to handle the signal and the consequent abrupt transition band. Another important question is in relation to the inverse processing, that is, how to obtain the reconstructed signal $\hat{x}[n]$ from the components $v_k[n], k = 0, 1, 2, \ldots, K - 1$ as illustrated in Figure 9.18b. This leads to the concept of the perfect reconstruction (PR) filter bank. A filter bank is said to be PR if $\hat{x}[n] = cx[n - l]$, where l is an integer constant and c is real. This means that the reconstructed signal is a shifted and scaled version of the input signal.

There are several textbooks on signal processing that address the theory of multirate filter banks in detail, such as references [1,11]. This section will cover the concepts that are considered of greatest importance to power system applications. In particular, we focus on topics that help to link the theory of filter banks and the wavelet transform.

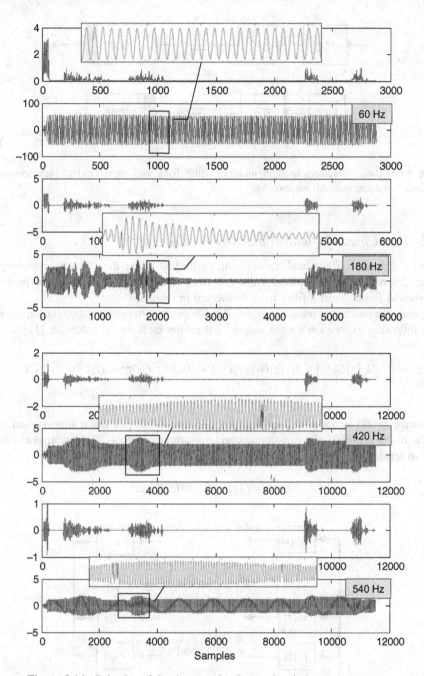

Figure 9.16 Behavior of the time-varying harmonics during system events.

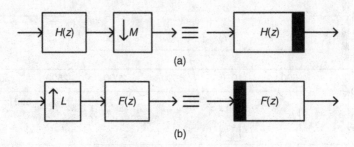

Figure 9.17 Basic structures used in multirate filter bank and its equivalent representation for $L = M = 2$: (a) decimator; (b) interpolator.

9.4.1 Two-Channel Quadrature-Mirror Filter Bank

Several concepts in filter bank theory can be studied using the structure represented in Figure 9.19, referred to as a quadrature-mirror filter (QMF) bank. The QMF bank presents an analysis and synthesis of a filter bank connected in cascade.

By using the multirate concepts presented in Chapter 8, after some manipulation we arrives at the following expression for the output of the filter bank in the z domain [11]:

$$Y(z) = \frac{1}{2}\{[H_0(z)G_0(z) + H_1(z)G_1(z)]X(z) + [H_0(-z)G_0(z) + H_1(-z)G_1(z)]X(-z)\}.$$

$$(9.29)$$

Equation (9.29) is not a transfer function since the QMF is a linear time-variant system because of the presence of down-sampler and up-sampler devices. This equation can be split into two terms

$$Y(z) = T(z)X(z) + A(z)X(-z) \tag{9.30}$$

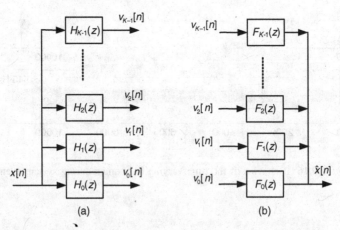

Figure 9.18 (a) Analysis filter bank and (b) synthesis filter bank.

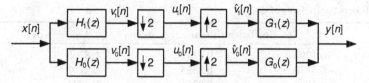

Figure 9.19 Quadrature-mirror filter bank.

where

$$T(z) = \frac{1}{2}[H_0(z)G_0(z) + H_1(z)G_1(z)] \tag{9.31}$$

and

$$A(z) = \frac{1}{2}[H_0(-z)G_0(z) + H_1(-z)G_1(z)]. \tag{9.32}$$

$T(z)$ is referred to as a *distortion transfer function* and $A(z)$ as the *aliasing term*. Spectrum aliasing may occur due to a down-sampler operation, both in low-pass and in high-pass paths. Note however that it is possible to choose the analysis and synthesis filters such that $A(z) = 0$ leads to a linear time-invariant system. The *aliasing-free* filter bank is obtained by zeroing Equation (9.32), or setting

$$H_0(-z)G_0(z) + H_1(-z)G_1(z) = 0 \text{ aliasing free condition.} \tag{9.33}$$

The *perfect reconstruction* condition is obtained if the system is aliasing-free and the distortion transfer function is:

$$T(z) = \frac{1}{2}[H_0(z)G_0(z) + H_1(z)G_1(z)] = cz^{-l} \quad \text{PR condition.} \tag{9.34}$$

Under those circumstances the output is a shifted and scaled version of the input, that is, $y[n] = cx[n - l]$.

9.4.1.1 Numerical Example

Filter banks and wavelet theory have many commonalities. In fact, the MATLAB® implementation as shown in next section is a filter bank with PR characteristics. The next example will verify if the filter used to implement the discrete wavelet transform obeys the PR and the aliasing-free conditions.

The filter related to the mother wavelet *Dabauchie order 2* (db2) can be obtained using the MATLAB® function [Lo_D,Hi_D,Lo_R,Hi_R] = wfilters('db2') where Lo_D, Hi_D, Lo_R and Hi_R are the coefficients of the filters $H_0(z)$, $H_1(z)$, $G_0(z)$ and $G_1(z)$ respectively. The filter coefficients are listed in Table 9.2.

Table 9.2 Filters coefficients of db2.

Filter	Coefficients $[a_0, a_1, \ldots, a_N]\, a_0 + a_1 z^{-1} + \cdots + a_N z^{-N}$			
$H_0(z)$	−0.1294	0.2241	0.8365	0.4830
$H_1(z)$	−0.4830	0.8365	−0.2241	−0.1294
$G_0(z)$	0.4830	0.8365	0.2241	−0.1294
$G_1(z)$	−0.1294	−0.2241	0.8365	−0.4830

The PR condition (Equation (9.34)) can be evaluated manually or by using the MATLAB® commands:

```
X1=conv(Lo_D,Lo_R);
X2=conv(Hi_D,Hi_R);
T=(X1+X2)/2;
T=[0 0.0000 0 1.0000 0 0.0000 0];
```

Note that $T(z) = z^{-3}$. To evaluate the alias-free condition using MATLAB®, firstly $H_0(-z)$ and $H_1(-z)$ must be obtained. This can be achieved by changing the signals of the odd coefficients:

```
change_sig=(-1).^(0:length(Lo_D)-1);
H0_1=change_sig.*Lo_D;
H1_1=change_sig.*Hi_D;
X1=conv(H0_1,Lo_R);
X2=conv(H1_1,Hi_R);
A=(X1+X2)/2;
A=[0 0 0 0 0 0 0];
```

We then have $A[z] = 0$.

9.4.2 An Alias-Free Realization

The alias-free condition can be obtained if the following filter relationships [11] are selected:

$$\begin{aligned}
G_0(z) &= H_1(-z) \\
G_1(z) &= -H_0(-z).
\end{aligned} \tag{9.35}$$

If this relationship is used in Equation (9.33), it can be easily verified that $A[z] = 0$.

9.4.3 A PR Condition

Equation (9.34) shows that the PR condition requires the product of two low-pass and two high-pass filters. Let these products be defined:

$$P_0(z) = H_0(z)G_0(z) \tag{9.36}$$

$$P_1(z) = H_1(z)G_1(z). \tag{9.37}$$

Table 9.3   Special properties of filter $P_0(z)$

1	All even powers of z are canceled
2	$P_0(z)$ must have only one term with an odd power of z (the term of order l)
3	l must be an odd integer

We then have

$$T(z) = \frac{1}{2}(P_0(z) + P_1(z)) = cz^{-l}. \tag{9.38}$$

where c is a constant and l is an integer. By using the alias-free relationship expressed in Equation (9.35) and setting $c = 1$, we obtain

$$P_0(z) - P_0(-z) = 2z^{-l} \tag{9.39}$$

which leads to the condition expressed in Table 9.3. Equation (9.39) can be rewritten more conveniently by multiplying both sides by z^l:

$$P(z) - P(-z) = 2 \tag{9.40}$$

where

$$P(z) = z^l P_0(z) \tag{9.41}$$

by using the information in Table 9.3 and the previous equation. This implies that $P(z)$ must be a zero-phase half-band low-pass filter [11]. Table 9.4 lists the properties of the polynomial $P(z)$, which is of the form:

$$P(z) = 1 + p_1(z + z^{-1}) + p_3(z^3 + z^{-3}) + \ldots \tag{9.42}$$

The problem is then to find the coefficients $p_1, p_3, \ldots, p_{4K+2}$. Note that the order of $P(z)$ has been assigned $4K + 2$, where K is an integer value. In order to obtain extra characteristics, additional conditions are imposed on the polynomial. In the compression application it is convenient to choose the maximum number of zeros at $z = -1$ for $H_0(z)$. One general form of $P(z)$ that allows to choose m zeroes at $z = -1$ for $H_0(z)$ is given by [13]:

$$P(z) = \left(1 + z^{-1}\right)^m (1 + z)^m R(z) \tag{9.43}$$

Table 9.4   Special condition of filter $P(z)$.

1	Zero-phase half-band low-pass filter
2	The power zero coefficient of $P(z)$ must be 1
3	As l is odd, the $P(z)$ must have only odd power of z, except the power zero coefficient that must be 1

where

$$R(z) = r_0 + \sum_{s=1}^{m-1} r_s(z^s + z^{-s}).$$ (9.44)

Example 9.1

Find $P(z)$ using $m = 2$. In this case,

$$R(z) = az + b + az^{-1} \quad \text{and} \quad P(z) = (1 + z^{-1})^2 (1 + z)^2 (az + b + az^{-1}).$$

By using the properties listed in Table 9.3 and simplifying, we arrive at the following restriction for a and b:

$$\begin{cases} 4a + b = 0 \\ 8a + 6b = 1 \end{cases}$$

which results in $a = -1/16$ and $b = 1/4$.

9.4.4 Finding the Filters from $P(z)$

Equation (9.43) represents an interesting method of finding the $P(z)$ polynomial that uses the characteristics of having zeros at $z = -1$. The problem now lies in how to factor $P(z)$ to obtain $H_0(z)$ and $G_0(z)$ individually. $H_1(z)$ and $G_1(z)$ follow from Equation (9.35). There are different methods of factoring $P(z)$; one of those is presented in [11]:

$$H_0(z) = \frac{1}{8}\left(1 + 3z^{-1} + 3z^{-2} + z^{-3}\right)$$

$$G_0(z) = \frac{1}{2}\left(-1 + 3z^{-1} + 3z^{-2} - z^{-3}\right)$$

$$H_1(z) = \frac{1}{2}\left(-1 - 3z^{-1} + 3z^{-2} + z^{-3}\right)$$

$$G_1(z) = \frac{1}{8}\left(-1 + 3z^{-1} - 3z^{-2} + z^{-3}\right).$$

Note that the previous filters are linear phase FIRs. The frequency response of $H_0(z)$ and $H_1(z)$ is shown in Figure 9.20.

Despite the fact that the filters have the PR property, observe that the low-pass and high-pass filters are not power complementary. In fact, it is possible to demonstrate that power complementarity filter cannot be obtained when the linear phase is required for the decomposition filters. As such, in several situations it is preferred to relax the linear phase characteristics in order to obtain power complementary filters. The procedure is summarized in Table 9.5.

Note that the magnitude response of $H_0(z)$ and $G_0(z)$ are the same. MATLAB® function *irpr2chfb* designs a PR QMF filter bank, returning the analysis and synthesis filters which are

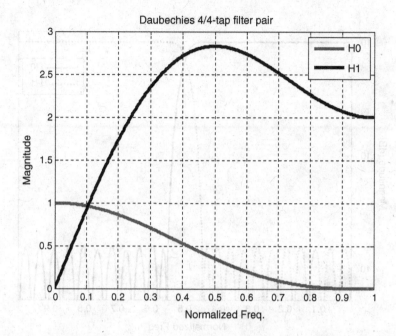

Figure 9.20   Daubechies 4/4-tap frequency response.

orthonormal and power-symmetric filter banks. The function is of the form: $[h0, h1, g0, g1] = \text{firpr2chfb}(N, fp)$ where N is the order of all four filters and must be an odd number and $h0, h1, g0$ and $g1$ are the coefficients (the impulse response) of the FIR filters. The parameter fp is the pass-band edge for the low-pass filters $h0$ and $g0$ and must be less than 0.5; $h1$ and $g1$ are high-pass filters with pass-band edge given by $1 - fp$.

Example 7.2

Design a filter bank with filters of order 45 and pass-band edges of 0.45.

```
N = 45;
     [h0,h1,g0,g1] = firpr2chfb(N,.45);
     fvtool(h0,1,h1,1,g0,1,g1,1);
```

Table 9.5   Finding the power complementary filters from $P(z)$.

1	Design $P(z)$ with order $2N$, by using Equation (9.43)
2	$P(z)$ is a factor form of zeros inside the unity circle; half of them are assigned to $H_0(z)$ and the others are assigned to $G_0(z)$
3	This leads to $G_0(z) = z^{-N}H_0(z^{-1})$ (the mirror image of $H_0(z)$)
4	The other filters are found using Equation (9.35)

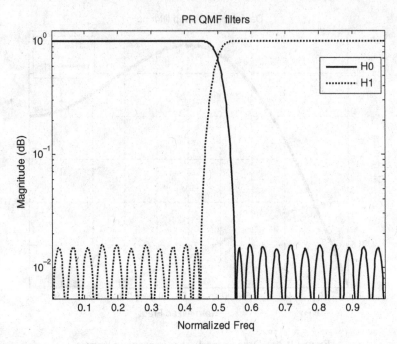

Figure 9.21 Magnitude response of PR filters using firpr2chfb Matlab function.

Figure 9.21 shows the magnitude response of $H0(z)$ and $H1(z)$, where $Hi(z)$ is the z-transform of the filter hi, $i = 1, 2$.

9.4.5 General Filter Banks

The previous section dealt with QMF banks, the PR principle and some methods of designing the filters to obtain PR QMF banks. The concept of QMF banks can be extended for a generic filter bank or, in other words, a generic PR filter bank can be designed using the basic structure of a QMF. To illustrate this idea, consider the filter bank structure presented in Figure 9.22; note the highlighted PR QMF bank. Since this section of the figure is a PR filter bank, it can be substituted by a direct connection between the down-sampler A and the up-sampler B (in reality the branch has a delay that can be easily moved outside the rectangle). As such, the

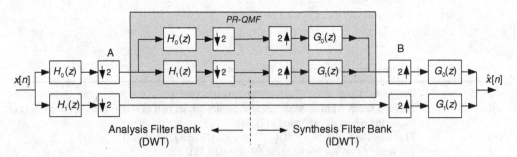

Figure 9.22 A generic PR filter banks.

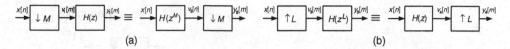

Figure 9.23 Nobles identities: (a) I and (b) II.

reader can verify that a second PR QMF is formed. Consequently, the original structure is also a PR filter bank.

The filter bank presented in Figure 9.22 is well known as a binary tree or octave-band. The next section will show that the analysis filter bank corresponds to the discrete wavelet transform (DWT) implementation and the synthesis filter bank to the inverse discrete wavelet transform (IDWT) implementation. Using the noble identities presented in Section 6.3.2, repeated in Figure 9.23 for convenience, Figure 9.22 can be redrawn as shown in Figure 9.24.

By grouping the cascade filters in each branch we arrive at structures such as those depicted in Figure 9.3. In fact, any configuration of a filter bank can be implemented using the basic procedure of a PR QMF bank. The problem is reduced to choosing four filters of the QMF structures and, through the noble properties, obtaining the equivalent filter banks structure. The following section explores the problem of building a filter bank for a harmonic decomposition.

9.4.6 Harmonic Decomposition Using PR Filter Banks

The structure in Figure 9.18a demonstrates how to build an analysis filter bank directly, and how to divide the input signal into its odd harmonic components. In this case the transfer function in the z domain of the kth band-pass filter is centered at the kth harmonic and must be designed to have 3 dB bandwidth lower than $2f_1$, where f_1 is the fundamental frequency. If only odd harmonics are assumed present in the input signal, the 3 dB bandwidth can be relaxed to be lower than $4f_1$. As mentioned before, Figure 9.18 is not a multirate system because it does not includes a sampling rate alternation device.

In this situation it is best to construct an equivalent filter bank using the multirate technique with the PR QMF bank, as mentioned in the previous section. Figure 9.25 shows how a structure equivalent to Figure 9.18 can be obtained by using the multirate approach. For the sake of simplicity, the filters $H_0(z)$ and $H_1(z)$ are renamed as H0 and H1 and the cascade of these filters with the down-sampler are redrawn by using the equivalent blocks of Figure 9.17. The output of the harmonic decomposition is from lower to higher frequencies: 1, 3, 7, 5, 15, 13, 9 and 11. This sequence seems to be abnormal, but it is correct and explained in the following.

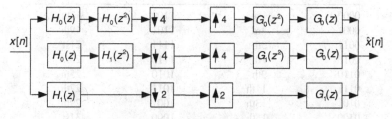

Figure 9.24 Equivalent filter bank.

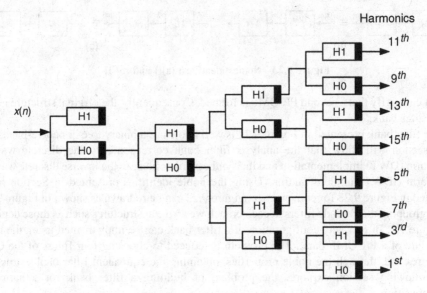

Figure 9.25 Analysis filter banks for harmonic decomposition.

To understand the decomposition sequence presented in Figure 9.25, we have to use the noble identities. Moving all down-samplers from left to right, the following equivalent filters centered around the first up to the seventh harmonics are generated:

$$\mathcal{H}_1(z) = H_0(z)H_0(z^2)H_0(z^4)H_0(z^8)$$
$$\mathcal{H}_3(z) = H_0(z)H_0(z^2)H_0(z^4)H_1(z^8)$$
$$\mathcal{H}_5(z) = H_0(z)H_0(z^2)H_1(z^4)H_1(z^8)$$
$$\mathcal{H}_7(z) = H_0(z)H_0(z^2)H_1(z^4)H_0(z^8).$$

Note that the order of each harmonic in the bank corresponds to the well-known digital *Gray code*. The Gray code is a binary numeral system where two successive values differ in only one digit. Table 9.6 shows the Gray code for four bits (equivalent to the four levels in the

Table 9.6 The Gray Code and corresponding harmonic outputs.

Gray code	Harmonic	Gray code	Harmonic
0000	1st	1100	17th
0001	3rd	1101	19th
0011	5th	1111	21st
0010	7th	1110	23rd
0110	9th	1010	25th
0111	11th	1011	27th
0101	13th	1001	29th
0100	15th	1000	31st

filter banks of our example). Considering that in this table '0' corresponds to H0(z) and '1' corresponds to H1(z), it is easy to identify the location of each harmonic output in Figure 9.25. Despite the fact that Figure 9.5 shows the output up to the 15th harmonic, the extension for up to the 30th odd harmonic is straightforward using Table 9.6.

Figure 9.26 shows the magnitude response for the filter

$$\mathcal{H}_5(z) = H_0(z)H_0(z^2)H_1(z^4)H_1(z^8).$$

This figure was obtained using a sampling rate equal to 64 samples per cycle and QMF FIR filters of 69th order (that contain 70 coefficients). The M-function used for generating this figure is [H,w,Hc]=freq_bank(br,H0,H1,a_s), where br contains the 0s and 1s corresponding to the sequence of filters H0 and H1 from left to right. As for example, for the 5th harmonic br = [0 0 1 1]. Note from the Gray code in Table 9.6 that this defines the filter centered at the 5th harmonic; the parameter a_s = 0 for analysis filter banks and a_s = 1 for synthesis filter banks. The output H is the magnitude response in dB, Hc is the magnitude in linear scale and w is the vector frequency in rad (0–π radians). Figure 9.26 shows the frequency response for the 5th harmonic obtained using the basic MATLAB® code:

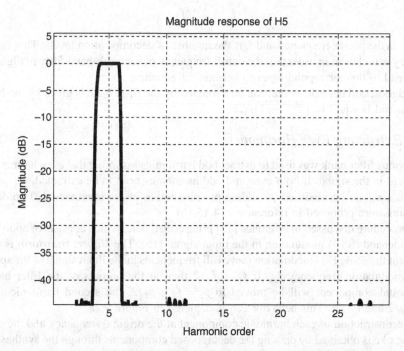

Figure 9.26 Magnitude response of filter centered at 5th harmonic.

```
br=[0 0 1 1];           % Gray code for H5
load H_pr_69b;          % H0, H1, F0 and F1 filters H=[H0; H1; F0; F1]
h0=H(1,:);
h1=H(2,:);
br=[0 0 1 1];           % Gray code for H5
[ H,w,Hc] =freq_bank(br,h0,h1,0); % analysis filter
f=w* 64* 60/(2*pi);
plot(f/60,H,'k','LineWidth',3)
grid on
title({ 'Magnitude response of H5'} ,'FontSize',16,'FontName','Verdana')
xlabel({ 'Harmonic order'} ,'FontSize',14,'FontName','Verdana')
ylabel({ 'Magnitude (dB)'} ,'FontSize',14,'FontName','Verdana')
```

The response for all filter banks is presented in Figure 9.27. Note that the central frequency corresponds to an odd harmonic. Furthermore, note the poor rejection of the even harmonics. This means that if they are present in the input signal they will be mixed with the odd harmonics. As such, further processing is needed in order to eliminate the spillover of the even harmonics towards the odd band filter. This will be discussed in Section 9.4.8.

9.4.7 The Sampling Frequency

To obtain the frequency response shown in Figure 9.27 a sampling rate of 64 samples per cycle is used. This rate is related to the number of decomposition levels at the filter banks (4 levels in Figure 9.27) through Equation (9.45):

$$f_s = 2^{L+1} f_1 \tag{9.45}$$

where f_1 is the power frequency and L is the number of decomposition levels. This sampling frequency was chosen in order that the cutoff frequency of each sub-band filter in Figure 9.27 should lead to the corresponding even harmonic frequency.

The highest odd harmonics that can be extracted by the filter bank is given by the Nyquist Theorem and is equal to $(2^{L+1} - 1) f_1$.

9.4.8 Extracting Even Harmonics

The previous filter bank was used to extract odd harmonics assuming that even harmonics are not present in the signal. If both even and odd harmonics need to be extracted, the previous approach needs to be modified by adding pre-processing of the input signal. Two different procedures were proposed in references [14,15,16].

The procedure discussed in reference [15] is described here, which consists of applying the single-sideband (SSB) modulation in the input signal [15]. The *Hilbert transform* is used to implement the SSB. This modulation moves all frequencies in the input signal of the spectrum by the modulation frequency f_m. If $f_m = f_1/2$ then at the outputs of the filter bank the fundamental component will be moved to $f_1 = f_1 + f_1/2$, the second harmonic to $f_2 = 2f_1 + f_1/2$ and so on. This behavior can is depicted in Figure 9.28.

The reconstruction of each harmonic component at the original frequency and the original sampling rate is obtained by passing the decomposed components through the synthesis filter bank and applying the inverse SSB, as illustrated in Figure 9.29.

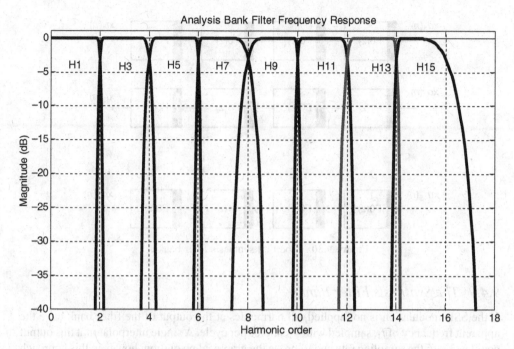

Figure 9.27 Analysis bank filter frequency response.

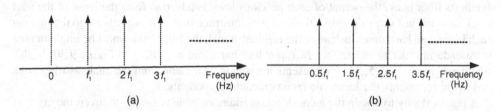

Figure 9.28 Harmonics behavior: (a) before SSB modulation and (b) after the SSB modulation.

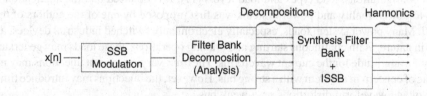

Figure 9.29 Proposed system overview.

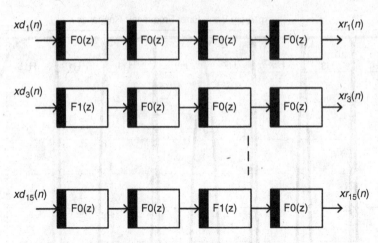

Figure 9.30   Modified synthesis filter banks.

9.4.9 The Synthesis Filter Banks

If the SSB modulation is not applied, all harmonics at the output of the filter bank have the apparent frequency of f_1, sampled with 4 samples per cycle. A single interpolator at this output could increase the sampling rate and improve the graphical resolution; however, this approach will not change the frequency back to the original harmonic frequency. To reconstruct each harmonic into its original frequency it is necessary to use the synthesis filter bank structure followed by the inverse SSB modulation if the inverse SSB is used in the pre-processing as depicted by Figure 9.29.

The synthesis filter bank used for this purpose is not a conventional one. In the conventional synthesis filter bank, the output of each previous level is added to form the input of the next level, so in the last stage the original signal is reconstructed. As the objective is to reconstruct each individual harmonics instead of the original signal, the filter banks must be implemented in cascade in order to obtain the corresponding harmonic according to Figure 9.30. In this figure $xd_i[n]$, $i = 1, 3, 5, \ldots, k$ represents the signals that came from the analysis filter bank and $xr_i[n]$ represents the harmonic reconstructed components.

A practical way to obtain the reconstruction filters cascade is simply to invert the order of the Gray code in Table 9.1 and consider 0 as $F_0(z)$ and 1 as $F_1(z)$.

9.5 Wavelet

Waveform distortions in power systems are inherently time-varying due to continuous changes in system configurations, load types and load levels [17]. It is believed that the application of wavelets to power quality and power systems was first proposed by one of the authors of this book [18]. Many power system loads, especially electronically switched industrial devices, are dynamic in nature. With continuous starting and braking operations, these loads may generate a time-varying amplitude for the current waveforms. The time-varying current injections may not be an issue of concern in a system with a strong bus. However, the injections may introduce time-varying voltage waveform distortions on a weak bus.

The time-varying nature of waveform distortions requires a precise tool for analysis and its visualization is essential for power quality studies, such as evaluation of the effects of harmonics on electrical devices and setting limits for the time-varying harmonics [19,20]. Identification of the harmonic time variations is important in estimating the temperature rise and the associated aging of equipment. As well as waveform distortions, the power system signals are subject to disturbances of several other types and need to be detected and classified to identify the possible causes and solutions. These types of signals are evidently non-stationary in nature and a time-frequency analysis is used for detecting and classifying the disturbances [21,22].

The previous sections have already considered some techniques that have a joint time-frequency representation (TFR) of the waveform. The STFT, SWRDFT and filter banks were considered and their applications to signal decomposition were discussed. The STFT was presented using a filter banks structure approach, where each filter in the bank had the same bandwidth and was generated from a basic prototype filter (the 'mother filter'), using modulation operation. The SWRDFT was presented as a variation of STFT, where the individual harmonic components were obtained using the generator approach. A general filter bank theory was presented in Section 9.4. Some important aspects of filter banks were presented and the similarity between the filter banks and wavelets was mentioned.

This section explores wavelet transforms and their applications in power systems. As mentioned, wavelet transforms are closely related to the filter bank theory. Even before the wavelet transform was formally introduced [23], the non-uniform filter bank had been used in the speech processing literature since 1972.

The literature on wavelet transform is extensive, but most of it requires a mathematical background that is beyond the majority of both power systems engineers and signal processing experts. As such, this section will present wavelets using the filter banks approach described previously whenever possible, which is easily understood by power system engineers.

9.5.1 Continuous Wavelet Transform

The continuous wavelet transform (CWT) was developed as an alternative approach to the STFT to overcome the resolution problem described in Section 9.2.3. The wavelet transform is a multi-resolution analysis (MRA) that analyzes the signal at different frequencies (scales) with different resolutions. MRA gives good time resolutions at high frequencies and poor time resolution at low frequencies. The progressive resolution characteristic of the wavelet transform can provide adequate time resolution for different components simultaneously.

In the traditional Fourier transform, a continuous-time function $x(t)$ is written in terms of the basic functions $e^{j\Omega t}$. The basic functions in a Fourier transform are parameterized by the frequency variable Ω. There are an infinite number of terms in the basis and each function has infinite time duration.

The basic philosophy of wavelet transform is to decompose a signal into a series of dilations and translations of a mother wavelet denoted $\psi(t)$. That is, all functions in the basis are generated by dilations and a shift of a single function, the *mother wavelet*. The mother wavelet is a fast-decaying oscillating waveform with a zero mean. Thus, instead of representing $x(t)$ as a linear combination of functions $e^{j\Omega t}$ (as in the Fourier transform), the wavelet attempts to represent the time function as a linear combination of

$$\psi_{\tau,s}(t) = |s|^{-1/2}\psi\left(\frac{t-\tau}{s}\right), \tau, s \in R; s \neq 0 \tag{9.46}$$

where s and τ are the dilation and translation factors, respectively. The functions $\psi_{\tau,s}(t)$ are the wavelet basis functions that are generated from the mother wavelet $\psi(t)$ by dilations and translations. A mother wavelet is a small wave that is oscillatory and obeys some specific properties, such as the *Morlet wavelet* in Figure 9.31. The Morlet wavelet is obtained from [2]:

$$\psi(t) = e^{-t^2/2} \cos(5t). \tag{9.47}$$

In the mathematical sense it has infinite support because its time duration is infinite, but for a practical reason its effective support is -4 to 4 s. The term *time support* of the wavelet is related to the duration in time of the wavelet. Note that the Morlet is symmetric, but this is not a mandatory property. The main properties that the wavelet function must obey are that of finite energy:

$$E = \int_{-\infty}^{\infty} |\psi(t)|^2 dt$$

and the admissibility condition (the mean value of the wavelet must be zero):

$$\int_{-\infty}^{\infty} \psi(t) dt = 0.$$

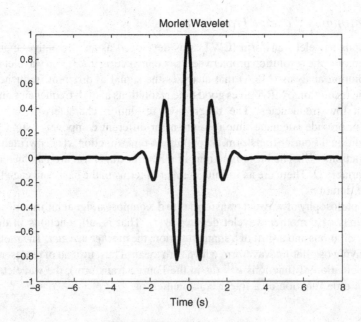

Figure 9.31 Morlet wavelet.

This is equivalent to the Fourier transform of the wavelet and has no DC component, i.e.

$$\Psi(j\Omega) = \int_{-\infty}^{\infty} \psi(t)e^{j\Omega t}\,dt \quad \text{Fourier Transform}$$

$$\Psi(0) = \int_{-\infty}^{\infty} \psi(t)dt = 0.$$

There are a number of other properties that are required for wavelets when the wavelet's basis is orthogonal or biorthogonal (see Section 9.5.4 for an explanation of these terms). However, only the useful properties for power system applications are presented in this section.

The CWT of $x(t)$ is defined as [1–4]:

$$X_{CWT}(\tau, s) = \int_{-\infty}^{\infty} x(t)\psi_{\tau,s}^*(t)dt \tag{9.48}$$

or

$$X_{CWT}(\tau, s) = |s|^{-1/2} \int_{-\infty}^{\infty} x(t)\psi^*\left(\frac{t-\tau}{s}\right)dt, \tag{9.49}$$

where $\psi_{\tau,s}(t) = |s|^{-1/2}\psi\left(\frac{t-\tau}{s}\right)$.

The continuous wavelet transform is a function of the translation parameter τ and the scale parameter s. Equation (9.48) shows how the function $x(t)$ can be expanded as a linear combination of the wavelet basis. It is very similar to the Fourier transform, except that the basis functions are not exponential and do not need to have infinite support. Equation (9.48) is known as an analysis equation and $X_{CWT}(\tau, s)$ is the continuous wavelet coefficient.

Equations (9.48) and (9.49) can alternatively be interpreted as the output of an infinite bank of linear filters described by the impulse response:

$$h_s(t) = |s|^{-1/2}\psi^*(-t/s), \tag{9.50}$$

$$X_{CWT}(\tau, s) = \int_{-\infty}^{\infty} x(t)h_s(\tau - t)dt \tag{9.51}$$

over a continuous range of scales $s \neq 0$. If the basic filter (the *mother* filter) is assumed to be a band-pass filter and the scale factor is assumed to be higher than 1, then as s increases the impulse responses will be expanding overtime. The corresponding frequency bandwidth consequently contracts, moving towards the zero frequency. The filter in the bank with the lowest s factor therefore corresponds to the filter that responds to a higher frequency, and vice versa.

MATLAB® uses the factor a instead of s to represent the scale factor. Figure 9.32 shows an example of a continuous wavelet transform of a synthetic chirp signal. In the chirp signal the frequency is increased as the time increases. This figure was generated using the MATLAB® wavelet toolbox called by the command *wavemenu*. This command opens a graphical user interface (GUI) that provides several wavelet tools such as the continuous wavelet transform, the discrete wavelet transform and the wavelet packet. In Figure 9.32 the upper graphics are the chirp signal and the lowest plot is the scalogram. The vertical axis in the scalogram is the scale factor. As mentioned, the higher scale factor corresponds to lower frequencies; the darker the shade, the higher the energy of the corresponding wavelet coefficient.

The continuous wavelet transform coefficient plots are the time-scale view of the signal. A time-scale plot is similar to a time-frequency Fourier view. However, while the electrical engineer is tempted to always correlate a time-scale with time-frequencies, it is important to highlight that the time-scale is a very natural way of viewing data derived from a great number of natural phenomena other than electrical phenomena.

Any signal processing performed on a computer must be performed on a discrete signal. If the processing is to be performed on a continuous-time signal, the signal first needs to be digitalized. What then is the application of the continuous wavelet transform? The main difference between the CWT and the DWT (discussed further in the following section) is the set of scales and translation that can be used at the transform. Unlike the DWT which imposes limits on the values of the scale and translation factors, the CWT can operate at every scale and translation according to the requirement for detail.

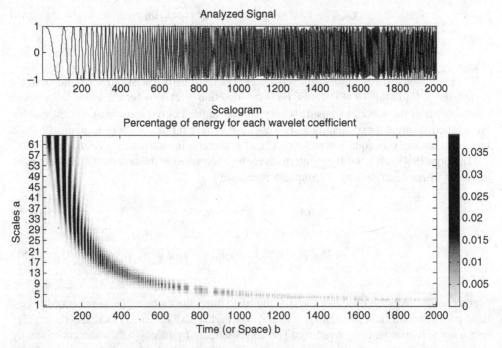

Figure 9.32 The continuous wavelet transform: (top) chirp signal; and (bottom) corresponding scalogram.

9.5.2 The Inverse Continuous Wavelet Transform

The inversibility of the CWT is an important feature of wavelet theory. All good transformations must have an analysis equation (CWT) and a synthesis equation (inverse CWT or ICWT). The condition for the existence of an ICWT is the admissibility condition (mean of wavelet must be zero) for the mother wavelet. However, this is of less importance for power system applications and, as such, the ICWT is not considered further here. Interested readers should consult specialized texts on wavelet transforms such as references [2,3,4,23].

9.5.3 Discrete Wavelet Transform (DWT)

The CWT is analyzed by computer using dense grids, both in time and scale factors. However, these grids need to be sufficiently spaced in order to obtain a coherent and accurate representation and not overload the computer. The grid used is generally known as *dyadic*, where the time is spaced at $\tau = 2^k n T_s$ and the scale is $s = 2^k$, $k \in Z$. This choice is equivalent to evaluating Equation (9.48) at a discrete set of points in the time-frequency grid:

$$X_{\text{CWT}}(2^k n T_s, 2^k) = 2^{-k/2} \int_{-\infty}^{\infty} x(t) \psi^*(2^{-k} t - n T_s) dt. \tag{9.52}$$

Using the dyadic grid in Equation (9.50) results in

$$h_k(t) = 2^{-k/2} \psi^*(-2^{-k} t) \tag{9.53}$$

which, using Equation (9.52), becomes

$$X_{\text{CWT}}(2^k n T_s, 2^k) = \int_{-\infty}^{\infty} x(t) h_k(2^k n T_s - t) dt. \tag{9.54}$$

The above integral represents the convolution between the input signal $x(t)$ and the $h_k(t)$ evaluated at a discrete set of points $2^k n T_s$. Figure 9.33 illustrates the convolution processing of Equation (9.54). The blocks allocated after the filters are the sampling functions that retain every $2^k n T_s$ samples. The output of this block is denoted $X_{\text{DWT}}(n, k)$, $n, k \in Z$ where n and k are related to the shifting and scale operators according to Equation (9.54).

At this point it is useful to discuss the frequency response of the filters given by Equation (9.53). Note that for $k = 0$, the filter impulse response is

$$h(t) = h_0(t) = \psi^*(-t). \tag{9.55}$$

The impulse response of the filter is equal to the conjugate of the time-reversed mother wavelet. It is easy to verify that

$$h_k(t) = 2^{-k/2} h(2^{-k} t). \tag{9.56}$$

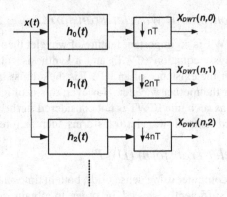

Figure 9.33 Filter banks interpretation of Equation (9.33).

In the frequency domain, we have

$$H_k(j\Omega) = 2^{k/2}H(j2^k\Omega). \tag{9.57}$$

All frequency responses of each filter in the bank are therefore obtained by frequency-scaling of a prototype filter $H(j\Omega) = H_0(j\Omega)$. The scale factor $2^{-k/2}$ in Equation (9.56) ensures that the energy of the filters will be independent of k, according to reference [1].

If the mother wavelet is chosen so that the correspondent prototype filter is a band-pass filter with cutoff frequency of α and β, then Figure 9.34 illustrates the frequency response of the filter banks of Figure 9.32. For the sake of clarity, the factor $2^{k/2}$ has been omitted.

Note that the filter $H_0(j\Omega)$ is centered at the highest band of the spectrum and the output signal of this filter has a higher frequency component. Consequently, the sampling factor applied at the output of the corresponding filter must be lower, as can be seen in Figure 9.33. The second filter in the bank is a band-pass filter, with smaller bandwidth (half size) and centered in a smaller frequency than $H_0(j\Omega)$. The sampling factor at this output is twice the previous one. This is repeated for the following filter in the filter banks. To summarize, the scale factor s is substituted by k. The higher the value of k, the lower the frequency component at the corresponding filter output and the higher its resampling factor.

The time-frequency diagram is illustrated in Figure 9.35. As can be observed, the time-frequency division is not as uniform as in STFT. Note also that the frequency spacing is smaller for the lower frequency and the corresponding time-space is larger. For high frequency, the spacing is larger while the time spacing is smaller. To understand that this time-frequency grid is superior to the STFT grid, recall that frequency resolution is directly related to the number of cycles of the corresponding signal, as is observed inside the window.

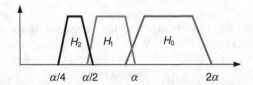

Figure 9.34 Frequency response of filter bank of Figure 9.33.

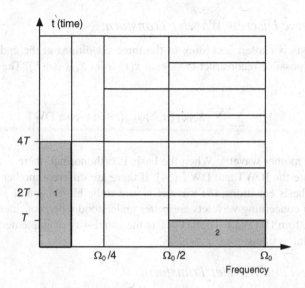

Figure 9.35 Time-frequency spacing for wavelet transform.

If the signal is of low frequency, a larger observation window is necessary to improve the resolution. Alternatively, if the processed signal is of high frequency, a short observation window can be used.

In Figure 9.35 two regions are highlighted. Region 1 corresponds to a low frequency, which means there is a narrowband filter of bandwidth $\Omega_0/8$. As mentioned before, the signal at the output of this filter has a low-frequency component. It is therefore necessary to use a larger observation window to achieve a good frequency resolution. At the same time, the samples can be kept at lower rates. Region 2 contains a filter for a high-frequency signal; as the signal at the filter output has high-frequency components, a lower observation window can be used as the sampling rate must be high. The horizontal division gives an idea of the length of the time observation window.

When the mother wavelet is adequately chosen in order to generate an orthonormal (or biorthogonal) wavelet basis, and the time-scale of the discrete sampling is a dyadic grid, this wavelet is referred to in literature as a discrete wavelet transform (DWT). Equation (9.54) is rewritten:

$$X_{\mathrm{DWT}}(n, k) = \int_{-\infty}^{\infty} x(t) h_k(2^k n T_s - t)\mathrm{d}t. \tag{9.58}$$

In summary, the term DWT is used to designate a special category of CWT sampling that satisfies the conditions: (1) the time-scale grid must be the dyadic grid; (2) the family of wavelets must form an orthogonal or a biorthogonal basis; and (3) the analysis wavelet must be compactly supported.

9.5.4 The Inverse Discrete Wavelet Transform

If the wavelet basis is chosen according to the three conditions at the end of the previous section, then it is possible reconstruct the signal $x(t)$ from $X_{\mathrm{DWT}}(n, k)$. The inverse is of the form

$$\hat{x}(t) = \sum_k \sum_n X_{\mathrm{DWT}}(n, k)\psi_{k,n}(t) \quad \text{Inverse DWT} \tag{9.59}$$

where $\psi(t)$ is the mother wavelet. When the basis is orthonormal, there is a unique mother wavelet to compute the IDWT and DWT [24]. If there are different mother wavelets for the analysis and synthesis equations, the wavelet is said to be biorthogonal.

Several aspects concerning wavelets are better understood when considering the discrete-time wavelet transfom (DTWT). The DTWT is the usual way of implementing the wavelet transform in digital systems.

9.5.5 Discrete-Time Wavelet Transform

Until now, only the continuous-time signal $x(t)$ has been considered and the output of the continuous-time filter bank was sampled to generate the DWT. However, the signals to be processed in the digital world are discrete-time signals; the connection between the DWT and the discrete-time wavelet transform (DTWT) needs to be made. The DTWT is the name given to a wavelet transform when the signal is represented by its samples (discrete-time signals) and the filters are discrete-time filters [1]. This denomination is not universally accepted; in fact, the MATLAB® wavelet toolbox and some authors use DWT for what other authors call DTWT. To be consistent with MATLAB®, after the differences between DWT and DTWT are explained we will adopt the term DWT for both signals and filters considered in discrete time.

To link DTWT and DWT the filter bank approach is followed again [1]. The filters in Figure 9.34 were generated using an analog prototype filter (or mother wavelet). The others filters in the bank were generated using the scaling property of the Fourier transform. The property cannot be used when using a digital filter, where the frequencies of interest are limited to the interval of $-\pi$ to π and the frequency response of the filters are periodic. If the same frequency division as in Figure 9.34 is to be mimicked using digital filters, we need to refer to Section 9.4.5 on general filter banks where several band-pass filters were generated by the cascade of $H_0(z)$ and $H_1(z)$.

Figure 9.24 is repeated below in Figure 9.36a and b . Figure 9.36a is known as the *wavelet decomposition tree* or *binary tree decomposition*. For many signals, the low-frequency

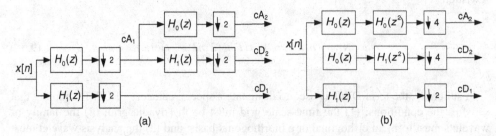

Figure 9.36 DTWT: (a) wavelet decomposition tree; and (b) equivalent implementation.

content carries the most important information about the signal. In general, the low-frequency content carries the signal identity. In Figure 9.36a the low-frequency is denoted cA_2 in accordance with multi-resolution analysis. The high-frequency content, on the other hand, gives details of the signal. This branch is denoted cD_k, where $k = 1, 2$. As a consequence, the approximations will be the high-scale or low-frequency components of the signal. The details are the low-scale, high-frequency components.

Figure 9.37 presents the frequency response of the equivalent filters. In the figure $H0 = H_0(z) \cdot H_0(z^2)$, $H1 = H_0(z) \cdot H_1(z^2)$ and $H2 = H_1(z)$. As can be observed from the figure, H2 is the higher band in the bank and the down-sampler is the smallest in the bank. H1 is the intermediary band and H0 is the lowest band in the bank.

Additional branches can be included in the previous wavelet decomposition tree of Figure 9.36a. For example, the approximation cA_2 coefficient can be decomposed at an added level as represented in Figure 9.38a. Figure 9.38b shows the new frequency response of the equivalent wavelet decomposition tree.

The multi-resolution analysis makes it possible to analyze a signal at different frequencies with different resolutions. For high frequencies (low scales), which generally correspond to power system phenomena of a short duration in time, high sampling rates are needed. For low-frequency phenomena (high scales) a long period of time observation window is required (the approximation coefficient and high-order details must be obtained). This multi-resolution is obtained using the previous filter banks. The low-pass and high-pass filtering branches of the filter bank retrieve the approximations and details of the signal $x[n]$, respectively. The filter banks shown in Figure 9.38a can be expanded to an arbitrary number of levels, depending on the desired details. In Figure 9.36a the number of levels is 2 and in Figure 9.38a it is 3. The coefficients cA_3 represent the lowest half of the frequencies in $x[n]$. The number of samples in

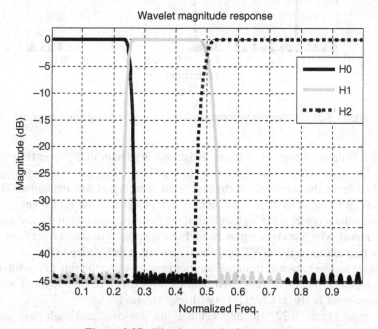

Figure 9.37 Wavelet magnitude response.

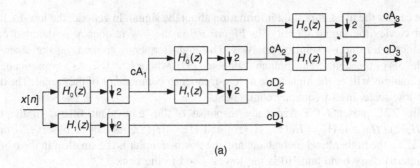

(a)

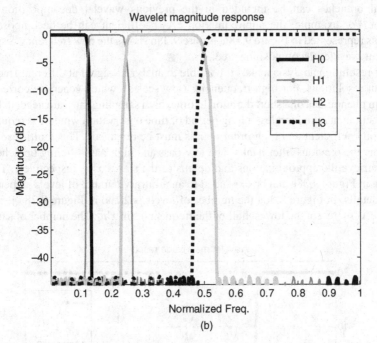

(b)

Figure 9.38 Wavelet (a) decomposition tree; and (b) magnitude response.

cA_L, where L is the total number of levels in the tree, is a factor of 2^L lower than the original signal.

Still in the L level, the output cD_L is the first detail in the signal decomposition. The number of samples is also a factor of 2^L lower than the original signal. After each level, the output of the high-pass filter represents the highest half of the frequency content of the low-pass filter of the previous level, which leads to a pass-band. For a special set of filters $H_0(z)$ and $H_1(z)$ this structure is referred to as the DWT, and the filters are called wavelet filters.

These filters can be obtained from the wavelet toolbox using the function wfilters.m. This function computes four filters associated with the orthogonal or biorthogonal wavelet. The complete command is [H0,H1,G0,G1] = wfilters('wname').

According to Figure 9.22, H0 and H1 are the low-pass and high-pass analysis (or decomposition) wavelet filters, respectively and G0 and G1 are the low-pass and high-pass synthesis (or reconstruction) filters, respectively. The string 'wname' is the respective wavelet.

The available wavelet names 'wname' are: (a) Daubechies: 'db1' or 'haar', 'db2', . . . , 'db45'; (b) coiflets: 'coif1', . . . , 'coif5'; (c) symlets: 'sym2', . . . , 'sym45'; (d) discrete Meyer wavelet: 'dmey'; (e) biorthogonal: 'bior1.1', 'bior1.3', 'bior1.5', . . . , 'bior6.8'; (f) reverse biorthogonal: 'rbio1.1', 'rbio1.3', 'rbio1.5', . . . , 'rbio6.8'.

Some wavelet families do not have discrete-time wavelet filters and consequently do not have DWT. For example, by typing *waveinfo('morlet')* in the MATLAB® command window the response is generated:

```
Definition:
morl(x) = exp(-x^2/2) * cos(5x)
Family                  Morlet
Short name              morl
Orthogonal              no
Biorthogonal            no
Compact support         no
DWT                     no
CWT                     possible
Support width           infinite
Effective support       [-4, 4]
Symmetry                yes
```

The two highlighted lines above show that the Morlet wavelet can be used only in the continuous form; there are no discrete-time filters (FIR) that can be used to implement the DWT.

On the other hand, if the command *waveinfo('db3')* is typed into the MATLAB® command line the response is generated:

```
Family                  Daubechies
Short name              db
Order N                 N strictly positive integer
Examples                db1 or haar, db4, db15

Orthogonal              yes
Biorthogonal            yes
Compact support         yes
DWT                     possible
CWT                     possible
Support width           2N-1
Filters length          2N
Regularity              about 0.2 N for large N
Symmetry                far from
Number of vanishing moments for psi          N
```

The two highlighted lines above show that the wavelet can be used as a continuous wavelet transform, or can be implemented as a filter bank. If the Daubechies is to be implemented as a CWT it needs the mother wavelet function $\psi(t)$ and the scale function. However, this wavelet has no explicit expression, except for db1, which is the *Haar wavelet*. In this case, when no explicit wavelet or scale functions are available but the wavelet filters are, the continuous wavelet transform can be generated from the filters though a recursive algorithm that uses the dilatation equation [4]:

$$\varphi^{(k+1)}(t) = \sqrt{2} \sum_{n=0}^{N-1} h_0[n]\varphi^{(k)}(2t - n). \tag{9.60}$$

In this algorithm the initial value for $\varphi^{(0)}(t)$ must be given. The wavelet function can be obtained using

$$\psi^{(k+1)}(t) = \sqrt{2} \sum_{n=0}^{N-1} h_1[n]\varphi^{(k)}(2t - n). \tag{9.61}$$

The wavelet function [phi,psi,xval] = wavefun('wname',ITER) returns approximations of wavelet function (psi) 'wname' and the associated scaling function (phi), if they exist, on the points grid XVAL. The positive integer ITER specifies the number of iteration in the algorithm. For example, the small program below plots the scaling and wavelet function for db3 depicted in Figure 9.39:

```
[phi,psi,xval] = wavefun('db3',10);
figure(1)
plot(xval,phi,'k','LineWidth',3);
title({'Approximations of the scaling function for
db3'},'FontSize',16,'FontName','Verdana')
xlabel({'Time (s)'},'FontSize',14,'FontName','Verdana')
figure(2)
plot(xval,psi,'k','LineWidth',3);
title({'Approximations of the wavelet funtion for
db3'},'FontSize',16,'FontName','Verdana')
xlabel({'Time (s)'},'FontSize',14,'FontName','Verdana')
```

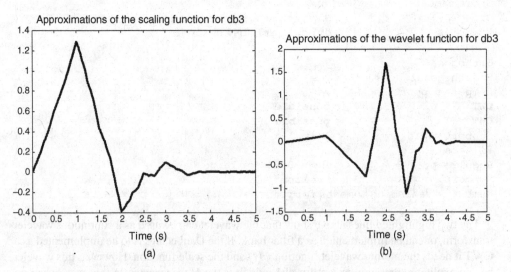

Figure 9.39 Daubechies wavelet: (a) scaling function; and (b) wavelet function.

9.5.6 Design Issues in Wavelet Transform

The choice of a mother wavelet for a given application is an important question to be addressed. To choose the appropriate wavelet, information about the wavelet properties vanishing moments, the size of support, support versus moments and regularity are needed. These properties are more strongly related to the practical use of these functions.

9.5.6.1 Vanishing Moments

The vanishing moment is a criterion that explains how a function decays toward infinity. The rate of decay can be estimated by the integration [3]:

$$m_k = \int\limits_{-\infty}^{\infty} t^k f(t) \mathrm{d}t,$$

where the parameter k indicates the rate of decay. The wavelet has p vanishing moments if

$$m_k = \int\limits_{-\infty}^{\infty} t^k \psi(t) \mathrm{d}t = 0, \quad \text{for} \quad 0 \leq k \leq p. \tag{9.62}$$

This result implies that all polynomial signals given by

$$s(t) = \sum_{0 \leq j \leq p} a_j t^j \tag{9.63}$$

have zero wavelet coefficients. As a consequence, the details are also zero. This property ensures the suppression of signals that are polynomials of a degree lower than or equal to p.

9.5.6.2 Size of Support

The size of support indicates the filter length. Note that the FIR filter is commonly used due to its stability and implementation issue. In reference [3] it is demonstrated that if $h_0[n]$ has a finite support both $\varphi(t)$ as $\psi(t)$ have finite support.

9.5.6.3 Regularity

The regularity of $\psi(t)$ has significant influence on the error introduced by thresholding or quantizing the wavelet coefficients. The concept of regularity is in relation to the differentiation of a function at a given point. If the signal is r-times continuously differentiable at t_0 and r is an integer ($r \geq 0$), then the regularity is r. It is said that a wavelet function is extra 'smooth' or regular if the regularity is higher. When reconstructing a signal from wavelet coefficients,

$$f(t) = \sum_j \sum_n \langle f(t), \psi_{j,n}(t) \rangle \psi_{j,n}(t) \tag{9.64}$$

Table 9.7 Regularity for Daubechies wavelets.

$\psi(t)$	db1 = Haar	db2	db3	db4	db5	db7	db10
Regularity	Discontinuous	0.5	0.91	1.27	1.59	2.15	2.90

where $\langle f(t), \psi_{j,n}(t) \rangle$ is the inner product and the reconstruction error is related to the mother wavelet chosen. If the signal is smooth, then a good wavelet choice is the one with higher regularity. The regularity of certain wavelets is known; Table 9.7 lists some information for Daubechies wavelets [25].

When the command waveinfo(db) is used it returns the information that the regularity factor is equal to $0.2N$ for large N.

9.5.6.4 Connecting Scale and Frequency

The common question that arises when working with wavelet transforms regards the relationship between scale and frequency. This question arises as one is strongly linked to the concept of frequency, brought to us by Fourier. Figure 9.38 shows the equivalent frequency response of a wavelet tree with 3 levels. This figure is redrawn in Figure 9.40 in a more appropriate shape, labeled with the details and the approximation coefficients. As the filters are broad bands, the signals at the output of each filter (the wavelet coefficients) are mixed in frequency. Strictly speaking, the relationship between a single frequency and each coefficient cannot be made. One natural relationship links the wavelet coefficients with the central frequency of the band-pass filter in the respective tree. Table 9.8 shows this relationship, remembering the correspondence between digital frequency (ω in rad) with analog frequency (f in Hz) given by

$$f = \frac{\omega F_s}{2\pi} \tag{9.65}$$

where F_s is the sampling rate.

For a wavelet tree with L levels, Table 9.9 lists the mathematical expression for each coefficient. The previous manner of connecting scale and frequency can only be used for DWT. For CWT, when the scale factor can assume any real value, another procedure has to be used. In [25] the following expression (used in MATLAB®) is proposed:

$$F_a = \frac{F_c}{aT_s} \tag{9.66}$$

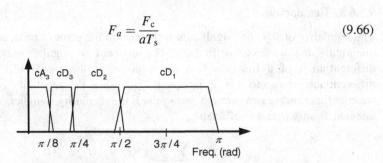

Figure 9.40 The equivalent frequency response of a wavelet tree of 3 levels.

Table 9.8 The relationship between the scale and the central frequency.

Coefficient	Bandwidth (Hz)	Center band (Hz)
cD_1	$F_s/4 - F_s/2$	$3F_s/8$
cD_2	$F_s/8 - F_s/4$	$3F_s/16$
cD_3	$F_s/16 - F_s/8$	$3F_s/32$
cA_3	$0 - F_s/16$	$F_s/32$

Table 9.9 The relationship between the coefficients and the central frequency.

Coefficient	Bandwidth (Hz)	Center band (Hz)
$cD_k, k = 1, \ldots, L$	$F_s/2^{k+1} - F_s/2^k$	$3F_s/2^{k+2}$
cA_L	$0 - F_s/2^{L+1}$	$F_s/2^{L+2}$

where a is the scale, T_s is the sampling period, F_c is the center frequency and F_a is the pseudo-frequency (in Hz) corresponding [7,26]. to the scale a. The idea developed in [25] is to associate a given wavelet with a purely periodic signal of frequency F_c. The central frequency F_c is defined as the one that maximizes the FFT of the wavelet modulus. The function centfrq.m can be used to compute the center frequency of a given mother wavelet. Figure 9.41 depicts the plot of the db10 together with sinusoid signal with central frequency $F_c = 0.69231$.

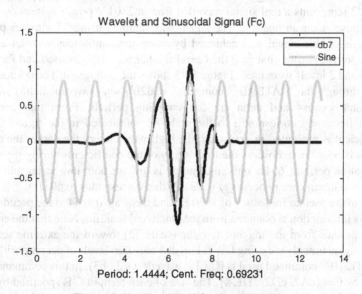

Figure 9.41 The central frequency concept.

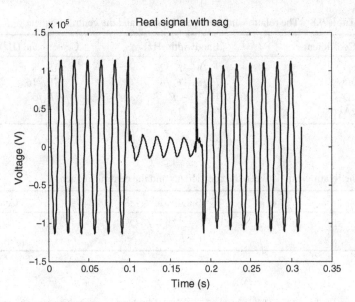

Figure 9.42   A sag recorded from a real system 230 kV.

9.5.7 Power System Application of Wavelet Transform

The wavelet transform has been used in power systems in many areas with emphasis in protection, power quality and diagnosis [7,26]. This section will present applications of wavelet transform for the analyses of power quality disturbances. Power quality encompasses numerous kinds of electrical disturbances (voltage sags, transients, overvoltages, harmonic distortions, flicker, imbalances, etc.). The sag disturbance is chosen as an illustrative example for this section.

Figure 9.42 represents a real signal recorded from a 230 kV power system during a fault. The figure shows a sag in phase a. The signal was sampled using 32 samples per cycle.

The analysis of this signal was achieved by using the continuous wavelet and discrete wavelet transforms. For the first case the Gaus4 mother wavelet was used and for the second case the db3 and 2 levels were used. Figure 9.43 shows the scalogram. The pseudo frequency is obtained through the MATLAB® command scal2frq(scales,wname,delta) where scales = 1:64, wname = gaus4 and delta are the sampling periods. From that command the fundamental frequency corresponds to the scale 32, as labeled in the figure. The energy of the coefficient is presented in grayscale; the higher the energy the darker the color. From this figure, it is possible to observe the low energy of the coefficients during the sag. During normal operation periods, 60 Hz only, the energy is low for both low scales and high scales because the maximum energy occurs at scale 32 (that corresponds to 60 Hz). The scale $a = 1$ corresponds to the pseudo frequency of 1920 Hz and the scale $a = 64$ to the pseudo frequency of 30 Hz (this information is obtained from the command scal2frq. Note that the energy of the coefficient vanishes from the maximum value (scale 32) toward the extreme scale values.

The same signal is analyzed using DWT. For, db3 only two levels of decomposition are used and the MATLAB® command used is [C,L] = wavedec(x,2,db3). In this command the vector C is formed by C = [cA2, cD2, cD1, x]. The size of each term in C is specified by the vector L = [size(cA2), size(cD2),size(cD1), size(x)]. The plot is represented in Figure 9.44, where

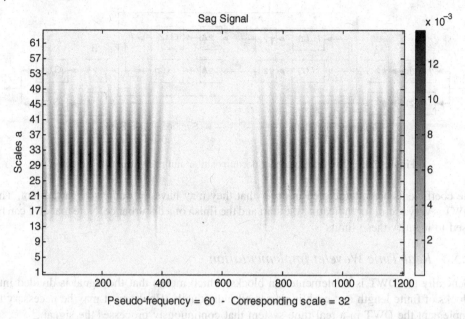

Figure 9.43 Scalogram of the sag signal using the Gaus4 as mother wavelet.

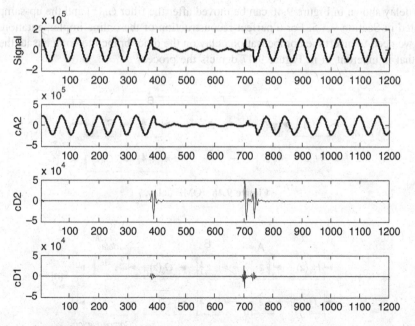

Figure 9.44 DWT decomposition of a sag signal. From top to bottom: the original signal, the approximation coefficient and two detail coefficients.

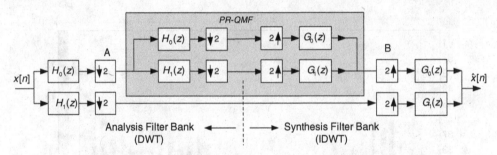

Figure 9.45 Two-level wavelet decomposition and reconstruction structure.

the coefficients are interpolated in order that they may have the same size in the plot. The DWT is very useful for indicating the start and the finish of a disturbance. Note that cD2 can be used to identify these limits.

9.5.8 Real-Time Wavelet Implementation

Generally the DWT is implemented in blocks, which means that the signal is divided into blocks of finite length and each is processed individually. However, it may be necessary to implement the DWT in a real-time system that continuously processes the signal.

Figure 9.45 shows a two-level DWT. This structure can be seen as a filter bank, and as seen before the highlighted block is a PR QMF and can be substituted by a delay as shown on Figure 9.46. This delay affects the signal reconstruction unless it is compensated on the other branch.

The delay shown in Figure 9.46 can be moved after the filter $G_0(z)$ and the up-sampler, as presented in Section 9.4.5. For a perfect reconstruction of the online implementation of the DWT, we need to insert the compensation delay on the other branch to compensate the QMF delay that is inherent in it. Figure 9.47 depicts the process.

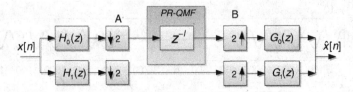

Figure 9.46 QMF delay.

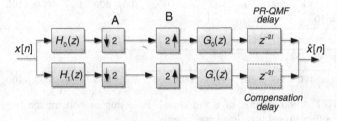

Figure 9.47 Block diagram of the process.

9.5.8.1 Numerical Example

Consider the example of Section 9.4.1 where the filters were designed based on a Daubechie2 function and $T(z) = z^{-3}$. The decimation can be done by two ways: throwing out the even samples of the signal or throwing out the odd samples. In the first case we can compensate the QMF delay by adding a delay block of six samples ($2l = 2 * 3 = 6$). For the second case, if we throw out the odd samples it means that we are advancing the signal one sample; $T(z) = z^{-2}$ and our delay block is of four samples ($2l = 2 * 2 = 4$).

9.6 Conclusions

This chapter presented the basic concepts related to signal time-frequency decomposition, emphasizing the applications for power system signals. Considering the increased complexity of the electrical network associated with the development of smart grids and the computational power available, time-frequency decomposition methods will become one of the main methods to determine the performance and state of the electric grid. Time-frequency decomposition can be used for dealing with phenomena ranging from millisecond transients to load and generation profiles. Additional applications are presented and discussed in Chapter 11.

References

1. Vaidyanathan, P.P. (1993) *Multirate Systems and Filter Banks*, Prentice Hall, Englewood Cliffs, New Jersey.
2. Daubechies, I. (1992) *Ten Lectures on Wavelets*, SIAM, Philadelphia, PA, pp. 24–54.
3. Mallat, S. (2009) *A Wavelet Tour of Signal Processing: The Sparse Way*, 3rd edition, Academic Press–Elsevier, Burlington, MA.
4. Sodney Burrus, C., Gopinath, R.A. and Guo, H. (1998) *Introduction to Wavelets and Wavelet Transforms: A Primer*, Prentice Hall, New Jersey.
5. Hartley, R. and Welles, K.II (1990) Recursive computation of the Fourier transform. IEEE International Symposium on Circuits and Systems. May 1990. pp. 1792–1795.
6. Silveira, P.M., Duque, C.A., Baldwin, T. and Ribeiro, P.F. (2008) Time-varying power harmonic decomposition using the sliding-window DFT. The 13th IEEE International Conference on Harmonics and Quality of Power. Woolongong, Australia.
7. Bollen, M.H.J. and Gu, I.Y.H. (2006) *Signal Processing of Power Quality Disturbances*, IEEE Press and Wiley-Interscience, NJ.
8. Qian, S. and Chen, D. (1999) Joint time-frequency analysis. *IEEE Signal Processing Magazine*, **16** (2), 53–67.
9. Cho, S.-H., Jang, G. and Kwon, S.-H. (2010) Time-frequency analysis of power-quality disturbances via the Gabor–Wigner transform. *IEEE Transactions on Power Delivery*, **25** (1), 494–499.
10. Vilbé, L.C.C. and Pierre (1992) On the uncertainty principle in discrete signals. *IEEE Transactions on Circuits-II*, **39** (6), 394–395.
11. Mitra, S.K. (2006) *Digital Signal Processing: A Computer Approach*. McGraw–Hill Publishing.
12. Phadke, J.S. and Thorpe, A.G. (1988) *Computer Relaying for Power Systems*, Research Studies Press Ltd., New Orleans, Louisiana.
13. Vetterli, M. and Kovacevic, J. (1995) *Wavelets and Subband Coding*, Prentice Hall, NJ.
14. Fabri, D.F., Martins, C.H.N., Silva, L.R.M., Duque, C.A., Ribeiro, P.F. and Cerqueira, A.S. (2010) Time-varying harmonic analyzer prototype. The 14th IEEE International Conference on Harmonics and Quality of Power, Bergamo. IEEE Conference Publication, Italy.
15. Fabri, D.F., Martins, C.H., Duque, C.A., da Silveira, P.M. and Ribeiro, P.F. (2011) Improved filter banks for time-varying power harmonic decomposition. In *Proceedings of IEEE PES Meeting*, Detroit. pp. 1–6.
16. Duque, C.A., Silveira, P.M. and Ribeiro, P.F. (2011) Visualizing time-varying harmonics using filter banks. *Electrical Power Systems Research*, **81**, 974–983.

17. Baghzouz, Y., Burch, R.F., Capasso, A. *et al.* (1998) (Committee, Probabilistic Aspects Task Force of the Harmonics Working Group Subcommittee of the Transmission and Distribution) Time varying harmonics. Part I: Characterizing measured data. *IEEE Transactions on Power Delivery*, **13**, 938–944.

18. Ribeiro, P.F. (1994) Wavelet transform: an advanced tool for analyzing non-stationary distortions in power systems. In *Proceedings of ICHPS VI*, Italy. pp. 365–369.

19. Baghzouz, Y., Burch, R.F., Capasso, A. *et al.* (2002) Time varying harmonics: Part II—Harmonic summation and propagation. *IEEE T. Power Syst.*, **17**, 279–285.

20. Wang, J., Ren, Q., Wang, F. and Ji, Y. (1998) Time-varying transient harmomics measurement based on wavelet transform. In *Proceedings of International Conference on Power Systems Technology*, **2**, 1556–1559.

21. Santoso, S., Powers, E.J., Grady, W.M. and Hofmann, P. (1996) Power quality assessment via wavelet transform. *IEEE Transactions on Power Delivery*, **11**, 924–930.

22. Wang, C. and Lin, C. (2006) Adaptive wavelet networks for power-quality detection and discrimination in a power system. *IEEE Transactions on Power Delivery*, **21**, 1106–1113.

23. Grossman, A. and Morlet, J. (1984) Decomposition of Hardy functions into square integrable wavelets of constant shape. *SIAM Journal of Mathematical Analysis*, **15** (4), 723–736.

24. Teolis, A. (1998) *Computational Signal Processing with Wavelets*, Birkhauser, Boston.

25. Misiti, M., Misiti, Y., Oppenheim, G. and Poggi, J.-M. (2013) *Wavelet Toolbox User Guide*. R2013a. Mathworks, Inc., MA, USA.

26. Gu, Y.H. and Bollen, M.H.J. (2000) Time-frequency and time-scale domain analysis of voltage disturbances. *IEEE Transactions on Power Delivery*, **15** (4), 1279–1284.

10

Pattern Recognition

10.1 Introduction

The fast growth of computational power and applications has helped to cross-fertilize many areas of science and technology, producing a great number of new methods, theories and models. For example, signal processing can be used in combination with other theories (statistical, geometry and applied mathematics) to find relationships which can be used for pattern observation and identification.

The complexity of the future grid will require not only advanced signal processing that can identify specific parameters, but also intelligent methods for identifying particular patterns of behavior. This recognition is directly related to the design of machines capable of classifying objects into a number of categories or classes, a fairly natural task for humans. Even a small child is capable of learning to recognize particular objects such as a ball, and can then recognize any similarly classified objects hence generalizing the learned concept of a ball. This illustrates one of the most important properties of a pattern recognition machine: the learning by examples and the capacity for generalization.

Pattern recognition applications received a boost in the last four decades due to the increasing need for automation, both in industry and at home. This demand has been met by the evolution of computers, digital signal processing and processors concurrently with the exponential growth of digital information available on the internet. Applications of pattern recognition can be found in a machine's vision, robotics, biometrics, information retrieval, instrumentation and also in power systems.

Examples of the application of pattern recognition in power systems include fault identification [1], power quality [2], consumer profile identification [3] and protection [4].

This pattern recognition will be very much required in future power systems due to the variability of electrical signals from diverse generators and loads, to aid the system operator to properly identify problems and to control the grid's power delivery process. As grid complexity increases these may include distributed and renewable generation, customers that consume and produce energy (prosumers), active distributed networks and, for example, new types of loads and generators that can respond to market signals. All of these are creating the complex smart grid of the future where pattern recognition is an important enabling tool for operation and control.

Power Systems Signal Processing for Smart Grids, First Edition. Paulo Fernando Ribeiro, Carlos Augusto Duque, Paulo Márcio da Silveira and Augusto Santiago Cerqueira.
© 2014 John Wiley & Sons, Ltd. Published 2014 by John Wiley & Sons, Ltd.
Companion Website: http://www.wiley.com/go/signal_processing/

10.2   The Basics of Pattern Recognition

In the design of a non-invasive automatic system to identify a power system consumer profile, we consider two classes: residential and commercial consumers.

Initially, the designer needs specific knowledge of the problem in order to decide what measured quantities of the consumers' information he or she will use for the analysis. For the commercial consumer, the power consumption is usually equally distributed during commercial hours. For the residential consumer however, the power consumption increases during non-commercial hours, excluding periods of sleep. The class or category of load used for commercial units also differs widely from the residential units. With such knowledge at hand, a designer may decide to use the daily readings of power consumption throughout the day and may add specific windows of acquired current waveforms for particular periods.

After the measured quantities are defined, the designer then needs to select (a) the inputs of the automatic classification system, and the hourly measurements for both (b) instantaneous power consumption and (c) current harmonic distortion. When these features are selected and extracted they are used as inputs for the classification algorithms, in this case a feature vector of size 2×24. It is important to stress that, to obtain the current harmonic distortion, the Discrete Fourier transform (DFT) could be used as a feature extraction technique.

Once the features are selected, the classifier system has to be designed. A classifier design can basically be viewed as the implementation of a separation surface between the two classes (in this case, a separation surface on a 48-dimension space). Figure 10.1 illustrates a two-dimensional problem where it is possible to see the separation surface implemented by a classifier.

Finally, the classifier performance should be correctly evaluated and, if the achieved performance is below design requirements, the system design should be reviewed. The

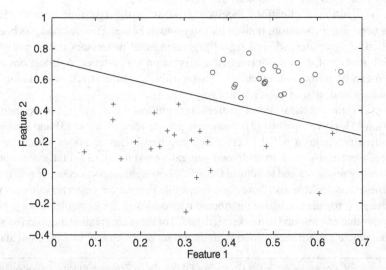

Figure 10.1 Illustration of a two-dimensional feature space for a two-class problem, where one class is represented by crosses and the other class by circles. The separation surface implemented by a linear classifier is also shown (dashed line).

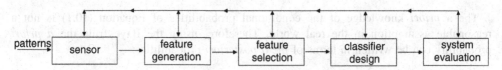

Figure 10.2 The basic stages involved in the design of a classification system [5].

revision can be made at any stage of the design, from the definition of the measured variables to the classifier design.

In reference [5], the design of a classification system is divided into five basic stages as depicted in Figure 10.2, which is a good illustration of the procedure of the previous example.

10.2.1 Datasets

In order to design and test a pattern recognition system, sets of samples from the patterns are required. Usually three datasets are used: for training, validation and testing. The training dataset is used for system design during the classifier learning phase. The validation dataset is used in order to verify the classifier generalization performance. Finally, the test dataset is used to verify the system performance.

10.2.2 Supervised and Unsupervised Learning

One of the most important characteristics of a classifier algorithm is the ability to learn from examples. This can be achieved through supervised or unsupervised learning during the design (training) of the classifier algorithm.

In a supervised learning algorithm (also called learning with a teacher), the training dataset is composed of samples (events) of all classes where the relationship between each sample and the respective class is known. The number of classes and the relationship between each sample and the respective class are therefore known.

In unsupervised learning, the relationship between each sample of the training dataset and the respective class is unknown; in some cases, the number of classes is not known either. The complexity of the problem grows in an unsupervised learning in comparison with a supervised learning; more complex models are therefore obtained through unsupervised learning.

10.3 Bayes Decision Theory

The classification problem can be stated in terms of conditional probabilities in the framework of Bayes decision theory. This approach leads to optimum classifiers but requires *a priori* information about the problem that is not usually available and could lead to very complex classifiers. Despite these problems, several interesting properties and results can be gleaned from the Bayes classifier.

Given a pattern with an unknown class label from measurements, with $\mathbf{x} = [x_1, x_2, \ldots, x_N]$ as the corresponding feature vector, the correct class feature vector needs to be established. Let M be the number of unknown classes $\omega_1, \omega_2, \ldots, \omega_M$. By considering the feature vector as a random vector, then according to the Bayes decision theory $\mathbf{x}$ is assigned to the class ω_i if

$$P(\omega_i/\mathbf{x}) > P(\omega_j/\mathbf{x}), \quad \forall j \neq i. \tag{10.1}$$

The a *priori* knowledge of the conditional probabilities of Equation (10.1) is not a reasonable assumption in the real world. Therefore, using the Bayes rule the a *priori* probability can be written in terms of the a *posteriori* probabilities:

$$P(\omega_i/\mathbf{x}) = \frac{f(\mathbf{x}/\omega_i)P(\omega_i)}{f(\mathbf{x})}, \tag{10.2}$$

where $f(\mathbf{x}/\omega_i)$ is the probability density function of the observed feature vector given that it belongs to class ω_i and $f(\mathbf{x})$ is the feature vector of the probability density function. Equation (10.1) can therefore be rewritten

$$f(\mathbf{x}/\omega_i)P(\omega_i) > f(\mathbf{x}/\omega_j)P(\omega_j), \quad \forall j \neq i. \tag{10.3}$$

Now, the Bayes classifier requires the knowledge of the a *priori* probability of each class and the a *posteriori* conditional probability density functions. If these quantities are not known, they can be estimated through the training feature vectors. If these quantities are known, the Bayes classifier leads to the optimum classifier in the sense of minimizing the error probability.

For the reasons listed above, several techniques and algorithms do not follow the Bayes approach to the classification problem. The goal of these techniques is to design directly the decision surface based on the minimization (or maximization) of a defined cost function and using the training feature vectors. Among these techniques, some of them are only capable of implementing linear separation surfaces on the feature space, usually called linear classifiers, while non-linear classifiers are capable of implementing non-linear separation surfaces.

Before presenting some classifier techniques, the following section introduces some important aspects related to feature extraction (a very important stage in pattern recognition system design).

10.4 Feature Extraction on the Power Signal

One of the most important stages in the design of pattern recognition systems is the extraction of features envisaging the improvement of the different 'classes' separation. If the extracted features result in a feature space where the classes are well separated, there is a high probability that the designed classifier will demonstrate a good performance. If not, the classifier will perform poorly no matter the technique used for the analysis.

Another requirement of a good feature extraction technique is the ability to provide the required separation in a low-dimensional space. In pattern recognition problems, we aim to find a reduced number of features in order to make the classifier design easier. If the feature extraction technique results in a high-dimensional space, the feature selection will play an important role in the system design. In order to select a feature extraction technique, good knowledge of the classification issues is required.

10.4.1 Effective Value (RMS)

The root mean square (RMS) is used to define the effective value of a signal (a time function or a set of samples) and is frequently used to express the voltage and current in a power system.

The RMS value is defined:

$$f_{\text{RMS}} = \lim_{T \to \infty} \sqrt{\frac{1}{2T} \int_{-T}^{T} f(t)^2 \, dt}, \tag{10.4}$$

where $f(t)$ is a continuous-time function. For periodic waves, the RMS value can be calculated over one period. For a discrete-time sequence $x[n]$, the RMS is given by:

$$x_{\text{RMS}} = \sqrt{\frac{1}{N} \sum_{n=1}^{N} x[n]^2}. \tag{10.5}$$

Due to its extensive use in power systems, the RMS value is often used as a feature extraction tool because of its capacity to represent the effective waveform values and its inherent ability to compress information carried by the waveform.

10.4.2 Discrete Fourier Transform

See Chapter 6 for a definition of the DFT. Due to its capacity to represent the spectral content of a sequence, it is also widely used to extract features from the power system signal. It is very useful when the power system signal frequency content carries relevant information about the separation of each class. For example, the DFT could be used in order to estimate the voltage of a fundamental component or to discriminate between normal operation, undervoltage and overvoltage.

10.4.3 Wavelet Transform

The Wavelet transform is widely used as a preprocessing technique for the power system signal, due to its time-frequency domain representation and the nature of the power system signal. The wavelet transform was discussed in Chapter 9.

A wavelet transform is a good choice of preprocessing technique for the location of time disturbances as it decomposes the signal into several frequency bands, preserving the time information. The wavelet transform is also a good preprocessing choice for the classification of power quality disturbances.

10.4.4 Cumulants of Higher-Order Statistics

The higher-order statistics are applicable when the process is non-Gaussian; several real-world applications are truly non-Gaussian [6]. Under such circumstances the cumulants reveal important properties that cannot be revealed through low-order statistics. Cumulants have been used in power system applications of pattern recognition [7].

For a zero-mean stationary random process $x(t)$ [8], the second-, third- and fourth-order cumulants are given by [6]:

$$C_{2,x}(\tau) = E\{x(t)x(t + \tau)\} \tag{10.6}$$

$$C_{3,x}(\tau_1 \tau_2) = E\{x(t)x(t + \tau_1)x(t + \tau_2)\} \tag{10.7}$$

$$C_{4,x}(\tau_1, \tau_2, \tau_3) = E\{x(t)x(t + \tau_1)x(t + \tau_2)x(t + \tau_3)\} - C_{2,x}(\tau_1)C_{2,x}(\tau_2 - \tau_3)$$
$$-C_{2,x}(\tau_2)C_{2,x}(\tau_3 - \tau_1) - C_{2,x}(\tau_3)C_{2,x}(\tau_1 - \tau_2) \tag{10.8}$$

where $E\{\cdot\}$ is the expected value [8] and τ is the lag.

For a zero-mean stationary random sequence $x[n]$ with a finite number of samples, an estimation of the second-, third- and fourth-order cumulants is:

$$\hat{C}_{2,x}[k] = \frac{2}{N} \sum_{n=0}^{N/2-1} x[n]x[n + k] \tag{10.9}$$

$$\hat{C}_{3,x}[k_1, k_2] = \frac{2}{N} \sum_{n=0}^{N/2-1} x[n]x[n + k_1]x[n + k_2] \tag{10.10}$$

$$\hat{C}_{4,x}[k_1, k_2, k_3] = \frac{2}{N} \sum_{n=0}^{N/2-1} x[n]x[n + k_1]x[n + k_2]x[n + k_3] - \hat{C}_{2,x}[k_1]\hat{C}_{2,x}[k_2 - k_3]$$
$$-\hat{C}_{2,x}[k_2]\hat{C}_{2,x}[k_3 - k_1] - \hat{C}_{2,x}[k_3]\hat{C}_{2,x}[k_1 - k_2] \tag{10.11}$$

where $k = 0, 1, \ldots, N/2 - 1$ is the kth delay.

The cumulant is a function of the delay k, resulting in a high-dimension feature space. A feature selection technique should therefore be used during the design phase in order to reduce the dimension of the feature space.

10.4.5 Principal Component Analysis

The principal component analysis (PCA) is very often used in pattern recognition applications as an additional pre-processing step during the feature generation stage. The main goal of the PCA is to find a new basis in the feature space, where the features are uncorrelated and organized by decreasing variance. Uncorrelated features are a very desirable property for the input classifier data, and the organization by decreasing variance could lead to the use of fewer features at the classifier input (feature selection).

More precisely, the PCA goal is to find a new basis to represent a multidimensional dataset which is a linear combination of the original basis that best represents the dataset. Let matrix $\mathbf{X}$ be the extracted features of a training dataset with M lines (each line represents one feature) and N columns (each column represents one event) and

$$\mathbf{Y} = \mathbf{PX}, \tag{10.12}$$

where $\mathbf{Y}$ is the new uncorrelated and ordered dataset matrix ($M \times N$ size) and $\mathbf{P}$ the linear transformation matrix (of size $M \times M$).

From the requirements imposed on the new dataset $\mathbf{Y}$, its covariance matrix $\mathbf{C_Y}$ must be diagonal. One way to find the linear transform matrix $\mathbf{P}$ is therefore to transform the covariance matrix of $\mathbf{X}$ into a diagonal matrix using the property [8]:

$$\mathbf{C_y} = \mathbf{D}^t \mathbf{C_X} \mathbf{D}, \quad \text{for } \mathbf{Y} = \mathbf{DX}. \tag{10.13}$$

D is therefore the linear transform that diagonalizes C_X and is formed by the eigenvectors of C_X [8]. Let e_i be the ith eigenvector of C_X and E its eigenvectors matrix formed by e_i rows, then the resulting linear transformation could be written as $D = E$.

One step of the PCA transformation is already achieved through D. However, another step related to the organizing of the features by decreasing variance value is missing. In order to complete this step, the eigenvectors e_i must be sorted by the decreasing value of its corresponding eigenvalues λ_i, forming the linear transformation E_S. The PCA transformation is therefore given by:

$$P = E_S. \tag{10.14}$$

10.4.6 Normalization

The normalization of extracted features is also an important requirement in the correct design of the classifier technique. It is important to adjust the input data values to a finite and small range (if possible) in order to avoid numerical problems with the design and implementation of the classifier system. The normalization could also enhance or weaken the classifier performance; it should therefore be carefully chosen and its impact on the classifier performance evaluated.

The normalization could be performed for each sample (event), for each feature or by considering both. Several normalization techniques can be used; in the following sections we describe some of methods commonly applied in pattern recognition problems.

10.4.6.1 Dividing by the Maximum Value

A simple way to normalize data is to divide each sample by its maximum absolute value

$$x_N^i = \frac{x^i}{x_{max}^i}, \tag{10.15}$$

where x_N^i is the ith normalized feature vector, x^i is the ith feature vector (sample) and x_{max}^i is the maximum absolute value of the ith feature vector considering all features. This is an example of a sample normalization.

The same procedure could be applied for each feature or by considering the maximum value of the total dataset. The sample approach is described by Equation (10.15), where the feature-by-feature approach could affect the class separation of the feature space. Furthermore, dividing the input data by the maximum value of the dataset will not affect the class separation.

10.4.6.2 Energy Normalization

A very common method used for data normalization is the energy normalization. In this case, each sample of the dataset is divided by its energy:

$$x_N^i = \frac{x^i}{x_{RMS}^i} \tag{10.16}$$

where x_N^i is the ith normalized feature vector and x^i is the ith feature. This technique is very useful when classifier immunity to the feature energy variation is required.

10.4.6.3   Mapping the Minimum and Maximum Values

This normalization is usually applied to features and its goal is to ensure that all selected features represent the same dynamic range. This is a desirable property when considering implementation issues. The algorithm is defined:

$$x_N^j(i) = \frac{\left[x_\mathbf{N}(i)_{\max} - x_\mathbf{N}(i)_{\min}\right]\left[x^j(i) - x(i)_{\min}\right]}{x(i)_{\max} - x(i)_{\min}} + x_N(i)_{\min} \qquad (10.17)$$

where $x_N^j(i)$ is the normalized jth sample of the ith feature, $x_N(i)_{\max}$ and $x_N(i)_{\min}$ are the required maximum and minimum values for the ith feature, $x^j(i)$ is the jth sample of the ith feature, $x(i)_{\max}$ and $x(i)_{\min}$ are the maximum and minimum values of the ith feature considering all samples.

10.4.7   Feature Selection

Once the features have been extracted from the acquired signals, they can rarely be directly applied to the classifier algorithm. This is because the dimension of the feature space can be very high, increasing the computational complexity of the problem. Another important reason for avoiding a high-dimension feature space is the higher the ratio of training samples N to the number of free classifier parameters (which is directly related to the number of features), the higher the quality of the generalization properties of the resulting classifier [9].

The goal of feature selection techniques is to select the most important features with respect to the class separation and reduce the number of features that will be used as inputs of the classifier, preserving as much as possible or even improving the classifier performance.

10.4.7.1   Outlier Removal

An outlier is a point that lies far away from the mean value of the corresponding random variable; such points with values distant from the rest of the data may cause large errors during the classifier design. This is not desirable, especially when the outliers are the result of noisy measurements. For normally distributed data, a threshold of 1, 2 or 3 times the standard deviation is used to define outliers. Points that lie further from the mean than this threshold are removed. However, for non-normal distributions, further measurements should be considered.

Outlier removal is performed feature by feature, comparing the feature value of a certain event from the training sample with the mean value of this feature over the entire training sample. The events considered as outliers should be removed from the data if the number of training samples remains high after their removal. If the number of events in the training dataset is not high, it would be preferable to keep the outlier event on the dataset but the value of the feature that is far from the mean should be reset to the mean value.

10.4.7.2   Confusion Area

The confusion area (CA) between the pdfs of each feature from two different classes could be used as a feature selection technique. In this case the goal is to remove some features from the input vector, reducing the feature space dimension (as opposed to the outlier removal technique that removes some events from the training dataset).

The area under a pdf is always 1; for two equal pdfs the confusion area is therefore 1. On the other hand, if there is no overlap between two pdfs the confusion area will be 0. If the CA between class one and class two could be calculated for each feature of the training dataset, features that represent the small CA values could be selected.

As the involved pdfs are rarely available, some pdf estimation techniques should be used. A simple way to estimate the required pdfs is through normalized histograms (area equal to 1). Furthermore, in order to measure the CA for a certain feature, the normalized histogram for one class (HN1) and for the other class (HN2) should be built and the CA between HN1 and HN2 calculated.

10.5 Classifiers

The Bayes classifier minimizes the error probability, resulting in the best separation surface on the feature space. It therefore provides the best generalization among the family of classifiers. Nevertheless, the design of the Bayes classifier is not always possible because the involved pdfs are complicated and their estimation is not an easy task. To overcome this problem, alternative classifiers can be designed by directly computing the decision surfaces by means of alternative cost functions. This approach leads to suboptimal classifiers but, in several practical cases, the performance of this kind of classifier is better than that achieved by the Bayes classifier designed using estimations of the involved pdfs.

10.5.1 Minimum Distance Classifiers

Several classifiers are based on distance measurements from the training data, described in the following sections.

10.5.1.1 Euclidian Distance Classifier

This is one of the most popular distance-based classifiers, computed from the Euclidian distance between the incoming unknown feature vector $\mathbf{x}$ and the mean value $\mathbf{m}$ of each class ω on the training dataset. The minimum distance indicates which class the unknown feature vector is assigned to. The unknown feature vector $\mathbf{x}$ is assigned to class ω_i if

$$\| \mathbf{x} - \mathbf{m_i} \| = \sqrt{(\mathbf{x} - \mathbf{m_i})^T (\mathbf{x} - \mathbf{m_i})} < \| \mathbf{x} - \mathbf{m_i} \|, \quad \forall i \neq j. \tag{10.18}$$

10.5.1.2 Mahalanobis Distance Classifier

The unknown feature vector $\mathbf{x}$ is assigned to class ω_i if

$$\sqrt{(\mathbf{x} - \mathbf{m_i})^T \mathbf{S}^{-1} (\mathbf{x} - \mathbf{m_i})} < \sqrt{(\mathbf{x} - \mathbf{m_j})^T \mathbf{S}^{-1} (\mathbf{x} - \mathbf{m_j})}, \quad \forall i \neq j \tag{10.19}$$

where $\mathbf{m}_i$ is the mean value of class ω_i on the training dataset and $\mathbf{S}$ is the covariance matrix [10] of the training dataset.

10.5.2 Nearest Neighbor Classifier

This is one of the most popular classifiers and is a very intuitive approach to the classifier problem. Basically, the distance between the unknown feature vector $\mathbf{x}$ and all samples from the training dataset are computed and the unknown feature vector is assigned to the class with k_i out of k nearest neighbors. The nearest neighbor rule can be summarized in the following steps [5].

1. Among the N training points, search for the k neighbors closest to $\mathbf{x}$ using a distance measure (e.g. Euclidean). The parameter k is user-defined. Note that it should not be a multiple of the number of classes. That is, for two classes k should be an odd number.
2. Out of the k closest neighbors, identify the number k_i of the points that belong to class ω_i.
3. Assign $\mathbf{x}$ to class ω_i for which $k_i > k_j, j \neq i$. $\mathbf{x}$ is therefore assigned to the class to which the majority of the closest neighbors belong.

10.5.3 The Perceptron

The basic concept behind the perceptron was invented by psychologist Rosenblatt in the 1950s. The perceptron is the simplest form of a neural network and has an important place in the historical development of neural networks.

The perceptron can be used to classify linearly separable patterns with only two classes. In order to classify more classes, additional perceptrons are required. Rosenblatt's perceptron is built around a non-linear neuron model, called McCulloch–Pitts model of a neuron. This model consists of a linear combination of the inputs followed by a step function (from -1 to 1) that is responsible for the saturated behavior of the perceptron. An illustration of the non-linear neuron model is depicted in Figure 10.3. The summing junction of the neural network computes the linear combination of each input signal, multiplied by the synaptic weights, and the bias. The resulting sum v is applied to the activation function. The perceptron output value is 1 if v is positive and -1 if it is negative.

From Figure 10.3, it can be seen that the perceptron inputs are represented by x_i, the synapses by w_i (for $i = 1, 2, \ldots, m$) and the bias is b. The induced local field is represented by v and can be written:

$$v = \sum_{i=1}^{m} w_i x_i + b. \tag{10.20}$$

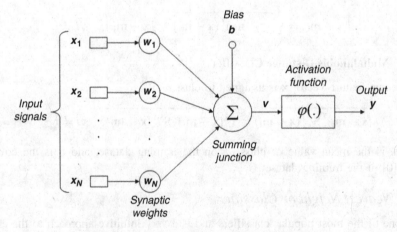

Figure 10.3 The McCulloach–Pitts Neuron.

The goal of the perceptron is to correctly classify the set of inputs into one of two classes, ω_1 and ω_2. The decision rule for the classification is to assign the point represented by the input to class ω_1 if the perceptron output is 1 and to class ω_2 if it is -1 [11].

The perceptron is capable to generate a hyperplane on the input space (feature space) that divides the space into two regions. The hyperplane is defined:

$$\sum_{i=1}^{m} w_i x_i + b = 0. \tag{10.21}$$

The induced local field can also be represented in a matrix form, including the bias b as an additional input with a fixed value 1 multiplied by an additional synaptic weight w_0, resulting in

$$v = \mathbf{w}^t \mathbf{x} \tag{10.22}$$

where $\mathbf{x} = [1 \quad x_1 \quad x_2 \cdots \quad x_m]^t$ and $\mathbf{w} = [w_0 \quad w_1 \quad w_2 \cdots \quad w_m]^t$.

The perceptron synaptic weights can be adapted on an iteration-by-iteration basis using the training dataset. The algorithm starts with an initial estimate of the hyperplane in the input space and converges to a solution in a *finite* number of iteration steps. The solution represented by $\mathbf{w}^P$ correctly classifies all the training points (assuming of course that they stem from linearly separable classes). Note that the perceptron algorithm converges to one of an infinite number of possible solutions. Different hyperplanes result by starting from different initial conditions. The update at the nth iteration step has the simple form:

$$\mathbf{w}(n+1) = \mathbf{w}(n) - \eta(n)\delta_x \mathbf{x_i}, \tag{10.23}$$

where $\mathbf{w}(n+1)$ is the synaptic weight vector at iteration $n+1$, $\mathbf{w}(n)$ is the weight vector at iteration n and $\eta(n)$ is a user-defined parameter referred to as the learning rate. $\delta_x = 0$ if the feature $\mathbf{x_i}$ was correctly assigned to its class; $\delta_x = 1$ if $\mathbf{x_i}$ was wrongly assigned to class ω_1; and $\delta_x = -1$ if $\mathbf{x_i}$ was wrongly assigned to class ω_2.

If $\eta(n) = \eta > 0$, where η is a constant, then we have a fixed-increment rule for the perceptron. The value of η is not important as long as it is positive [11]. It is important to stress that $\eta(n)$ controls the convergence behavior. The algorithm converges when the number of assigned vectors to the wrong class is zero. The classes must therefore be linearly separable on the feature space in order to ensure the convergence of the perceptron algorithm.

To solve problems with more than 2 classes, additional neurons must be used. For example, in a 3-class problem, 3 neurons must be used with the following decision rule: the input feature $\mathbf{x_i}$ is assigned to class ω_1 if the output of neuron 1 is 1 and the outputs of the others are -1; the input feature $\mathbf{x_i}$ is assigned to class ω_2 if the output of neuron 2 is 1 and the others are -1; the input feature $\mathbf{x_i}$ is assigned to class ω_3 if the output of neuron 3 is 1 and the others are -1. It is important to note that the training algorithm should also be modified in order to accommodate the additional neurons.

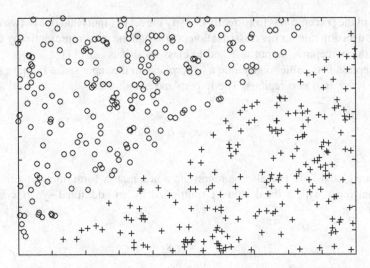

Figure 10.4 Linear separable classes in a two-dimensional feature space.

Figure 10.4 shows a linear separable classification problem in a two-dimensional feature space. This dataset was generated using the following commands in MATLAB®:

```
N=1000;
X=rand(2,N);
k=1;
j=1;
for i=1:N
    if X(2,i)>(X(1,i)+0.05);
        C1(:,j)=X(:,i);
        j=j+1;
    elseif X(2,i)<(X(1,i)-0.05);
        C2(:,k)=X(:,i);
        k=k+1;
    end
end
plot(C1(1,1:200),C1(2,1:200),'o')
hold
plot(C2(1,1:200),C2(2,1:200),'+')
```

The perceptron algorithm could be used in order to find a linear separation surface. Figure 10.5 shows the linear separation surface obtained using the perceptron algorithm with a constant $\eta = 0.5$. The perceptron algorithm was implemented for the classes depicted in Figure 10.4 using the code

```
%initializing the value of the weights
w0 =[ randn(1,3)];
BiasC1 = ones(1,length(C1));
BiasC2 = ones(1,length(C2));
xC1 =[ C1; BiasC1]';
```

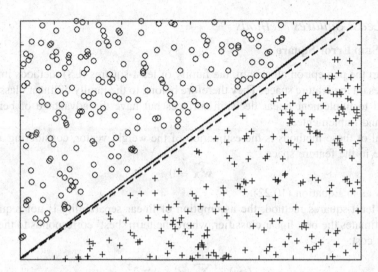

Figure 10.5 Linear separation surface obtained using the perceptron algorithm (solid line) and the ideal separation surface (dashed line).

```
xC2 =[ C2; BiasC2]';
%testing the initial performance
testC1 = w0*xC1';
testC2 = w0*xC2';
misC1 = find(testC1>0);
misC2 = find(testC2<0);
%weights update loop
cont=0;
newW = w0;
while ( length(misC1) ~= 0 || length(misC2) ~= 0);
      if cont == 100000;
           break;
      end
      newW = newW + 0.5*sum(xC1(misC1,:)) - 0.5*sum(xC2(misC2,:));
      testC1 = newW*xC1';
      testC2 = newW*xC2';
      misC1 = find(testC1<0);
      misC2 = find(testC2>0);
      cont=cont+1;
end
%ploting the separation surface
x = 0:1/10:1;
   y = -(newW(1)*x + newW(3))/newW(2);
   plot(x,y,'LineWidth',2)
```

It is possible to see from Figure 10.5 that the perceptron algorithm found a linear separation surface (solid line) that correctly classifies all training samples, although it is far from the optimal separation surface (dashed line) for this problem.

10.5.4 Least-Squares Methods

10.5.4.1 Sum Error Square

Similarly to the perceptron algorithm, the family of least-squares (LS) methods implements hyperplanes on the feature space. They therefore belong to the family of linear classifiers that are limited to implement linear decision surfaces but have the advantage of being easily designed and implemented.

The goal of these methods is the estimation of the weight vector, considering a decision hyperplane in the feature space

$$\mathbf{w}^t\mathbf{x} = 0 \tag{10.24}$$

where $\mathbf{x}$ is as in Equation (10.22).

For the least-squares method, the assumption of linear separability is not required. The method estimates the best linear classifier, where the term 'best' corresponds to the $\mathbf{w}^{LS}$ that minimizes cost:

$$J(\mathbf{w}) = \sum_{i=1}^{N} (y_i - \mathbf{w}^t\mathbf{x}_i)^2 \tag{10.25}$$

which is the sum of the squared error cost function and where y_i is the known class label for the feature vector $\mathbf{x}_i$, $i = 1, 2, \ldots, N$ and N is the number of training samples.

It is straightforward to show that the LS estimate is given by [9]:

$$\mathbf{w}^{LS} = (\mathbf{X}^t\mathbf{X})^{-1}\mathbf{X}^t\mathbf{y}, \tag{10.26}$$

where

$$\mathbf{X} = \begin{bmatrix} \mathbf{x}_1^t \\ \mathbf{x}_2^t \\ \vdots \\ \mathbf{x}_N^t \end{bmatrix} \quad \text{and} \quad \mathbf{y} = \begin{bmatrix} y_1 \\ y_2 \\ \vdots \\ y_N \end{bmatrix}. \tag{10.27}$$

Figure 10.6 shows the linear separation surface obtained by the sum squares error algorithm for the dataset generated in Section 10.5.3. It can be seen that the algorithm could find a linear separation surface capable of correctly classifying all data from the training dataset. The algorithm was implemented in MATLAB® using

```
%Obtaining the weights according to the sum error square algorithm
%building the input data matrix (xC1 and xC2 from previous example)
XSQR=[ xC1; xC2] ;
%building the output vector
out=[ ones (1,length (xC1)) -ones (1,length (xC2))]' ;
%sum square error algorithm to find the weights
w = inv (XSQR'*XSQR)*XSQR'*out;
%Coeficients of the linear separation surface
a = -w (1)/w (2);
b = -w (3)/w (2);
x = 0:1/10:1;
    y = a*x + b;
    plot (x,y,'LineWidth',2)
```

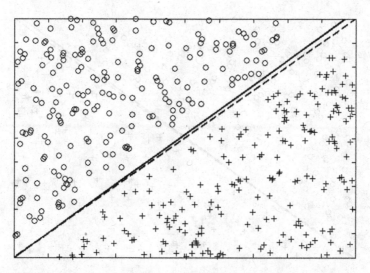

Figure 10.6 Linear separation surface obtained using the sum square error algorithm (solid line) and the ideal separation surface (dashed line).

10.5.4.2 LMS Algorithm

Some problems may arise from the LS method described before, mainly concerned with the matrix inversion of Equation (10.26). A stochastic approach to the problem can instead be used and the cost function written:

$$J(\mathbf{w}) = E\{|y - \mathbf{w}^t\mathbf{x}|^2\}. \tag{10.28}$$

In order to minimize Equation (10.28) a stochastic approximation [12] can be used and the resulting iterative scheme is given by [13]:

$$\mathbf{w}(n+1) = \mathbf{w}(n) - \eta(n)\mathbf{x_i}(y_i - \mathbf{w}(n)^t\mathbf{x_i}) \tag{10.29}$$

where $\mathbf{x_i}$ is an input vector, y_i its corresponding output on the training dataset and $\eta(n)$ is the learning rate that must satisfy two conditions in order to ensure the algorithm convergence:

$$\sum_{n=1}^{\infty} \eta(n) \to \infty \tag{10.30}$$

$$\sum_{n=1}^{\infty} \eta(n)^2 < \infty \tag{10.31}$$

Equation (10.29) is known as the least mean squares (LMS) algorithm.

The LMS algorithm was implemented for the dataset used in the previous examples and the separation surface obtained is depicted by Figure 10.7. It can be seen that the LMS for this case has a slightly better performance than the sum square error algorithm. This is due to the

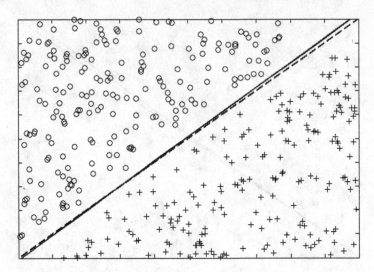

Figure 10.7 Linear separation surface obtained using the LMS algorithm (solid line) and the ideal separation surface (dashed line).

fact that the LMS separation surface is closer to the ideal separation surface. The LMS was implemented in MATLAB® using

```
%LMS
IND=randperm(length(XSQR));
%from the previous example
XLMS=XSQR;
outLMS=out;
%learning rate
n = 0.05;
%number of epochs
epoch = 10;
w = randn(1,3);
i = 1;
%LMS algorithm
while i < epoch
    i = i + 1;
    for j = 1:length(XLMS);
        A = outLMS(IND(j)) - XLMS(IND(j),:)*w';
        w = w + ((n/i)*XLMS(IND(j),:)*A);
    end
    w_final = w;
end
%linear separation surface coeficients
a = -w_final(1)/w_final(2);
b = -w_final(3)/w_final(2);
%ploting the separation surface
y = a*x + b;
plot(x,y,'LineWidth',2)
```

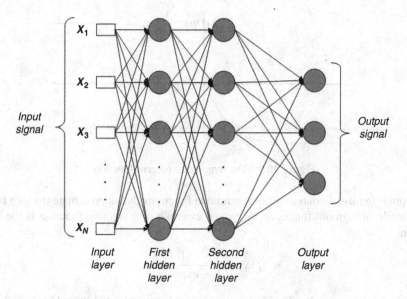

Figure 10.8 An illustration of a multilayer perceptron.

10.5.5 Multilayer Perceptron

The multilayer perceptron (MLP) is an example of an artificial neural network (ANN) that is used extensively for the solution of a number of different problems, including pattern recognition. It is a development of the single-layer perceptron (Section 10.5.3) and overcomes its practical limitations. Figure 10.8 shows a three-layer fully connected feed-forward MLP. This network has two hidden layers and the output layer. The MLPs must have at least one hidden layer of neurons and their non-linear activation functions must be differentiable due to the learning algorithm.

The hidden neurons play a critical role in the operation of a multilayer perceptron. They perform a non-linear transformation of the input data into a new space, where the classes of interest in a pattern recognition task may be more easily separated from each other than in the original input space [11].

The main challenge of the MLP is the development of an efficient learning algorithm capable of adjusting the synaptic weights associated with the hidden neurons, since the error is not directly available at the hidden neuron outputs. The back-propagation algorithm solves this problem, providing an efficient method for the training of MLPs.

10.5.5.1 Back-Propagation Algorithm

The back-propagation algorithm is a supervised learning technique that propagates the output errors to the hidden neurons, making it possible to adjust all synaptic weights from a fixed-architecture MLP by the minimization of an appropriate cost function. In order to propagate the errors to the hidden neurons the gradient is applied; the neuron activation functions should therefore be differentiable.

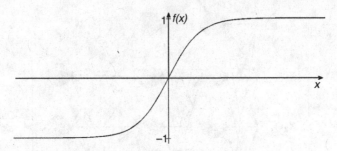

Figure 10.9 The hyperbolic tangent function.

A popular family of continuous differentiable functions that approximate the step function is the family of sigmoid functions. A typical example of a sigmoid function is the logistic function

$$f(x) = \frac{1}{1 + \exp(-ax)} \tag{10.32}$$

where a is a slope parameter. The hyperbolic tangent functions are also used because they vary between 1 and -1:

$$f(x) = c\frac{1 - \exp(-ax)}{1 + \exp(-ax)} = c\tanh\left(\frac{ax}{2}\right). \tag{10.33}$$

Figure 10.9 depicts the hyperbolic tangent function. In order to describe the algorithm, consider Figure 10.10 where the neuron and its main quantities are shown.

The back-propagation is an iterative algorithm similar to the LMS that applies a correction $\Delta w_{ji}(n)$ to the synaptic weight $w_{ji}(n)$. This is proportional to the gradient of the cost function $\varepsilon(n)$ with respect to the synaptic weight $w_{ji}(n)$, as shown in Equation (10.34).

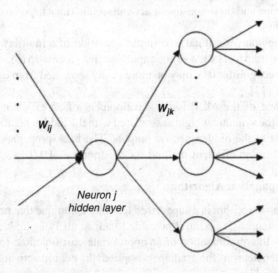

Figure 10.10 Illustration of a neuron located in a hidden layer.

$$\frac{\partial \varepsilon(n)}{\partial w_{ij}(n)} = \frac{\partial \varepsilon(n)}{\partial e_j(n)} \frac{\partial e_j(n)}{\partial y_j(n)} \frac{\partial y_j(n)}{\partial v_j(n)} \frac{\partial v_j(n)}{\partial w_{ji}(n)}.$$ (10.34)

Equation (10.34) can be used for neurons located at the output layer where the error is available. If the cost function is the sum square error

$$\varepsilon(n) = \frac{1}{2} \sum_{j \in C} e_j^2(n)$$ (10.35)

where the set C includes all neurons in the output layer, we have

$$\frac{\partial \varepsilon(n)}{\partial e_j(n)} = e_j(n).$$ (10.36)

The error signal is defined as $e(n) = d_j(n) - y_j(n)$, where $d_j(n)$ is the desired response. The second term on the right-hand side of Equation (10.34) is given by

$$\frac{\partial e_j(n)}{\partial y_j(n)} = -1.$$ (10.37)

The derivative of the neuron output with respect to the activation field is given by

$$\frac{\partial y_j(n)}{\partial v_j(n)} = \varphi_j'(v_j(n)).$$ (10.38)

Differentiating the last term of the right-hand side of Equation (10.34) results in

$$\frac{\partial v_j(n)}{\partial w_{ji}(n)} = y_i(n)$$ (10.39)

and substituting Equations (10.35)–(10.39) into Equation (10.34) yields

$$\frac{\partial \varepsilon(n)}{\partial w_{ji}(n)} = -e_j(n)\varphi_j'(v_j(n))y_i(n).$$ (10.40)

Using the same stochastic approximation as that used on the LMS algorithm, the weight correction is obtained as

$$\Delta w_{ji}(n) = -\eta \frac{\partial \varepsilon(n)}{\partial w_{ji}(n)}$$ (10.41)

where η is the learning rate. Substituting Equation (10.40) into Equation (10.41) results in

$$\Delta w_{ji}(n) = \eta e_j(n)\varphi_j'(v_j(n))y_i(n).$$ (10.42)

It is convenient to define the local gradient $\delta_j(n)$ as

$$\delta_j(n) = \frac{\partial \varepsilon(n)}{\partial v_j(n)} = e_j(n)\varphi_j'(v_j(n))$$ (10.43)

and Equation (10.42) can therefore be rewritten as

$$\Delta w_{ji}(n) = \eta \delta_j(n) y_i(n). \tag{10.44}$$

For neurons located at the output layer, the error can be directly calculated and Equation (10.44) can be used for weight correction. For neurons located in a hidden layer there is no desired response available and the computation of the local gradient is more complicated and is given by [11]:

$$\delta_j(n) = \varphi_j'(v_j(n)) \sum_k \delta_k(n) w_{kj}(n). \tag{10.45}$$

The term $\varphi_j'(v_j(n))$ depends only on the activation function associated with the hidden neuron j. $\sum_k \delta_k(n) w_{kj}$ is the sum of all local gradients from the neurons located on the layer immediately to the right of the hidden neuron j which are directly connect to neuron j, multiplied by the respective synaptic weights that connects the neurons in layer k to the neuron j, according to Figure 10.10.

The algorithm can be summarized by the following steps (recall that the network topology should be previously defined).

1. Initialization, where the synaptic weights should be randomly selected.
2. Using the training dataset, directly propagating a training sample and calculating the error at the network output.
3. Propagate backwards the error from the output layer up to the first layer, obtaining the local gradient for each neuron.
4. Using the error correction formula (10.44), update all weights.
5. Repeat the process from step 2 until the required performance is achieved.

The algorithm presented above describes the sample by sample weight update, but the neuron network could use a batch learning algorithm. Through the batch learning, all training samples are forward propagated and the mean square error over all training samples calculated. This error should then be backward propagated and used for the weights update to define a training epoch. Through the batch learning algorithm, the weights are therefore updated epoch by epoch. The forward and backward propagation through the network during the back-propagation algorithm is illustrated in Figure 10.11.

Figure 10.12 shows a two-dimensional feature space with non-linear separable classes; Class 1 is represented by circles and Class 2 by crosses. The classes can be generated by

```
N=1000;
X=rand(2,N);
k=1;
j=1;
for i=1:N
    if X(2,i)>((sin(2*pi*X(1,i))+5*X(1,i))/5+0.02);
        C1(:,j)=X(:,i);
        j=j+1;
    elseif X(2,i)<((sin(2*pi*X(1,i))+5*X(1,i))/5-0.02)
        C2(:,k)=X(:,i);
        k=k+1;
    end
end
```

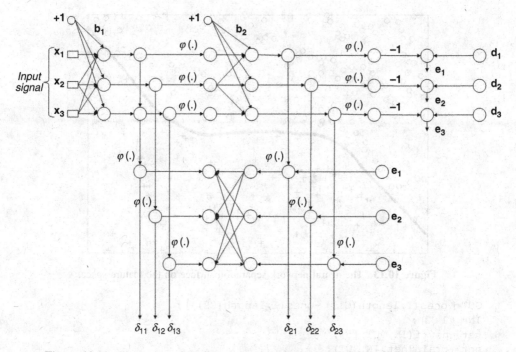

Figure 10.11 Illustration of the forward and backward propagation of a MLP network.

A multilayer perceptron with one hidden layer with 15 neurons using a hyperbolic tangent as an activation function was designed and the separation surface can be seen in Figure 10.13. The figure illustrates the ability of the neural network to solve non-linear classification problems and it can be seen that all features were classified correctly. The neural network was designed in MATLAB® by

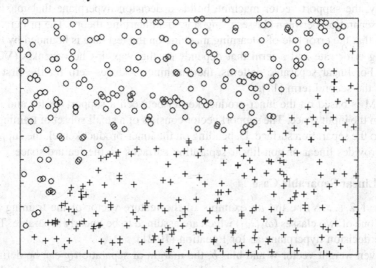

Figure 10.12 Two-dimensional feature space with non-linear separable classes.

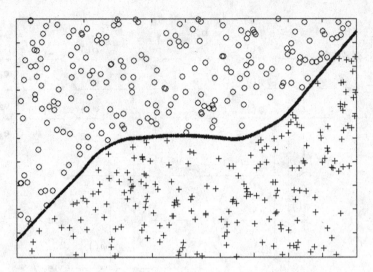

Figure 10.13 The neural network separation surface on the feature space.

```
OUT=[ ones(1,length(C1)) -ones(1,length(C2))];
IN=[ C1 C2];
net = newff(IN,OUT,15);
net = train(net,IN,OUT);
```

10.5.6 Support Vector Machines

A support vector machine (SVM) can be viewed as a universal feed-forward network which in the same way as an MLP can solve pattern recognition and non-linear regression problems. It was proposed in the late 1970s [14] and is based on statistical learning theory.

Basically, the support vector machine builds a decision hyperplane that maximizes the margin of separation between classes using statistical learning theory. The principle is based on the fact that the error rate of a learning machine on the test data is bounded by the sum of the training error rate and a term that depends on the Vapnik-Chervonenkis (VC) dimension [14]. For linear separable patterns, the machine produces zero for the first term and minimizes the second term [11].

The SVM is based on the inner-product kernel between a support vector and a vector $\mathbf{x}$ drawn from the input space. The support vector consists of a small subset of training data that is closest to the separation surface. Depending on the inner-product kernel, the support vector machine provides linear or non-linear separation surfaces on the feature space.

10.5.6.1 Linear Separable Case

Let $\mathbf{x_i}$ ($i = 1, 2, \dots, N$) be the ith example of the feature vector of the training set which belongs to one of two classes (ω_1, ω_2) which are assumed to be linearly separable. The goal is to find the decision hyperplane as for Equation (10.21).

For a given weight vector $\mathbf{w}$ and bias b, the margin of separation ρ can be defined as the separation between the hyperplane and the closest point in the dataset. The goal of the support vector machine is to maximize this margin of separation of both classes. Given that $\mathbf{w_o}$ and b_o denote the optimum values of the weight vector and bias, respectively, the linear discriminant function can be written as

$$g(\mathbf{x}) = \mathbf{w}_o^t \mathbf{x} + b_o. \tag{10.46}$$

The distance D from a given feature vector $\mathbf{x}$ from the separation hyperplane is given by

$$D = \frac{|g(\mathbf{x})|}{\| \mathbf{w} \|}. \tag{10.47}$$

The margin of separation ρ is therefore calculated using Equation (10.47) when the closest points in the dataset for both classes are used. An illustration is given in Figure 10.14. For a two-class problem of linearly separable patterns, we can write

$$\begin{aligned} \mathbf{w}_o^t \mathbf{x_i} + b_o &\geq 0 \quad \text{for } d_i = +1 \\ \mathbf{w}_o^t \mathbf{x_i} + b_o &< 0 \quad \text{for } d_i = -1 \end{aligned} \tag{10.48}$$

where $d_i = +1$ for class ω_1 and $d_i = -1$ for class ω_2. The goal is therefore to find the parameters $\mathbf{w_o}$ and b_o for the optimal hyperplane, given the training set $\{(\mathbf{x_i}, d_i)\}$. Rescaling Equation (10.21), the problem can be stated as

$$\begin{aligned} \mathbf{w}_o^t \mathbf{x_i} + b_o &\geq 1 \quad \text{for } d_i = +1 \\ \mathbf{w}_o^t \mathbf{x_i} + b_o &\leq -1 \quad \text{for } d_i = -1. \end{aligned} \tag{10.49}$$

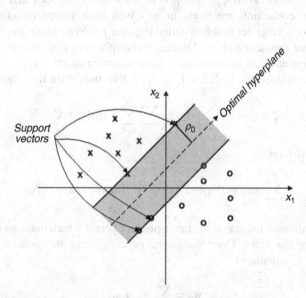

Figure 10.14 Illustration of an optimal hyperplane; support vectors and the optimal margin of separation are also shown.

The support vectors are the particular data points on the training set for which Equation (10.49) is equal to $+1$ (top) and -1 (bottom). From Equation (10.47) the distance from a given support vector $\mathbf{x}^S$ to the optimal hyperplane is

$$r = \frac{|g(\mathbf{x}^S)|}{\|\mathbf{w_o}\|} = \begin{cases} \dfrac{|+1|}{\|\mathbf{w_o}\|} & \text{if } d^S = +1 \\[3mm] \dfrac{|-1|}{\|\mathbf{w_o}\|} & \text{if } d^S = -1. \end{cases} \tag{10.50}$$

The optimum value of the margin of separation is the sum of both distances in Equation (10.50), resulting in

$$\rho = 2r = \frac{2}{\|\mathbf{w_o}\|}. \tag{10.51}$$

From Equation (10.51) it is possible to verify that, in order to maximize the margin of separation, the Euclidean norm of the weight vector should be minimized.

Given the training set $\{(\mathbf{x_i}, d_i)\}, i = 1, 2, \ldots, N$, find the optimal values of the weight vector and bias such that they satisfy the constraints

$$d_i(\mathbf{w}^t \mathbf{x_i} + b) \geq 1 \quad \text{for } i = 1, 2, \ldots N \tag{10.52}$$

and the weight vector $\mathbf{w}$ minimizes the cost function

$$J(\mathbf{w}) = \frac{1}{2} \mathbf{w}^t \mathbf{w} \tag{10.53}$$

where the factor of $1/2$ is included for convenience of presentation. This constrained optimization problem is called a *primal problem* and it can be seen that the cost function is convex and the constraints are linear in $\mathbf{w}$. Given such an optimization problem, it is preferable to construct another problem called the *dual problem* which has the same optimal value as the *primal problem* and use Lagrange multipliers to find the optimal solution.

The *dual problem* can now be formulated. Given the training set $\{(\mathbf{x_i}, d_i)\}, i = 1, 2, \ldots, N$, find the Lagrange multipliers $\{\alpha_i\}, i = 1, 2, \ldots, N$ that maximize the objective function

$$Q(\alpha) = \sum_{i=1}^{N} \alpha_i - \frac{1}{2} \sum_{i=1}^{N} \sum_{j=1}^{N} \alpha_i \alpha_j d_i d_j \mathbf{x_i} \mathbf{x}_j^t \tag{10.54}$$

subject to the constraints

$$\sum_{i=1}^{N} \alpha_i d_i = 0 \quad \text{and} \quad \alpha_i \geq 0 \quad \text{for } i = 1, 2, \ldots, N. \tag{10.55}$$

As already mentioned, the $\mathbf{x_i} \mathbf{x_j}$ is the inner-product kernel which plays an important role in the support vector machines. Once the Lagrange multipliers are found $\alpha_{o,i}$, the optimum weight vector $\mathbf{w_o}$ is calculated

$$\mathbf{w_o} = \sum_{i=1}^{N} \alpha_{o,i} d_i \mathbf{x_i}. \tag{10.56}$$

As for a given support vector $\mathbf{x}^S$ from the class ω_1, Equation (10.49) is equal to $+1$ and the bias can be found as

$$b_o = 1 - \mathbf{w}_o^t \mathbf{x}^S. \tag{10.57}$$

For practical cases, considering the subset of all support vectors $\{\mathbf{x}_i^S\}, i = 1, \ldots, N_S$ the bias can be found as

$$b_o = \frac{1}{N_S} \sum_{i=1}^{N_S} \left(d_i - \mathbf{w}_o^t \mathbf{x}_i^S \right). \tag{10.58}$$

10.5.6.2 Linear Non-Separable Case

For the non-separable case, vectors inside the margin of separation and the classification errors should be accounted in the optimization problems. The separation hyperplane can then be formulated as

$$d_i(\mathbf{w}^t \mathbf{x}_i + b) \geq 1 - \xi_i, \quad i = 1, 2, \ldots, N \tag{10.59}$$

where $\{\xi_i\}, i = 1, 2, \ldots, N$ are non-negative scalar variables called *slack variables*, that measure the deviation of a data point from the ideal condition of linear separation. From the margin of separation definition, the data points inside the margin are correctly classified and $0 \leq \xi_i \leq 1$; for misclassified data points $\xi_i > 1$.

The optimization task is now more complex, resulting in a non-convex problem that is not tractable. An approximation is therefore made and a new term is included in the cost function of Equation (10.53), resulting in

$$J(\mathbf{w}, \xi) = \frac{1}{2} \mathbf{w}^t \mathbf{w} + C \sum_{i=1}^{N} \xi_i. \tag{10.60}$$

The parameter C is a positive constant that controls the number of non-separable points allowed in the problem. Its value should be selected by the user experimentally or by the estimation of the VC dimension [11].

Therefore, given the training set $\{(\mathbf{x}_i, d_i)\}, i = 1, 2, \ldots, N$, we find the Lagrange multipliers $\{\alpha_i\}, i = 1, 2, \ldots, N$ that maximize the objective function

$$Q(\alpha) = \sum_{i=1}^{N} \alpha_i - \frac{1}{2} \sum_{i=1}^{N} \sum_{j=1}^{N} \alpha_i \alpha_j d_i d_j \mathbf{x}_i \mathbf{x}_j^t \tag{10.61}$$

subject to the constraints

$$\sum_{i=1}^{N} \alpha_i d_i = 0 \quad \text{and} \quad 0 \leq \alpha_i \leq C \quad \text{for} \quad i = 1, 2, \ldots, N. \tag{10.62}$$

It is important to note that if $C \to \infty$ the problem falls into the linear separable case.

10.5.6.3 Non-Linear Case

The non-linear SVM is a more general approach to solving pattern recognition issues. The basic idea is to perform a non-linear mapping of the input vectors into a high-dimensional feature space and then construct a linear optimal hyperplane (on the new feature space) for the separation of the features based on the maximization of the margin of separation.

This approach follows Cover's theorem on the separability of patterns. Cover's theorem states that a multidimensional feature space with non-linear separable patterns may be transformed into a new feature space where the patterns are linearly separable with high probability, providing that two conditions are satisfied: the transformation is non-linear and the dimensionality of the new feature space is sufficiently high [14].

The linear separation surface is built using the theory described in last section but with an important difference: the optimal linear hyperplane is built on the new feature space rather than in the original feature space.

Let $\varphi(\mathbf{x})$ denote the non-linear transformation from the input space to the new feature space. The linear discriminant function can be written

$$g(\mathbf{x}) = \sum_{j=1}^{m_1} w_j \varphi_j(\mathbf{x}) + b \tag{10.63}$$

where w_j and b_o are the weight vector and the bias, respectively, for the new data mapped into the new feature space by the non-linear transformation $\varphi_j(\mathbf{x})$ for $j = 1, \ldots, m_1$. The inner-product kernel $K(\mathbf{x}, \mathbf{x}_i)$ is defined

$$K(\mathbf{x}, \mathbf{x}_i) = \varphi(\mathbf{x})\varphi(\mathbf{x_i})^t = \sum_{j=0}^{m_1} \varphi_j(\mathbf{x})\varphi_j(\mathbf{x}_i) \tag{10.64}$$

where $\mathbf{x}_i$ is the ith example of the input vector, defined

$$\varphi(\mathbf{x}) = [\varphi_0(\mathbf{x}), \varphi_1(\mathbf{x}), \varphi_2(\mathbf{x}), \ldots, \varphi_{m_1}(\mathbf{x})] \tag{10.65}$$

where, by definition, $\varphi_0(\mathbf{x}) = 1$ for all $\mathbf{x}$ to incorporate the bias effect.

The optimization problem can be written similarly to Equation (10.61):

$$Q(\alpha) = \sum_{i=1}^{N} \alpha_i - \frac{1}{2} \sum_{i=1}^{N} \sum_{j=1}^{N} \alpha_i \alpha_j d_i d_j K(\mathbf{x}_i, \mathbf{x}_j) \tag{10.66}$$

with the constraints as defined in Equation (10.62). The optimal hyperplane is given by

$$\sum_{i=1}^{N} \alpha_{oi} d_i K(\mathbf{x}, \mathbf{x}_i) = 0 \tag{10.67}$$

where α_{oi} are the optimal values of the Lagrange multipliers.

An important property of the non-linear SVM is that the classification can be performed without knowledge of the non-linear mapping function $\varphi(\mathbf{x})$, since the optimization problem and the optimal hyperplane can be written in terms of the inner-product kernel.

The expansion of Equation (10.65) for the inner-product kernel is an important special case of *Mercer's theorem*. As such, the inner-product kernel should satisfy this theorem [15] and

Table 10.1 Summary of inner-product kernels [11].

Type of support vector machine	Inner-product kernel	Comments
Polynomial learning machine	$(\mathbf{x}\mathbf{x}_i^T + 1)^p$	The power p should be specified *a priori* by the user
Radial basis function network	$\exp\left(-\frac{1}{2\sigma^2}\|\mathbf{x} - \mathbf{x}_i\|^2\right)$	The width σ^2 is specified by the user
Two-layer perceptron	$\tanh(\beta_0\mathbf{x}^T\mathbf{x}_i + \beta_1)$	Mercer's theorem is satisfied only for some values of β_0 and β_1

the number of possible kernels to be used is reduced. Table 10.1 lists the most commonly used inner-product kernels.

Considering example described in Section 10.5.5 (classes C1 and C2 were defined), a non-linear SVM with radial basis kernel function was designed from the code:

```
OUT=[ ones(1,length(C1)) -ones(1,length(C2))];
IN=[ C1 C2];
IND = randperm(length(IN));
INsvm=IN(:,IND);
OUTsvm=OUT(1,IND);
f = -ones (length(IN),1);
b = 0;
LB = zeros(1,length(IN));
C = 100;
UB=C*ones(1,length(IN));
sigma=0.2;
Q=zeros(length(IN),length(IN));
        for i=1:length(IN)
            for j=1:length(IN)
            Q(i,j)=OUTsvm(i)*OUTsvm(j)*exp(-(norm(INsvm(:,i)-
            INsvm(:,j))^2)/(2*sigma^2));
            end
        end
OPTIONS = optimset('quadprog');
OPTIONS = optimset('MaxIter',10000);
% The Lagrange Multiplier are found
Lambda=quadprog(Q,f,OUTsvm,b,OUTsvm,b,LB,UB,[],OPTIONS);
MaxLambda=max(Lambda);
% Finding the index of the support vectors
Ind=find(Lambda>MaxLambda/100);
%Calculating the bias
bias=0;
    for y=1:size(Ind,1)
        for k=1:size(Ind,1)
            bias=-1/size(Ind,1)*OUTsvm(Ind(k))*Lambda(Ind(k))*(exp(-
            (norm(INsvm(:,Ind(y))-
            INsvm(:,Ind(k)))^2)/(2*sigma^2)))+bias;
        end
    end
```

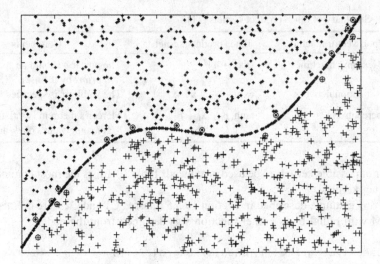

Figure 10.15 Support vectors and the separation surface obtained by the non-linear SVM.

```
%The output of the nonlinear SVM can be obtained using
out2=zeros(length(X1),1);
      for j=1:length(X1)
            for i=1:size(Ind,1)
            out2(j)=Lambda(Ind(i),1)*OUTsvm(1,Ind(i))*exp(-(norm(X1
            (:,j)-INsvm(:,Ind(i)))^2)/(2*sigma^2))+out2(j);
            end
      end
out2=out2+bias;
```

Figure 10.15 shows the support vectors (encircled data points on the feature space) and the separation surface found by the non-linear SVM; the ideal separation surface can be seen in Figure 10.16. It can be seen that the SVM separation surface is very similar to the ideal separation surface, illustrating the good generalization ability of the support vector machines. In the case of the separation surface obtained by a MLP depicted in Figure 10.13, it is not as close to the ideal separation surface as that of the non-linear SVM.

This example illustrates the better generalization ability of the non-linear SVM in comparison with the MLP. Nevertheless, the design of the neural network is much simpler and the SVM is more computationally complex than the MLP.

10.6 System Evaluation

The complete design of a pattern recognition system must include an evaluation of its performance, an important step which could lead to the complete redesign of the system. The goal is to estimate the probability classification error of the designed system with a finite dataset available for the system design.

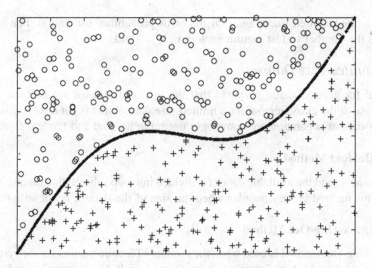

Figure 10.16 Samples of classes 1 (circles) and 2 (crosses) and ideal separation surface.

10.6.1 Estimation of the Classification Error Probability

The system error probability can be estimated from the test dataset using

$$\hat{P} = \sum_{i=1}^{M} P(\omega_i) \frac{k_i}{N_i} \tag{10.68}$$

where k_i is the number of misclassified vectors and N_i the total number of vectors, both from class ω_i. If not known, the class probability $P(\omega_i)$ can be estimated by

$$P(\hat{\omega}_i) = \frac{N_i}{N} \tag{10.69}$$

where N is the total number of feature vectors on the test dataset. The system error probability can therefore be estimated as

$$\hat{P} = \frac{1}{N} \sum_{i=1}^{M} k_i. \tag{10.70}$$

It can easily be shown that the error probability estimator is an unbiased estimator but only asymptotically consistent as $N \to \infty$. The number of feature vectors on the test dataset therefore cannot be small; otherwise, the classifier error probability estimation is not reliable.

A commonly used approach to define the number of feature vectors required to estimate the system error probability is

$$N \approx \frac{100}{E_P} \tag{10.71}$$

where E_P is the required uncertainty on the error probability estimation. For example, if $E_P = 0.1\%$ the number of test feature vectors is $N = 1000$.

10.6.2 Limited-Size Dataset

It is crucial for generalization purposes that the test dataset must not be used during the training phase of the classifier, but for a limited-size dataset this is not always possible. Two methods of dataset separation are commonly used: the holdout and the leave-one-out.

10.6.2.1 Holdout Method

In the holdout method the available dataset is divided into two subsets: the training and the test datasets. This method is very popular when the size of the dataset is not an issue.

10.6.2.2 Leave-One-Out Method

The Leave-One-Out method is used when the number of feature vectors for the system design is small. In this method, the training is performed with $N - 1$ samples and the test is performed with the excluded sample. If the test sample is misclassified it is counted as an error. This process is repeated N times, each time excluding a different sample. The total number of misclassified samples divided by N leads to the estimation of the system error probability.

In this method, the training is performed using practically all samples and the test itself is performed using all samples, although the independence between the training and test datasets is maintained. The problem of this technique is the high computational cost.

10.7 Pattern Recognition Examples in Power Systems

In this section, we present some examples of pattern recognition techniques to solve power system problems.

10.7.1 Power Quality Disturbance Classification

In the last decades, power quality (PQ) has become a major concern to both power utilities and power consumers as a result of deregulation, the widespread use of sensitive loads, power electronic devices and the complexity of industrial processes [16]. A good example of a pattern recognition application is the automatic classification of power quality disturbances. Several methods of PQ classification have been proposed in the literature [2,17,18].

A digital system for the detection and classification of power quality disturbances based on a wavelet transform with a support vector machine as the classifier algorithm is proposed in reference [19]. Six categories of disturbances are considered in this paper: voltage sags, voltage swells, harmonics, voltage fluctuations, voltage transients and short-time interruptions. The definitions of these disturbances can be found in reference [20]. Figures 10.17–10.19 illustrate a sag, a swell and harmonics, respectively.

The features used are generated by Fourier and wavelet transforms and the feature selection is based on expert knowledge. However, before the feature generation, the wavelet is also used as a de-noising technique. Five features are used as inputs of a non-linear SVM. The global efficiency achieved by the proposed method for the considered database was 95%.

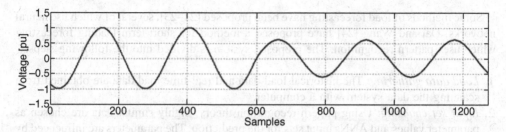

Figure 10.17 Illustration of a sag disturbance.

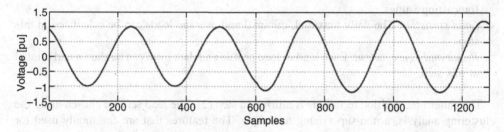

Figure 10.18 Illustration of a swell disturbance.

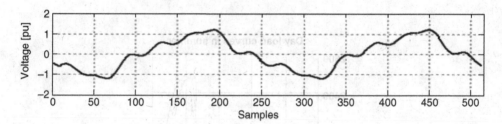

Figure 10.19 Illustration of harmonics.

10.7.2 Load Forecasting in Electric Power Systems

Electric power system load forecasting plays an important role in the energy management systems (EMS); it has a great influence on the operation, controlling and planning of electric power systems. Short-term load forecasting is useful in the safe and economic planning operation of an electrical power system. In normal working conditions, the system generating capacity should meet the load requirements at any time. If the system generating capacity is not large enough, essential measures should be taken such as adding generating units or importing some power from a neighboring network. On the other hand, if the system generating capacity is surplus to requirements, essential measures (e.g. shutting down some generating units or outputting some power to a neighboring network) should be taken. Load variation trends and feature forecasting are essential for the power dispatch, layout and design department of power systems.

Some methods of load forecasting have been proposed [21–23], several of which use neural networks. Dai and Wang [24] have proposed a method for short-term loading forecasting which uses pattern recognition. The proposed system is divided into the following steps.

1. *Load data gathering*: The historical load data and real-time load data are obtained from scanning the data system with a computer.
2. *Input set choosing*: Using pattern recognition theory, highly similar data are chosen as parameter values and ANNs input sets for the prediction. The parameters are influenced by factors such as temperature, humidity, rainfall and season.
3. *Load forecast*: The three-layered feed-forward artificial neural network is constructed. Its features are constructed and the back-propagation learning algorithm is used to obtain load forecasting values.
4. *Load statistics*: The daily maximal, minimal and average loads can be calculated in this step.
5. *Result and figure output*: Each kind of load profile and data form can be shown and printed at this step.

The pattern recognition technique is utilized in step (2). The load sets are chosen using the clustering analysis, a non-supervising technique. The features that are commonly used for short-term load forecasting (again expert knowledge is used) are the seasons, weather status and special events (e.g. political, cultural, etc.). An illustration of the achieved results by the proposed method for a 24-hour load forecasting is provided in Figure 10.20.

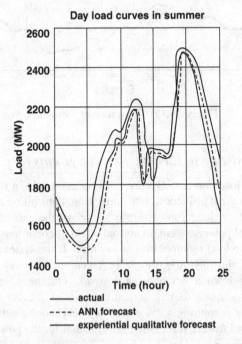

Figure 10.20 Illustration of a 24-hour forecast from [24].

10.7.3 Power System Security Assessment

The power system is subjected to many disturbances, and its security assessment involves determining whether it remains in a secure or insecure state. Traditional security assessment involved exhaustive steady-state load flow analysis, making it almost impossible to perform a real-time analysis.

Several methods based on pattern recognition techniques have been proposed to solve problems related to the power system security assessment [25,26]. Here, we describe a method based on artificial neural networks for static security assessment [25]. The goal is to predict the post-contingence system state using pre-contingence system variables.

The pre-contingence busbar power injection (P, Q) are the inputs of the classifier and the output is composed of the performance index (PI) values of the power system post-contingence state to a list of next contingencies. The main problem related to this approach is the dimension of the input feature space for practical systems. A good feature selection technique should therefore be applied in order to reduce the dimension of the feature space.

Several works in this field make use of relevance tests in order to select the best features. However, Luan *et al.* [25] propose the use of the regressional relief algorithm, claiming that it shows great advantage over the relevance test. The main characteristic of this method is that it models the problem as a non-linear regression, while others are based on a classification problem. The results for a simulated 77 busbar system show that the feature selection technique is capable of reducing the input feature space from 110 dimensions to 33 without noticeable performance loss [25].

10.8 Conclusions

The complexity and variability of electrical signals from variable generation and loads demand advanced signal processing tools to assist the system operators to properly identify problems and control the grid's power delivery process. Distributed, renewable generation and customers that consume and produce energy responding to market signals are creating a complex smart grid. Pattern recognition becomes an essential enabling tool for the identification and control of the upcoming electric smart-grid environment. This chapter has highlighted the main aspects of the necessary tools required to operate the grid of the future.

References

1. Ma, L. and Lee, K.Y. (2008) Fuzzy neural network approach for fault diagnosis of power plant thermal system under different operating points. In Proceedings of Power and Energy Society General Meeting: Conversion and Delivery of Electrical Energy in the 21st Century. IEEE, pp. 20–24.
2. Cerqueira, A.S., Ferreira, D., Ribeiro, M.V. and Duque, C.A. (2008) Power quality events recognition using a SVM-based method. *Electric Power Systems Research*, **78**, 1546–1552.
3. Gerbec, D., Gasperic, S., Smon, I. and Gubina, F. (2005) Allocation of the load profiles to consumers using probabilistic neural networks. *IEEE Transactions on Power Systems*, **20** (2), 548–555.
4. Kamwa, I., Grondin, R., Sood, V.K., Gagnon, C., Van Thich, N. and Mereb, J. (1996) Recurrent neural networks for phasor detection and adaptive identification in power system control and protection. *IEEE Transactions on Instrumentation and Measurements*, **45** (2), 657–664.
5. Theodoridis, S. and Koutroumbas, K. (2008) *Pattern Recognition*, 4th edn, Academic Press.
6. Mendel, J. (1991) Tutorial on higher-order statistics (spectra) in signal processing and system theory: theoretical results and some applications. *Proceedings of IEEE*, **79** (3), 378–305.

7. Ribeiro, M.V., Marques, C.A.G., Duque, C.A., Cerqueira, A.S. and Pereira, J.L.R. (2007) Detection of disturbances in voltage signals for power quality analysis using HOS. *EURASIP Journal on Advanced Signal Processing (Print)*, **2007**, 1–14.

8. Papoulis, A. (2002) *Probability, Random Variables and Stochastic Processes*, 4th edn, McGraw-Hill Europe.

9. Theodoridis, S., Pikrakis, A., Koutroumbas, K. and Cavouras, D. (2010) *Introduction to Pattern Recognition: A Matlab Approach*, 1st edn, Academic Press.

10. Kay, S. (2005) *Intuitive Probability and Random Processes using MATLAB®*, Springer.

11. Haykin, S. (2008) *Neural Networks and Learning Machines*, 3rd edn, Prentice Hall.

12. Robbins, H. and Monro, S. (1951) A stochastic approximation method. *Annals of Mathematical Statistics*, **22** (3), 400–407.

13. Widrow, B. (1960) Adaptive sample data systems. Proceedings of the First International Congress of the International Federation of Automatic Control, pp. 406–411.

14. Vapnik, V. (1979) *Estimation of Dependences Based on Empirical Data [in Russian]*, Nauka, Moscow, English translation: Springer Verlag, New York, 1982.

15. Mercer, J. (1908) Functions of positive and negative type, and their connection with the theory of integral equations. *Proceedings of Royal Society of London, Series A*, **83**, 69–70.

16. Bollen, M.H. and Gu, I. (2006) *Signal Processing of Power Quality Disturbances*, 1st edn, Wiley-IEEE Press.

17. Ferreira, D.D., Cerqueira, A.S., Duque, C.A. and Ribeiro, M.V. (2009) HOS-based method for classification of power quality disturbances. *Electronics Letters*, **45**, 183.

18. Jayasree, T., Devaraj, D. and Sukanesh, R. (2010) Power quality disturbance classification using Hilbert transform and RBF networks. *Neurocomputing*, **73** (7–9), 1451–1456.

19. Yu, X. and Wang, K. (2009) Digital system for detection and classification of power quality disturbance. Power and Energy Engineering Conference. APPEEC 2009. Asia-Pacific, pp. 1–4.

20. Bollen, M.H. (1999) *Understanding Power Quality Problems: Voltage Sags and Interruptions*, 1st edn, Wiley-IEEE Press.

21. Yamina, S.M., Shahidehpourb, Z. and Li, H.Y. (2004) Adaptive short-term electricity price forecasting using artificial neural networks in the restructured power markets. *Electrical Power & Energy Systems*, **26**, 571–581.

22. Kai, Z., Jianchun, P. and Feng, S. (2000) Short-term load forecast based pattern recognition. *Human Power Systems*, **20** (6), 1–12.

23. Chongqing, K., Xu, C., Qing, X., and Yu, S. (1999) A new unified approach to short-term load forecasting considering correlated factors. *Automation of Electric Power Systems*, **23** (18), 32–35.

24. Dai, W. and Wang, P. (2007) Application of pattern recognition and artificial neural network to load forecasting in electric power system. Third International Conference on Natural Computation, pp. 381–385.

25. Luan, W.P., Lo, K.L. and Yu, Y.X. (2000) ANN-based pattern recognition technique for power system security assessment. Proceedings of International Conference on Electric Utility Deregulation and Restructuring and Power Technologies, pp. 197–202.

26. Srinivasan, D., Chang, C.S., Liew, A.C. and Leong, K.C. (1998) Power system security assessment and enhancement using artificial neural network. Proceedings of International Conference on Energy Management and Power Delivery, vol. 2, pp. 582–587.

11

Detection

11.1 Introduction

In power systems, signal detection is extensively used for monitoring and protection. The basic theory relates to the random nature of nearly all physical signals. Statistical properties are used to design optimum detection systems.

A binary detection problem can be formulated with two hypotheses, one related to the event occurring (e.g. a short circuit between two phases of the power system) and another related to the non-occurrence of the event (e.g. system's normal operation). The statistical properties of the hypotheses are usually very complex or unknown. As such, sub-optimal detectors are used in practice.

For example, signal detection for power system protection can be performed using the three steps: (1) acquiring the voltage and current waveforms through voltage transformers (VTs) and current transformers (CTs); (2) estimation of the amplitude of the fundamental component for both waveforms through a simple DFT; and (3) event detection, where if the amplitude of the voltage and current exceed some pre-determined value the protection should act. In such a situation, the VT and CT are used for signal acquisition and the DFT is used for the feature extraction technique. The detection is performed using thresholds related to the statistical nature of the fault.

The binary nature of the detection problem can be classified as the occurrence or non-occurrence of the event. The steps required for the design of a detection system are similar to those presented in Chapter 10 for pattern recognition (Figure 10.2). The classical theory for both classification and detection is analogous and based on Bayes decision theory.

11.2 Why Signal Detection for Electric Power Systems?

In power systems, the detection of an event is one of the first steps for accomplishing several applications in protection, power quality monitoring, analysis and control. Any equipment designed for the detection and evaluation of disturbances and distortions in a system must be as selective as possible so that the alarm is given only for relevant events and not false alarms. Such an example is the need for differentiation between a fault or non-fault detection of any device. The fault detection activates other parts of the protection algorithms, such as the fault

Power Systems Signal Processing for Smart Grids, First Edition. Paulo Fernando Ribeiro, Carlos Augusto Duque, Paulo Márcio da Silveira and Augusto Santiago Cerqueira.
© 2014 John Wiley & Sons, Ltd. Published 2014 by John Wiley & Sons, Ltd.
Companion Website: http://www.wiley.com/go/signal_processing/

classification and the fault discriminator (in the case of transmission line protection). If a false alarm is generated by the detector, the protection would de-energize a healthy line. On the other hand, if the detector does not identify a fault, serious problems can lead to a total system collapse (a blackout could occur in both cases).

In power quality monitoring, the detector is also an important element; this could be an application that simply records the events for a future analysis, or it could be equipment such as a power quality event classifier that can be used in real time. In the first case a false alarm will lead to the recording of a normal waveform. Consequently, memory space can be quickly filled with non-useful information. As for the event classifier, if the detector does not perceive an event much information can be lost.

There are many other applications in power systems for event detection. Consider the CT saturation discussed in Chapter 2. One approach is the correction of the distorted waveform through CT saturation using a signal processing technique. However, the first step in correcting a waveform is the detection of the instant when the signal is saturated. This task is not easily accomplished in the presence of noise.

11.3 Detection Theory Basics

11.3.1 Detection on the Bayesian Framework

Classification and detection theories are based on the Bayes decision theory described in Section 10.3. This section will focus on its features as they relate to detection theory. Usually detection theory is formulated in the context of a communication problem when there is a transmitter and a receiver, with the goal of detecting the transmitted signal at the receiver end. For example, a transmitter could send two symbols, zero or one. If these are corrupted by the channel, the receiver end should still be able to determine which symbol was transmitted.

The observed data at the receiver end is represented by $\mathbf{r} = \{r_1, r_2, \ldots, r_N\}$ for dimensional observation space N. The hypotheses H_0 and H_1 are for a binary decision where the conditional probability densities at the receiver end could be represented by $f_{\mathbf{r}/H_0}(\mathbf{r}/H_0)$ and $f_{\mathbf{r}/H_1}(\mathbf{r}/H_1)$. The *a priori* probabilities are $P(H_0)$ and $P(H_1)$ at the transmitter side. The optimal detector can be found using the Bayes decision rule as in Equation (10.3), resulting in:

$$\begin{aligned} &\text{if } \ f_{\mathbf{r}/H_0}(\mathbf{r}/H_0)P(H_0) \geq f_{\mathbf{r}/H_1}(\mathbf{r}/H_1)P(H_1) \quad H_0 \text{ selected} \\ &\text{if } \ f_{\mathbf{r}/H_0}(\mathbf{r}/H_0)P(H_0) < f_{\mathbf{r}/H_1}(\mathbf{r}/H_1)P(H_1) \quad H_1 \text{ selected.} \end{aligned} \quad (11.1)$$

There are four possible results at the receiver: (1) H_0 transmitted; H_0 selected; (2) H_0 transmitted; H_1 selected; (3) H_1 transmitted; H_1 selected; and (4) H_1 transmitted; H_0 selected. The first and the third possibilities are the correct outcomes while the second and fourth are decision errors. Considering the four decision possibilities, costs could be assigned to each decision in order to increase and decrease the effect of each decision in the Bayes decision rule. If the costs are denoted C_{00}, C_{10}, C_{11} and C_{01} respectively, the Bayes decision rule becomes (for more details see reference [1]):

$$\begin{aligned} &\text{if } \ f_{\mathbf{r}/H_0}(\mathbf{r}/H_0)P(H_0)(C_{10} - C_{00}) \geq f_{\mathbf{r}/H_1}(\mathbf{r}/H_1)P(H_1)(C_{01} - C_{11}) \quad \text{select } H_0 \\ &\text{if } \ f_{\mathbf{r}/H_0}(\mathbf{r}/H_0)P(H_0)(C_{10} - C_{00}) < f_{\mathbf{r}/H_1}(\mathbf{r}/H_1)P(H_1)(C_{01} - C_{11}) \quad \text{select } H_1. \end{aligned} \quad (11.2)$$

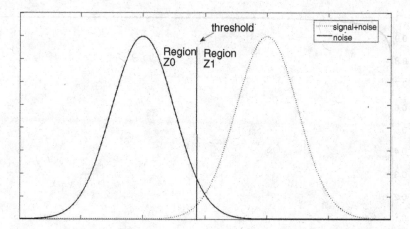

Figure 11.1 Conditional probabilities for a one-dimensional binary detection problem.

Equation (11.2) can be rewritten

$$\frac{f_{\mathbf{r}/H_1}(\mathbf{r}/H_1)}{f_{\mathbf{r}/H_0}(\mathbf{r}/H_0)} \underset{H_0}{\overset{H_1}{\gtrless}} \frac{P(H_0)(C_{10} - C_{00})}{P(H_1)(C_{01} - C_{11})}. \tag{11.3}$$

The quantity on the left side of Equation (11.3) is called the *likelihood ratio* and the quantity on the right side is the *threshold* (η).

Figure 11.1 shows the conditional probabilities considering Gaussian densities for $f_{\mathbf{r}/H_0}(\mathbf{r}/H_0)$ (noise: hypothesis H_0) and $f_{\mathbf{r}/H_1}(\mathbf{r}/H_1)$ (signal + noise: hypothesis H_1) for a one-dimensional binary detection problem. The probability of false alarm P_F, detection P_D and missing signal P_M can be calculated using Equation (11.4), which represent the probability of noise signals being incorrectly detected at the receiver end, the probability of signals correctly detected at the receiver end and the probability of signals being incorrectly assigned as noise at the receiver end, respectively. These are defined:

$$P_F = \int_{Z1} f_{r/H_0}(\mathbf{r}/H_0) d\mathbf{r}$$

$$P_D = \int_{Z1} f_{r/H_1}(\mathbf{r}/H_1) d\mathbf{r} \tag{11.4}$$

$$P_M = \int_{Z0} f_{r/H_1}(\mathbf{r}/H_1) d\mathbf{r}.$$

11.3.2 Newman-Pearson Criterion

In practical situations it is very difficult to assign costs or *a priori* probabilities to define the optimal threshold. A simple approach to solve this problem is to work with the conditional probabilities of false alarm P_F and detection P_D. The goal is usually to make the false alarm level as insignificant as possible and the detection probability as large as possible. Unfortunately in several practical cases when P_F is decreased, P_D also decreases.

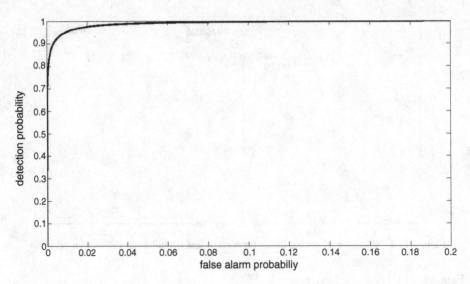

Figure 11.2 Receiving operating characteristics (ROC) for the distributions depicted in Figure 11.1.

The Newman-Pearson criterion constrains P_F and designs a test to maximize P_D. As such, the solution can be found using the Lagrange multipliers. See reference [1] for more details.

11.3.3 Receiving Operating Characteristics

A graphical way to evaluate the performance of a detector system is to use the receiver operating characteristics (ROC). This is a graphical visualization of $P_F \times P_D$ when the threshold is changed. For each threshold value, P_F and P_D are calculated according to Equation (11.4) in order to build the ROC. The ROC completely describes the performance of a detector system. Additionally, the threshold value could be evaluated using the ROC. Figure 11.2 shows the ROC curve for the conditional probabilities depicted in Figure 11.1. From Figure 11.2, it can be seen that for 6% of false alarms the detection probability is almost 100%.

Another example of a ROC curve can be seen in Figure 11.3b for the conditional probabilities showed in Figure 11.3a. In this case, the detector performance is worse than that in Figure 11.2 as the intersection (confusion area) between the conditional probabilities is larger than that in Figure 11.1. It can be seen in Figure 11.3b that for 10% of false alarm, the detection probability is around 80%.

11.3.4 Deterministic Signal Detection in White Gaussian Noise

Several practical problems can be formulated for the detection of a known deterministic signal in white Gaussian noise (WGN). This problem can be formulated in the discrete domain as

$$H_o : r[n] = w[n] \quad n = 0, 1, \ldots, N-1$$
$$H_1 : r[n] = s[n] + w[n] \quad n = 0, 1, \ldots, N-1 \tag{11.5}$$

where $r[n]$ is the observed signal, $s[n]$ is known and deterministic, and $w[n]$ is the WGN with a variance σ^2.

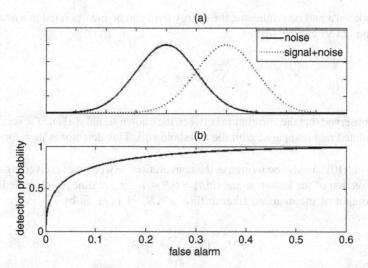

Figure 11.3 (a) Conditional probabilities of a detector system and (b) its ROC curve.

The detector can be designed using the likelihood ratio

$$L(\mathbf{r}) = \frac{f_{\mathbf{r}/H_1}(\mathbf{r}/H_1)}{f_{\mathbf{r}/H_0}(\mathbf{r}/H_o)} \overset{H_1}{\underset{H_0}{\gtrless}} \eta, \tag{11.6}$$

where $\mathbf{r} = [r[0] \quad r[1] \ldots \quad r[N-1]]^t$ and the conditional probabilities for $w[n]$ WGN can be written as

$$f_{\mathbf{r}/H_1}(\mathbf{r}/H_1) = \frac{1}{(2\pi\sigma^2)^{N/2}} \exp\left[-\frac{1}{2\sigma^2}\sum_{n=o}^{N-1}(r[n]-s[n])^2\right]$$

$$f_{\mathbf{r}/H_0}(\mathbf{r}/H_0) = \frac{1}{(2\pi\sigma^2)^{N/2}} \exp\left[-\frac{1}{2\sigma^2}\sum_{n=o}^{N-1}r^2[n]\right]. \tag{11.7}$$

The likelihood ratio therefore results in

$$L(\mathbf{r}) = \exp\left[-\frac{1}{2\sigma^2}\left(\sum_{n=0}^{N-1}(r[n]-s[n])^2 - \sum_{n=0}^{N-1}r^2[n]\right)\right] \overset{H_1}{\underset{H_0}{\gtrless}} \eta. \tag{11.8}$$

Applying logarithms to both sides of Equation (11.8), H_1 is chosen if

$$\sum_{n=o}^{N-1}r[n]s[n] > \sigma^2\ln\eta + \frac{1}{2}\sum_{n=o}^{N-1}s^2[n]. \tag{11.9}$$

As $s[n]$ is known and deterministic, the energy term can be incorporated in a new threshold and Equation (11.9) can be written

$$\sum_{n=o}^{N-1} r[n]s[n] > \eta'. \tag{11.10}$$

It can be observed that the correlation between the known signal $s[n]$ and the received signal $r[n]$ is calculated and compared with the threshold (η'). This detector is therefore known as *correlator*.

Equation (11.10) can also be written as the convolution between the received signal $r[n]$ and the flipped version of the known signal ($h[n] = s[N - 1 - n]$) at time $N - 1$, called a *matched filter*. The output of the matched filter at time $n = N - 1$ is given by

$$y[N - 1] = \sum_{n=0}^{N-1} r[n]s[n]. \tag{11.11}$$

Consider a deterministic sinusoidal signal $s[n]$ that is corrupted by an additive Gaussian white noise (AWGN) after transmission, which should be detected at the receiver end of a communication device. Figure 11.4 shows $s[n]$ and Figure 11.5 shows four realizations of the received signal $r[n]$, where the upper two are related to the signal while the lower two are related to noise. Visually it is very difficult to distinguish between signal and noise. Ten thousand realizations for both hypotheses can be generated by using the following coding in MATLAB®

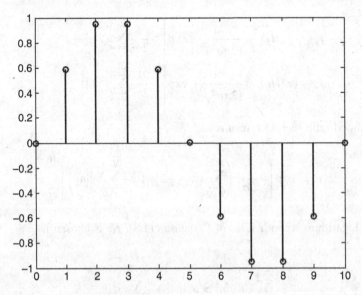

Figure 11.4 Example of a deterministic sinusoidal signal $s[n]$ that should be detected.

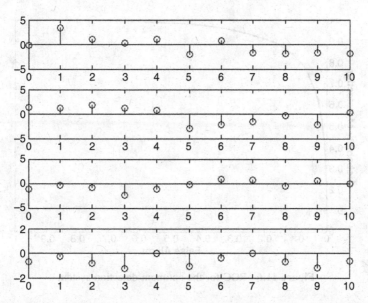

Figure 11.5 Samples of the received signal. The two realizations at the top are related to signal while the two at the bottom are related to noise.

```
%Detection Example
n=1:11;
%deterministic sinusoidal signal
s=sin(2*pi*1/10*(n-1));
t=0:10;
stem(t,s)
pause
%Gaussian White Noise
w=randn(10000,11);
w1=randn(10000,11);
sn=repmat(s,10000,1);
%received signals
r1=sn+w;
r0=w1;
%examples of signals and noise at receiver
figure
subplot(4,1,1),stem(t,r1(1,:))
subplot(4,1,2),stem(t,r1(2,:))
subplot(4,1,3),stem(t,r0(3,:))
subplot(4,1,4),stem(t,r0(4,:))
```

The optimum detector in this case is the correlator. Its estimated ROC curve can be seen in Figure 11.6. If both hypotheses are equiprobable, the optimum threshold (minimum probability error) can be calculated according to

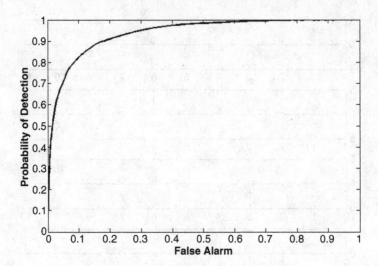

Figure 11.6 ROC for the optimum detection system.

$$\eta' = \sigma^2 \ln\left(\frac{P(H_1)}{P(H_0)}\right) + \frac{1}{2}\sum_{n=0}^{N-1} s^2[n]. \tag{11.12}$$

This results in a probability of detection of around 87% for a false alarm around 13%. The detection system can be implemented in MATLAB® using

```
%Correlator Detector
Det=s;
%Correlator outputs
rH1=r1*Det';
rH0=r0*Det';
%Histogram for rH0 (noise) and rH1 (signal+noise)
figure
hist(rH0,30)
hold
hist(rH1,30)
title('Histogram')
hold
%ideal threshold for equal a priori probabilities
TH=log(1)+1/2*sum(s.^2);
%probability of detection and false alarm for the optimum threshold
Pd=length(rH1(rH1>=TH))/10000
Pf=length(rH0(rH0>TH))/10000
%building the ROC curve
THRoc=[-5:17/200:12];
For i=1:201
PdRoc(i)=length(rH1(rH1>=THRoc(i)))/1000;
```

```
PfRoc(i)=length(rH0(rH0>THRoc(i)))/1000;
end
figure
plot(PfRoc,PdRoc)
title('ROC curve')
```

11.3.5 Deterministic Signals with Unknown Parameters

The power system fundamental component signal $f[n]$ can be modeled as a deterministic signal with unknown parameters as $f[n] = A_0 \cos(2\pi f_0/f_s n + \theta_0)$. We may not be sure of the values of amplitude A_0, fundamental frequency f_0 and the phase θ_0, but the sampling frequency f_s is usually known. There are two main approaches to solve this issue. The first considers the signal as deterministic and uses the generalized likelihood ratio test (GLRT). The second is a Bayesian approach, where the signal is modeled as a realization of a random process.

In the classical case of unknown deterministic signal parameters, an optimal detector usually does not exist. A suboptimal detector is therefore required and the GLRT will usually demonstrate good detector performance. If the Bayesian approach is adopted, the resulting detector can be said to be optimal. The difficulty of using the Bayesian approach is in specifying the prior probability density functions and carrying out the integration in practical scenarios.

In this section we discuss the case of unknown signal amplitude in order to illustrate the design of the detector test using both approaches. First, we discuss the importance of signal knowledge for the design of a detector system.

11.3.5.1 Unknown Signal

Consider a detection problem

$$
\begin{aligned}
H_0 &: r[n] = w[n] \quad n = 0, 1, \ldots, N-1 \\
H_1 &: r[n] = s[n] + w[n] \quad n = 0, 1, \ldots, N-1
\end{aligned}
\tag{11.13}
$$

where $s[n]$ is deterministic but completely unknown and $w[n]$ is the WGN with a variance of σ^2. A GLRT would select H_1 if

$$
\frac{f_{\mathbf{r}/\hat{s},H_1}(\mathbf{r}/\hat{s}[0], \ldots \hat{s}[N-1], H_1)}{f_{\mathbf{r}/H_0}(\mathbf{r}/H_0)} > \eta
\tag{11.14}
$$

where $\hat{s}[n]$ is the maximum likelihood estimation (MLE) under H_1. To determine the MLE, the likelihood function should be maximized over the signal samples

$$
f_{r/\hat{s},H_1}(\mathbf{r}/\hat{s}[0], \ldots \hat{s}[N-1], H_1) = \frac{1}{(2\pi\sigma^2)^{N/2}} \exp\left[-\frac{1}{2\sigma^2} \sum_{n=0}^{N-1} (r[n] - s[n])^2\right],
\tag{11.15}
$$

resulting in $\hat{s}[n] = r[n]$. Thus, from Equation (11.14):

$$\frac{\dfrac{1}{(2\pi\sigma^2)^{N/2}}}{\dfrac{1}{(2\pi\sigma^2)^{N/2}}\exp\left[-\dfrac{1}{2\sigma^2}\sum_{n=0}^{N-1}r^2[n]\right]} > \eta. \qquad (11.16)$$

After taking logarithms of both sides, this yields

$$\sum_{n=0}^{N-1}r^2[n] > \eta' \quad \text{select } H_1. \qquad (11.17)$$

It can be seen from Equation (11.17) that this is just an energy detector. The performance of the energy detector is worse than the correlator (optimal), which is due to the design of the energy detector where the signal was considered unknown. A comparison between both detectors can be performed using the example of Section 11.3.4, where a sinusoidal deterministic signal should be detected. Figure 11.7 shows the ROC curves for both detectors (correlator and energy detector), where it is possible to verify that the correlator presents a much better performance than the energy detector. These curves can be reproduced in MATLAB® using

```
%Detection Example
n=1:11;
%deterministic sinusoidal signal
s=sin(2*pi*1/10*(n-1));
t=0:10;
%Gaussian White Noise
```

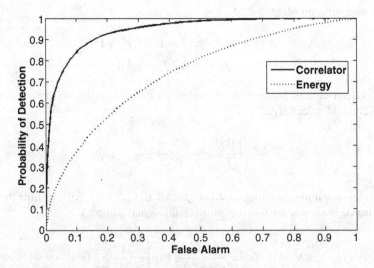

Figure 11.7 Comparison between the correlator and the energy detector for determinist signal detection under WGN.

```
w=randn(10000,11);
w1=randn(10000,11);
sn=repmat(s,10000,1);
%received signals
r1=sn+w;
r0=w1;
%Correlator Detector
Det=s;
%correlator outputs
rH1=r1*Det';
rH0=r0*Det';
%building the ROC curve
THRoc=[ -5:17/200:12];
For i=1:201
PdRoc(i)=length(rH1(rH1>=THRoc(i)))/10000;
PfRoc(i)=length(rH0(rH0>THRoc(i)))/10000;
end
figure
plot(PfRoc,PdRoc)
title('ROC curves')
%Energy Detector
r2H1=sum(r1.^2,2);
r2H0=sum(r0.^2,2);
THRoc=[ 0:55/200:55];
For i=1:201
Pd2Roc(i)=length(r2H1(r2H1>=THRoc(i)))/10000;
Pf2Roc(i)=length(r2H0(r2H0>THRoc(i)))/10000;
end
hold
plot(Pf2Roc,Pd2Roc,'r')
```

11.3.5.2 Unknown Amplitude

Consider the problem of detecting a known deterministic signal with unknown amplitude in WGN. The problem can be formulated:

$$
\begin{aligned}
H_0 &: r[n] = w[n] \quad n = 0, 1, \ldots, N-1 \\
H_1 &: r[n] = As[n] + w[n] \quad n = 0, 1, \ldots, N-1
\end{aligned}
\tag{11.18}
$$

where $s[n]$ is known and the amplitude A is unknown. The $w[n]$ is WGN with a variance of σ^2. In order to build the detector, the LRT

$$
\frac{\dfrac{1}{(2\pi\sigma^2)^{N/2}} \exp\left[-\dfrac{1}{2\sigma^2} \displaystyle\sum_{n=0}^{N-1} (r[n] - As[n])^2\right]}{\dfrac{1}{(2\pi\sigma^2)^{N/2}} \exp\left[-\dfrac{1}{2\sigma^2} \displaystyle\sum_{n=0}^{N-1} r^2[n]\right]} > \eta \quad \text{select } H_1
\tag{11.19}
$$

can be used. Taking logarithms of both sides results in

$$A \sum_{n=0}^{N-1} r[n]s[n] > \sigma^2 \ln \eta + \frac{A^2}{2} \sum_{n=0}^{N-1} s^2[n] = \eta'. \tag{11.20}$$

If the sign of A is known and $A > 0$, the test for whether H_1 should be selected is reduced to:

$$\sum_{n=0}^{N-1} r[n]s[n] > \frac{\eta'}{A} = \eta''. \tag{11.21}$$

If $A < 0$, we select H_1 if

$$\sum_{n=0}^{N-1} r[n]s[n] < \frac{\eta'}{A} = \eta''. \tag{11.22}$$

The tests are optimal and the detector reduces to the correlator. If the amplitude sign is unknown however, it is not possible to construct a unique test and the detector should be designed using the GLRT or the Bayesian approach.

11.3.5.3 GLRT Design

Using the generalized likelihood ratio test, H_1 is selected if

$$\frac{f_{\mathbf{r}/\hat{A},H_1}(\mathbf{r}/\hat{A}, H_1)}{f_{\mathbf{r}/H_0}(\mathbf{r}/H_0)} > \eta, \tag{11.23}$$

where $\hat{A}$ is the maximum likelihood estimator of A under the assumption of H_1, given by

$$\hat{A} = \frac{\sum_{n=0}^{N-1} r[n]s[n]}{\sum_{n=0}^{N-1} s^2[n]}. \tag{11.24}$$

From Equation (11.20), H_1 is selected if

$$\hat{A} \sum_{n=0}^{N-1} r[n]s[n] > \sigma^2 \ln \eta + \frac{\hat{A}^2}{2} \sum_{n=0}^{N-1} s^2[n]. \tag{11.25}$$

From Equation (11.24), we have

$$\hat{A}\hat{A} \sum_{n=0}^{N-1} s^2[n] - \frac{\hat{A}\hat{A}}{2} \sum_{n=0}^{N-1} s^2[n] > \sigma^2 \ln \eta. \tag{11.26}$$

The GLRT design therefore leads to the selection of H_1 if

$$|\hat{A}| > \sqrt{\frac{2\sigma^2 \ln \eta}{\sum_{n=0}^{N-1} s^2[n]}}. \tag{11.27}$$

Equivalently, Equation (11.27) can be written

$$\left|\sum_{n=0}^{N-1} r[n]s[n]\right| > \sqrt{2\sigma^2 \ln \eta \sum_{n=0}^{N-1} s^2[n]} \tag{11.28}$$

which is a correlator that accounts for the unknown amplitude sign. It is important to stress that the detector performance has deteriorated in comparison to the correlator (Equation (11.22)), but is much better than the energy detector (Equation (11.17)).

The performance of the GLRT detector in comparison to the correlator and the energy detector can be evaluated using the previous example (detection of a sinusoidal signal). Figure 11.8 shows the ROC curves for the three detectors. It is possible to see that the performance of the GLRT detector lies between that of the correlator and the energy detector. This behavior is expected since the amplitude sign must be known for the correlator detector, but not for the GRLT.

11.3.5.4 Bayesian Design

The Bayesian approach considers the signal $s[n]$ as a random process and requires knowledge of the prior probability density function of A. Assuming that A is a random variable following a Gaussian distribution with mean μ_A and variance σ_A^2, independent of $w[n]$, the decision to

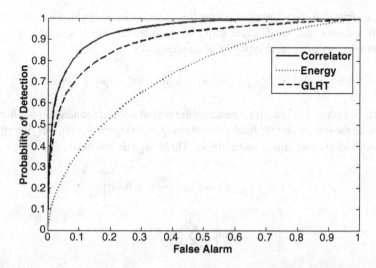

Figure 11.8 Comparison between the correlator, the GLRT and the energy detector.

select H_1 is determined by (see details in reference [2]):

$$\frac{\mu_A}{\sigma^2+\sigma_A^2\sum\limits_{n=0}^{N-1}s^2[n]}\sum_{n=0}^{N-1}r[n]s[n]+\frac{\sigma_A^2}{2\sigma^2\left(\sigma^2+\sigma_A^2\sum\limits_{n=0}^{N-1}s^2[n]\right)}\left(\sum_{n=0}^{N-1}r[n]s[n]\right)^2>\eta'. \qquad (11.29)$$

This detector is a combination of the correlator and the second-order correlator (GRLT). Its performance is optimal in the Newman-Person sense [2].

11.4   Detection of Disturbances in Power Systems

The increasing levels of harmonic distortion in power line signals is driving the development of signal processing tools capable of performing more detailed and precise detection of disturbances and events. One of the major requirements for power quality monitoring equipment is the detection of PQ disturbances fixed in time, with the ability to detect both the initiation and end of the fault.

The necessity for improved detection performance in continuous electric signals monitoring devices has motivated the development of several techniques that have a good tradeoff between computational complexity and performance [3–5].

Much research has been performed on wavelet-transform-based techniques for the purposes of detection. There are also several techniques that make use of second-order information about the error signal, which results from the subtraction of the fundamental component from the electric signal. Analysis of the error signal is an attractive and interesting solution to characterize the presence of disturbances [4]. The techniques proposed in references [3,6,7] are very similar in the sense that all of them make use of second-order statistics of the error signal for the detection of the occurrence of disturbances.

11.4.1 The Power System Signal

The discrete version of the monitored power line signals can be divided into non-overlapped frames of N samples and the discrete sequence in a frame. These can be expressed as an additive contribution of several types of phenomena:

$$x[n] = x(t)|_{t=nT_s} := f[n] + h[n] + i[n] + t[n] + v[n], \qquad (11.30)$$

where $n = 0, \ldots, N-1, T_s = 1/f_s$ is the sampling period and the sequences $f[n], h[n], i[n], t[n]$ and $v[n]$ denote the power system fundamental component signal, harmonics, interharmonics, transient and background noise, respectively. These signals are defined:

$$f[n] = A_0[n]\cos\left[2\pi\frac{f_0[n]}{f_s}n + \theta_0[n]\right], \qquad (11.31)$$

$$h[n] = \sum_{m=1}^{M} h_m[n], \qquad (11.32)$$

$$i[n] = \sum_{j=1}^{J} i_j[n], \tag{11.33}$$

$$t[n] = t_{\text{spi}}[n] + t_{\text{not}}[n] + t_{\text{cas}}[n] + t_{\text{dam}}[n], \tag{11.34}$$

and $v[n]$ is an independently and identically distributed (i.i.d.) noise as normal $N(0, \sigma_v^2)$ and independent of $f[n]$, $h[n]$, $i[n]$ and $t[n]$.

In Equation (11.31), $A_0[n]$, $f_0[n]$ and $\theta_0[n]$ refer to the magnitude, fundamental frequency and phase of the power system fundamental signal, respectively. In Equation (11.32) and (11.33), $h_m[n]$ and $i_j[n]$ are the mth harmonic and the jth interharmonic, respectively, which are defined

$$h_m[n] = A_m[n]\cos\left[2\pi m \frac{f_0[n]}{f_s} n + \theta_m[n]\right] \tag{11.35}$$

$$i_j[n] = A_{1,j}[n]\cos\left[2\pi \frac{f_{1,j}[n]}{f_s} n + \theta_{1,j}[n]\right] \tag{11.36}$$

where $A_m[n]$ is the magnitude and $\theta_m[n]$ is the phase of the mth harmonic and $A_{1,j}[n], f_{1,j}[n]$ and $\theta_{1,j}[n]$ are the magnitude, frequency and phase of the jth interharmonic, respectively. In Equation (11.34), $t_{\text{spi}}[n]$, $t_{\text{not}}[n]$, $t_{\text{cas}}[n]$ and $t_{\text{dam}}[n]$ are transients named spikes, notches, decaying oscillations, and damped exponentials.

11.4.2 Optimal Detection

The detection of an event or disturbance can be performed using the N samples frame that can be represented by vector $\mathbf{x}$. When the power system signal is in its normal operation,

$$\mathbf{x} = \mathbf{f}_n + \mathbf{v}, \tag{11.37}$$

where $\mathbf{f}_n$ is the fundamental component vector of the power system signal in its normal range of operation (amplitude and frequency) and $\mathbf{v}$ is the background noise vector. When a disturbance occurs, vector $\mathbf{x}$ can be expressed as

$$\mathbf{x} = \mathbf{f}_n + \Delta\mathbf{f}_n + \mathbf{h} + \mathbf{i} + \mathbf{t} + \mathbf{v}, \tag{11.38}$$

where $\Delta\mathbf{f}_n$ represents the abnormal conditions related to the fundamental component (sag, swell, interruptions or frequency variation), $\mathbf{h}$ is the vector related to the harmonics, $\mathbf{i}$ interharmonics and $\mathbf{t}$ transients. A disturbance occurs if any of these signals are present in $\mathbf{x}$.

The detection of disturbances can be formulated as a binary hypothesis test:

$$\begin{aligned} H_0 &: \mathbf{x} = \mathbf{f}_n + \mathbf{v} \\ H_1 &: \mathbf{x} = \mathbf{f}_n + \Delta\mathbf{f_n} + \mathbf{h} + \mathbf{i} + \mathbf{t} + \mathbf{v} \end{aligned} \tag{11.39}$$

where H_0 refers to the normal operation of the power system signal (absence of disturbance) and H_1 refers to the occurrence of a disturbance on the power system signal.

The first step in the design of a detection system is to generate and select representative features **p** from the signal frame **x** that can map the input data into a feature space where the separation between both hypotheses is increased. In order to design the optimum detector without assigning costs for the decisions, Equation (11.1) should be used. It requires knowledge of $f_{\mathbf{x}/H_0}(\mathbf{x}/H_0), f_{\mathbf{x}/H_1}(\mathbf{x}/H_1), P(H_0)$ and $P(H_1)$. These quantities are not normally known, but they can be estimated using the training dataset [7].

Probability density function estimation is not an easy task in a high-dimensional space. This is a common situation in several detection problems. Additionally, this approach can lead to a detection system of high computational complexity. Sub-optimum detection algorithms are therefore commonly used, for example the techniques presented in Chapter 10 including LS methods, distance-based classifiers, neural networks and SVM.

11.4.3 Feature Extraction

Several disturbance detection techniques make use of the wavelet transform and second-order statistics as a feature extraction technique due to the nature of the power system signal (see Equation (11.30)). The wavelet transform divides the input signal into different frequency bands. For each band the energy is measured (second-order statistics) and used as input to the detection algorithm.

Another common approach for feature extraction is the removal of the fundamental component $\mathbf{f}_n$ from the input signal vector **x**. As can be seen in Equation (11.39), the fundamental component is common to both hypotheses; its removal from **x** therefore increases the separation between the hypotheses. As such, instead of using **x** for the design of the detection system, the error signal $\mathbf{e} = \mathbf{x} - \hat{\mathbf{f}}_n$ is used where $\hat{\mathbf{f}}_n$ is the estimation of the fundamental component of the power system signal. The error signal can be generated using a notch filter (Chapter 5) or a fundamental component estimation technique. Some detection techniques directly use the error signal as an input for the detection algorithm while others measure the energy of the error signal, reducing the dimension of the feature space.

Another relevant approach for feature extraction is the use of higher-order statistics (HOS) cumulants, since these are blind to any kind of Gaussian process (background noise) whereas second-order information is not. Cumulant-based signal processing techniques can handle colored Gaussian noise automatically, whereas second-order techniques may not. Cumulant-based techniques therefore boost signal-to-noise ratio when electric signals are corrupted by Gaussian noise [8].

11.4.4 Commonly Used Detection Algorithms

Concerning detection algorithms, several techniques have been employed for power system disturbance detection. Techniques are usually suboptimal as the complexity of the problem impedes the development of the likelihood ratio detector designed using the Bayesian approach.

Several authors have made use of specialized knowledge to perform the feature extraction, and the detection threshold is chosen using the training dataset. Such an example is to detect transients; for this the cycle-by-cycle difference [9] could be used followed by an energy detector.

When several features are extracted from the power system signal in order to perform the disturbance detection, the training dataset should be used to design a detector based on distance (e.g. Euclidian distance), LS, ANN or SVM or a fuzzy-based detector, among other techniques.

11.5 Examples

11.5.1 Transmission Lines Protection

The first task in the protection of transmission lines is fault detection. This is very important for the design of the effective isolation of the electric power system that may be experiencing such a fault. Basically, a fault detection algorithm must differentiate the normal state from the faulted state. This need to be carried out within a few milliseconds and its performance (speed and accuracy) directly affects the performance of the protection system. In order to reduce the risk of a false alarm, a counter is used to confirm and differentiate between a fault and non-fault. A fault is detected only if a consecutive number of samples that overpass a threshold are higher than a predefined value. In reference [10], three consecutive samples are considered.

Fault detection algorithms can use the characteristic fault components to perform their task while maintaining their ability to identify an occurrence even in the presence of harmonic frequencies, white noise, notches, spikes and frequency variations.

One of the simplest algorithms used for fault detection is described in reference [10], called the short Fourier filter. In this case, a FIR filter with zeros located at each harmonic is used and its output $y[n]$ is expressed as the difference between two samples spaced by one cycle (that corresponds to N samples) of the current signal, that is

$$s[n] = |y[n] - y[n - N + 1]|. \tag{11.40}$$

An even simpler form is the detection based on the comparison of two consecutive samples of the signal $i[n]$ [11]. The equation in the discrete-time domain is:

$$s[n] = |i[n] - i[n - 1]|. \tag{11.41}$$

However, there is a more robust method of performing a detection task: a long Fourier filter [11]. The main difference between a long and a short Fourier filter is that the former has double zeros on the unit circle spaced every $2f_0$ (twice the fundamental frequency). This process makes it somewhat more immune to low-frequency variations, especially the frequencies around 60 Hz (or 50 Hz). It is implemented by the equation:

$$s[n] = \left| i[n] + i\left[n - \frac{N}{2} + 1\right] - i[n - N + 1] - i\left[n - \frac{3N}{2} + 1\right] \right|. \tag{11.42}$$

A very interesting class of parameters is the HOS or cumulants which has been successfully applied as a pre-processing tool in detection issues. In power quality, it provides an appropriate tool for detecting and classifying disturbances [12–14]. The remarkable results of [15–17] for detection purposes, parameter estimation and classification applications must be mentioned.

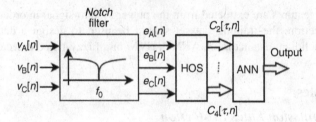

Figure 11.9 The block diagram structure for fault detection.

Figure 11.9 shows a detector stage as described in reference [17]. The detection process starts with the filtering process performed by an IIR notch filter (see Chapter 5), which has a transfer function

$$H(z) = \frac{1+\alpha}{2} \frac{1 - 2\beta z^{-1} + z^{-2}}{1 - \beta(1+\alpha)z^{-1} + \alpha z^{-2}}. \qquad (10.43)$$

This filter removes the fundamental frequency component f_0. Its use is based on the fact that various techniques of detection are performed using an error signal. The formulation based on hypotheses test:

$$\begin{aligned} H_0 &: e_v[n] = r_v[n] \\ H_1 &: e_v[n] = r_v[n] + t_v[n] \end{aligned} \qquad (11.44)$$

can be used for fault detection. The error signal is composed of a Gaussian noise $r_v[n]$ if the H_0 hypothesis is true. This situation is equivalent to normal operation of the electrical power system (EPS). On the other hand, the error signal is composed of a Gaussian noise component $r_v[n]$ added to a transient component $t_v[n]$, considering that the H_1 hypothesis is true. This situation is equivalent to the operation of the EPS under fault conditions. This is a reasonable formulation because, after the attenuation of the fundamental component, only a noise component is expected at the output of the filter (and possibly very small harmonics).

After the filtering stage, the HOS block is used to calculate the cumulants of the error signals. The cumulants are obtained at this stage and applied to the N-length sliding windows of signals $e_A[n]$, $e_B[n]$ and $e_C[n]$. Since the third-order cumulants of a symmetrically distributed random process are zero, only the second- and fourth-order cumulants are used. It is worthwhile mentioning that the detection is made in this approach using the voltage signal.

A combination of cumulants for phases A, B and C is the input of the third and last stage. This stage is composed of an artificial neural network.

Simulations were performed using a sampling frequency $f_s = N \times f_0$ with $N = 256$. The fault distances considered were $d = \{5, 10, 15, \ldots, 145\}$ with fault inception angles of $\{0°, 30°, 60°, \ldots, 330°\}$ and fault resistances between phase and ground $R_g = \{0.1, 1.0, 10, 100, 400\}$. The resistance between phases $R_p = \{0.1, 1, 10, 50\}$ in the case of faults involving phases. The system used to train the detector and all other functions of the relay system is depicted in

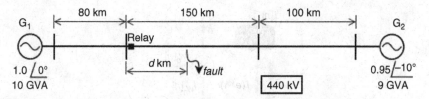

Figure 11.10 The simulated electrical power system.

Figure 11.10. For the detection, the data windows have a length of 1/8 (one-eighth) of the fundamental period.

Figure 11.11 shows the comparative results of the detection using the methodology discussed above and the traditional methodologies, which involve sample by sample (s_s) and cycle by cycle (s_c). For the example presented in Figure 11.11, the system frequency was varied over the range 59–61 Hz.

The main point to be highlighted in the detection of faults based on neural networks and cumulants is that after training the NN the threshold does not need to be adjusted as in the case of traditional methods; the threshold is established by the neural network optimization.

11.5.2 Detection Algorithms Based on Estimation

The information obtained from the estimation algorithms can be used for detection, despite the fact that this procedure is not commonly used in protection systems. However, if the estimation algorithm is able to indicate quickly that there is a significant deviation from the expected estimation value, it can be used to detect a fault. The estimation algorithm presented

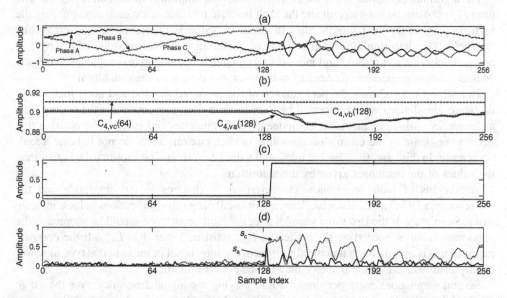

Figure 11.11 Performance of the detection method proposed in reference [17]: (a) fault signal; (b) cumulants used in detection process; (c) output of the proposed detector; (d) and output of the detection s_a and s_c.

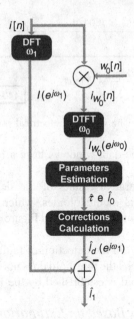

Figure 11.12 Detection method based on parameter estimation.

in Chapter 5 shows this property. Figure 11.12 shows the basic block diagram of the estimation and detection method.

The algorithm presented in Figure 11.12 estimates the amplitude of the current signal. The term $I_{w_0}(e^{j\omega_0})$ can be used to indicate the fault because its values change abruptly during the fault. In fact, at the first sample after the fault its value is high enough to overpass the threshold. The detection algorithm consists of obtaining the absolute value of $I_{w_0}(e^{j\omega_0})$ to compare this with a threshold and increment a counter for consecutive overpasses; when the counter reaches a specified number of overpasses, the detector issues an alarm.

To analyze and compare the performance of the proposed method and other methods in literature, simulations were performed by inserting non-fault components in the system such as frequency variations, harmonic components, noise, notches and spikes. The changes that occur in response to the estimation algorithm in such circumstances do not indicate a fault occurrence. In this case, the adjacent minimum values L_{min} should be chosen to be higher than the values of the responses given by the algorithms.

Simulations of fault occurrence, characterized as changes in the amplitude and the existence of a DC exponential with decaying or oscillating sub-synchronous values, indicate that the minimum of the first three samples after the fault occurrence should be adopted as the maximum value of borderline L_{max}. Clearly, L_{min} must be lower than L_{max}. If the opposite happens, this is an indication that the algorithm in question is not immune to that type of power quality disturbance and the detector may indicate a false alarm.

Several simulations were performed while varying the signal frequency over the range 48–72 Hz, the harmonic amplitude by 10–40% of the fundamental value, the SNR over the range 20–50 dB, the fundamental amplitude variation in fault condition 20–100%, the time

Table 11.1 The worst case for defining the threshold.

	L_{min}				L_{max}			
Frequency	Harmonics	Noise	Notches	Spikes	Fundamental amplitude	Exponential decaying amplitude	τ	Sub-synchronous oscillation
Proposed method	2.3040	1.1081	0.5998	0.6004	1.6125	32.2500×10^3	8.0768	13.5621	0.9017 $\times 10^6$

constant of an exponential from 0.5 up to 10 cycles, the amplitude of the exponential by 0–10 times the value of the nominal current and, finally, the sub-synchronous frequency from 10 to 50 Hz and its time constant from 0.01 to 0.1 s.

Table 11.1 presents the worst-case scenario used to define the threshold of the detection algorithm. It can be observed that any value higher than 2.304 and smaller than 8.0763 can be used as threshold.

The development of new algorithms for use in digital protection devices, monitoring and control requires validation in real time. In order to validate the system under conditions as close as possible to that of the power system where it is going to be installed, it is necessary that both the algorithm under test and the grid where it is going to be installed can operate in real time.

The real-time digital power system simulator (RTDS) is one of the most-used real-time electromagnetic digital simulators worldwide. It has traditionally been used to test protective relays and control equipment, but its main feature is the ability to interact in real time with external components and equipment that, through the I/O cards interface (analog to digital or A/D card and digital to analog or D/A card) become part of the simulation loop. This feature is referred to as hardware-in-the-loop (HIL) [18] and provides, among other features, the ability to validate prototypes of new equipment and components as well as the study of real-time transients in an integrated system.

On the other hand, when developing new algorithms for protection, control and monitoring, the implementation of these on dedicated hardware may require a long development time of the software or, in the case of a field-programmable gate array (FPGA), a long time to synthesize the digital circuit. An interesting approach that reduces the development time is the use of virtual hardware, such as dSPACE. dSPACE allows the algorithms previously tested in MATLAB®/Simulink to be compiled according to a real-time operating system, running on high-performance processors. The virtual instrument is then connected to the RTDS through the A/D and D/A cards of both instruments. Figure 11.13 shows the HIL implemented to test the detection algorithm described above. The estimation and detector algorithms were compiled from MATLAB® and transferred to the dSPACE board. The real time simulation was performed in the RTDS and the network is presented in Figure 11.14.

Figure 11.15 shows an example of a detection algorithm working in real time. The threshold used for this example was 3; the fault is applied at cycle 79 and then removed at cycle 84. Fault elimination is obtained directly by removing the fault resistance. The figure shows the current signal, the amplitude estimation, the output $I_{w_0} = I_{w_0}(e^{j\omega_0})$ and the trigger signal $I_{w_0} > 3$.

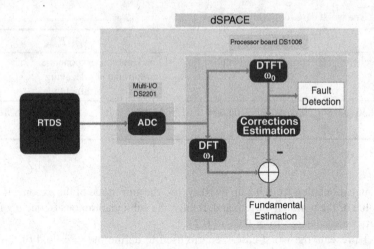

Figure 11.13 HIL to test the detection and estimation algorithm.

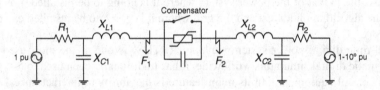

Figure 11.14 Network implemented in RTDS to test the detection and estimation algorithm.

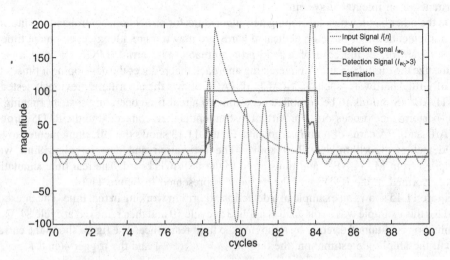

Figure 11.15 Example of the detection algorithm in real time.

11.5.3 Saturation Detection in Current Transformers

The importance of correcting the current distortion to prevent operation errors of protective devices connected to the current transformer (CT) secondary operating in saturation has been discussed extensively in the literature. The first step in making this correction is to know precisely where the saturation region starts and finishes. The correction methods known from literature are highly dependent on accurate knowledge of the saturated and unsaturated regions, due to the fact that in non-saturated areas the secondary current is a true copy (scale) of the primary fault current. On the other hand, the secondary current behaves differently in saturated regions depending on the non-linearity of the magnetizing inductance of the CT.

Several techniques exist for detecting the saturation of transformers. The main theory behind the detection algorithm is that the saturation causes a breakdown of the current waveform. Any method that then detects the breakdown in a signal would appear to be able to detect the saturation region. The most commonly used methods are based on wavelets [19] and derivatives [20]. Both work very well if there is no added noise in the signal. However, if a small noise equivalent to 50 dB of SNR is added to the system, the detection performance deteriorates completely.

Figure 11.16 shows the distorted secondary current and the respective detail coefficient d_1 when the wavelet db4 is used. Note that the detail coefficient is a good indicator of the saturation region; in this case, the signal does not contain noise. When noise is added, it is noted from Figure 11.17 that the first level of detail of the discrete wavelet transform (DWT) is seriously affected. There is no safe threshold level for the coefficient d_1 that could be used to detect the beginning and the end of the saturated regions. Further study of the second, third and fourth levels of detail (d_2, d_3 and d_4, respectively) shown in Figure 11.17 however indicate a possible way to detect this error. Since the coefficient values oscillate over a large number of samples around the start and end points of saturation, the simple choice of a threshold is not

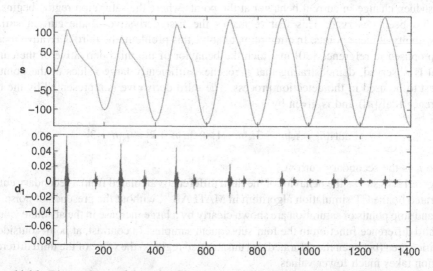

Figure 11.16 Distorted part of the secondary current revealed in the detail coefficient at level d_1. No noise added.

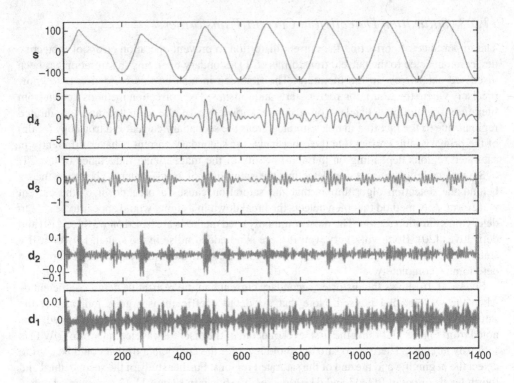

Figure 11.17 Distorted portion of the secondary current and detail coefficients with noise.

enough to identify saturated regions with accuracy and further processing is required to try to better identify such points.

A sudden change in current behavior at the points where the saturation region begins and ends has been observed. This feature makes the first derivative of the current suffer a discontinuity at these points. In attempting to solve this problem, the third derivative method was proposed in reference [20] in which the behavior of the third derivative of the current signal is observed, demonstrating that it reaches sufficiently large values at the points of interest to be used in the detection process. The third derivative is represented by the third difference ($\mathrm{del}_3[n]$) and is given by

$$\mathrm{del}_3[n] = i_s[n] - 3i_s[n-1] + 3i_s[n-2] - i_s[n-3] \tag{11.45}$$

where i_s is the secondary current.

Figure 11.18 shows the behavior of the third difference calculated from a secondary current generated by the CT simulation algorithm in MATLAB®, without the presence of noise. The start and stop points of saturation are shown clearly by a large increase in the absolute value of the third difference function in the four subsequent samples. In contrast, at points outside the transition area (between the saturated and unsaturated regions) the value of the third difference function takes much lower values.

The theoretical threshold value can be established by considering the output of Equation (11.45) in the unsaturated region. The secondary current in the unsaturated region

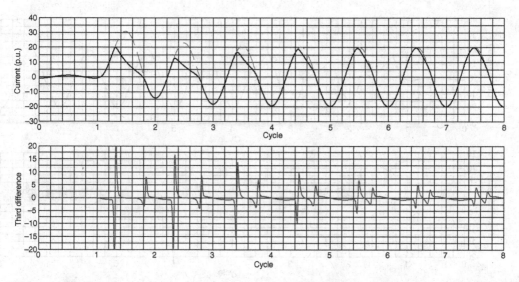

Figure 11.18 Distorted part of secondary current and the third derivative function when no noise is added to the signal.

can be written

$$i_s[n] = \frac{I_{pf\max}}{K_n}\left[\cos\left(\frac{2\pi}{N}n\right) - \exp(-nT_s/\tau)\right], \tag{11.46}$$

where $I_{pf\max}$ is the maximum current at the primary side, K_n is the transformer relation and τ is the time constant of the DC component. The maximum value of the third difference can be obtained by considering the hypotheses of zero DC component. The final value is defined:

$$\max(\text{del}_3[n]) = \frac{I_{pf\max}}{K_n}\left[2\sin\left(\frac{\pi}{N}\right)\right]^3, \tag{11.47}$$

which can be used to set the threshold for detecting the start and the stop of the saturation function of the third difference. In order to prevent filter inaccuracies of the inherent sensitivity of the algorithm, a margin factor k is adopted so that the threshold Tsh_{TD} (third difference) can be expressed:

$$\text{Tsh}_{\text{TD}} = k\frac{I_{pf\max}}{K_n}\left[2\sin\left(\frac{\pi}{N}\right)\right]^3. \tag{11.48}$$

Despite good results obtained by the previous method, for the detection of faults in noisy scenarios the above will be inaccurate. In reference [20], the authors used a first-order RC low-pass filter to smooth the effect of noise caused by the converter process (digital to analog and analog to digital) connected to the output of the ATP and input of the detector. The studies performed showed that the correct choice of cutoff frequency associated with the adjustment of the margin factor retains the efficiency of the method of the third difference. However, no

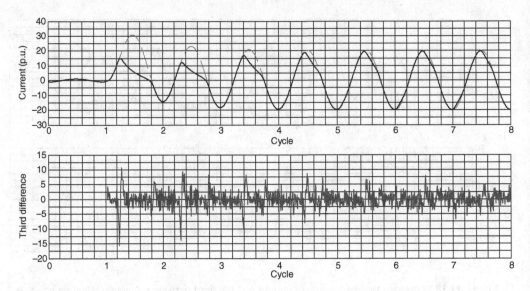

Figure 11.19 Distorted region of the secondary current and the third derivative function when noise is added to the signal and the signal is pre-filtered.

studies have been conducted regarding the background noise from the power system and electromagnetic interference in the CT secondary circuit. The derivative has the property of increasing high-frequency noise. For example, Figure 11.19 shows the same example of Figure 11.18 with noise of 50 dB added to the signal and filtered by a second-order low-pass Butterworth filter, with cutoff frequency of 500 Hz.

Further studies indicate that the use of more complex filters, such as the Chebchev [21] of 36 samples (for a sampling rate of 200 samples per cycle) can restore the effectiveness of the method of the third difference. However, these types of filters increase the computational effort and the delay in 18 samples.

11.6 Smart-Grid Context and Conclusions

Signal detection is used in power systems to monitor disturbances and protection activities. Detection theory is related to the random nature of physical signals and its statistical properties are used in order to design optimum detection systems. In power systems protection, detection can be performed with the adequate signal processing of waveforms that extract the information and detect parameters and thresholds related to the statistical nature of the event.

The higher complexity of smart grids will produce more intricate and compound signals that will make the detection process more difficult. This chapter introduced the basic aspects of detection theory using the Bayesian framework and discussed the deterministic signal detection in white Gaussian noise. We have also described example applications of power system signals in transmission line protection and saturation detection in current transforms. The more complex the smart grid of the future, the more detailed and advanced the detection methods will have to be in order to monitor the network and mitigate problems for the proper and continuous operation of the grid.

References

1. Van Trees, H.L. (2001) *Detection, Estimation and Modulation Theory, Part I: Detection, Estimation and Linear Modulation Theory*, John Wiley and Sons.
2. Kay, S.M. (1998) *Fundamentals of Statistical Signal Processing: Detection Theory*, Prentice Hall, Signal Processing Series.
3. Duque, C.A., Ribeiro, M.V., Ramos, F.R. and Szczupak, J. (2005) Power quality event detection based on the principle divided to conquer and innovation concept. *IEEE Transactions on Power Delivery*, **20** (4), 2361–2369.
4. Gu, I.Y.H., Ernberg, N., Styvaktakis, E. and Bollen, M.J.H. (2004) A statistical-based sequential method for fast online detection of fault-induced voltage dips. *IEEE Transactions on Power Delivery*, **19** (2), 497–504.
5. Ece, D.G. and Gerek, O.N. (2004) Power quality event detection using joint 2-D-wavelet subspaces. *IEEE Transactions on Instrumentation and Measurement*, **53** (4), 1040–1046.
6. Abdel-Galil, T.K., El-Saadany, E.F. and Salama, M.M.A. (2003) Power quality event detection using Adaline. *Electric Power Systems Research*, **64** (2), 137–144.
7. Theodoridis, S. and Koutroumbas, K. (2008) *Pattern Recognition*, 4th edn, Academic Press.
8. Mendel, J.M. (1991) Tutorial on higher-order statistics (spectra) in signal processing and system theory: theoretical results and some applications. *Proceedings of the IEEE*, **79** (3), 278–305.
9. Bollen, M.H.J. and Gu, I.Y.H. (2006) *Signal Processing of Power Quality Disturbances*, IEEE Press, Power Engineering Series.
10. Mohanty, S.R., Pradhan, A.K. and Routray, A. (2008) A cumulative sum-based fault detector for power system relaying application. *IEEE Transactions on Power Delivery*, **23** (1), 79–86.
11. Wiot, D. (2004) A new adaptive transient monitoring scheme for detection of power system events. *IEEE Transactions on Power Delivery*, **19** (1), 42–48.
12. Gerek, O.N. and Ece, D.G. (2006) Power-quality event analysis using higher order cumulants and quadratic classifiers. *IEEE Transactions on Power Delivery*, **21** (2), 883–889.
13. Ribeiro, M.V., Marques, C.A.G., Duque, C.A., Cerqueira, A.S. and Pereira, J.L.R. (2007) Detection of disturbances in voltage signals for power quality analysis using HOS. *EURASIP Journal on Advances in Signal Processing*, **1**, 177.
14. Ferreira, D.D., Cerqueira, A.S., Duque, C.A. and Ribeiro, M.V. (2009) HOS-based method for classification of power quality disturbances. *Electronics Letters*, **45** (3), 183–185.
15. Giannakis, G.B. and Tsatsanis, M.K. (1990) Signal detection and classification using matched filtering and higher order statistics. *IEEE Transactions on Acoustics, Speech and Signal Processing*, **38** (7), 1284–1296.
16. Colonnese, S. and Scarano, G. (1999) Transient signal detection using higher order moments. *IEEE Transactions on Signal Processing*, **47** (2), 515–520.
17. Carvalho, J.R., Coury, D.V., Duque, C.A. and Jorge, D.C. (2011) Development of detection and classification stages for a new distance protection approach based on cumulants and neural networks. *IEEE Power Energy Soc. Gen. Meet.*, 1–7.
18. Liu, Y., Steurer, M. and Ribeiro, P.F. (2009) Real-time simulation of time-varying harmonics. Chapter 18 in *Time-Varying Waveform Distortions in Power Systems* (Ribeiro, P.F., ed.), Wiley-IEEE Press.
19. Li, F., Li, Y. and Aggarwal, R.K. (2002) Combined wavelet transform and regression technique for secondary current compensation of current transformers. *IEE Proceedings: Generation, Transmission and Distribution*, **149** (4), 497–503.
20. Kang, Y.-C., Ok, S.-H. and Kang, S.-H. (2004) A CT saturation detection algorithm. *IEEE Transactions on Power Delivery*, **19** (1), 78–85.
21. Mitra, S.K. (2003) *Digital Signal Processing: A Computer-Based Approach*, McGraw Hill Publishing.

12

Wavelets Applied to Power Fluctuations

12.1 Introduction

As discussed in detail in Section 9.6 of Chapter 9 the wavelet transform is a mathematical tool that analyzes time-varying signals. Unlike conventional frequency analysis methods, wavelets give information about the range of frequency components of a signal as a function of time. Wavelets applications are of value when there are large time-varying discrepancies of the frequencies to be analyzed. This is especially true when the signal to be analyzed is time-varying and non-periodic. In power systems, wavelets are usually applied as a diagnostic tool for power quality identification; as for Fourier series, the flexibility of the tool allows application to many scientific and technological areas.

The advent of smart grids intensified the need for tools to support the balancing of loads both in generation and distribution. Under these circumstances, wavelets can be an enabling tool. In this chapter wavelets are applied to analyze power fluctuations in load and generation profiles. As the generation and load in AC power systems must always be balanced, wavelets can be applied to identify the required balancing capacity of the system by zooming into unforseen frequency variations. This is achieved by observing the time-varying nature of the dominant frequencies that are contained in the signals.

To guarantee a stable supply of an AC power system, a continuous and precise balance between the electric power generation and load demands is needed. To achieve this equilibrium, the frequency must continuously remain within very strict limits. In general, there are three reasons for imbalance: (a) deviation of the load from its predicted values; (b) the loss of generation or network failures; and (c) the deviation of the generation from its expected values.

With the advent of smart grids, defined as electricity networks that can intelligently integrate the behavior and actions of all users connected to it in order to efficiently deliver sustainable, economic and secure electricity supplies, the dynamics and complexity of an electricity grid has been substantially increased. Due to the introduction of renewable energy sources (RES) for a more distributed and renewable generation, unbalances can occur more

Power Systems Signal Processing for Smart Grids, First Edition. Paulo Fernando Ribeiro, Carlos Augusto Duque, Paulo Márcio da Silveira and Augusto Santiago Cerqueira.
© 2014 John Wiley & Sons, Ltd. Published 2014 by John Wiley & Sons, Ltd.
Companion Website: http://www.wiley.com/go/signal_processing/

frequently due to the increase in deviations between generation and load. Methods used to detect these variances are usually based on time-series analysis, a Fourier transform or fuzzy sets [1–3]. All deal with the uncertainty of generation and loads.

Traditionally, the need for the balancing capacity of a network [4–10] is defined with the help of probabilistic methods. These techniques specify a certain value for expected losses or the non-delivery of the specific load (energy), which are all based on a certain availability of balancing capacities of that load and its system [11–12]. The increase of distributed and renewable generators requires the development of new techniques. These should be able to define the fluctuation and variability of generation instead of its unavailability. Furthermore, these techniques should be applied to determine and size the specific requirements for generation and electricity storage characteristics.

In references [13–15], a method that uses a fast Fourier transform (FFT) to evaluate the needs of different classes of reserve capacity is established. One of the properties of FFT is the assumption of periodicity of a load; changes in the frequency domain over time will therefore not be reflected in the results.

On the other hand, wavelet theory (WT) can be applied to identify fluctuations of non-periodic generators. WT uses different window lengths to identify the different frequencies in a signal. Short windows are applied for high frequencies and long windows are used for low frequencies. The suggestion of using WT in power systems was first made during a meeting of the IEEE Working Group on Harmonic Modeling Simulation in 1993 and subsequently published in 1994 [16].

Wavelet transforms are used in different fields of expertise, including mathematics, quantum physics, seismic geology and electrical engineering [17]. In the field of electrical power engineering, wavelets are mainly applied to identify transients in power systems and for power system protection analysis [18]. Other research focuses on the application of wavelets in wind power forecasting [19,20].

In this chapter, we propose the application of wavelet theory to identify fluctuations in generation by RES. With the use of RES, the results can be used to define the needs for extra balancing capacity or for electricity storage concerns and provide the information needed to balance the power output of specific generators.

12.2 Basic Theory

As detailed in Section 9.6, wavelet theory is based on small (oscillatory) transitorily finite waves are used for the construction or reconstruction of time-varying signals. A wavelet is a function that must satisfy two basic conditions: it must be oscillatory and it must decay rapidly to zero [21]. The assumption is that any signal $P(t)$ can be represented by the superposition of scaled variations of a basic wavelet (the so-called mother wavelet). For this to happen, the mother wavelet is expanded in time (or space) and translated. A mother wavelet is defined as $\psi(t)$, where the original signal is a superposition such as [22]:

$$P(t) = \sum_k \sum_n P_{\text{DWT}}(n, k) \psi_{k,n}(t) \tag{12.1}$$

where $P(t)$ is the original signal converted to wavelets, $P_{\text{DWT}}(n, k)$ are the wavelet coefficients to be found using the transformed wavelets $W_{m,n}$ and the wavelets $\psi_{k,n}(t)$ are transformations of the mother wavelet $\psi(t)$ using [22–24]:

$$\psi_{\tau,s}(t) = |s|^{-1/2}\psi\left(\frac{t-\tau}{s}\right), \quad \tau, s \in R; s \neq 0. \tag{12.2}$$

Alternatively, assuming $s = 2^k$ and $\tau = 2^k n$ results in [23]

$$\psi_{n,k}(t) = (2)^{-k/2}\psi\left(\frac{t-2^k n}{2^k}\right), \quad \tau, s \in R; s \neq 0. \tag{12.3}$$

The wavelet coefficients $P_{\mathrm{DWT}}(n, k)$ can be found by taking the inner product of $P(t)$ and $\psi_{n,k}(t)$:

$$P_{\mathrm{DWT}}(n,k) = \left\langle P(t), \psi(t)_{k,n}\right\rangle. \tag{12.4}$$

The mother wavelet is dilated and translated. The dilation depends on k and is equal to 2^k. The translation depends on both k and n and is equal to $2^k n$ [25]. By dilating and translating the mother wavelet $\psi(t)$, the wavelet analysis allows for the breakdown of a signal according to scale (long window: low frequency; short window: high frequency).

By changing the scaling factor k the mother wavelet $\psi(t)$ is measured in time by a factor of 2^k, where different frequency ranges can be selected. To some extent the scaling factor can be regarded as a filter in the frequency domain [25]. The relation between the frequency range F_k and the scale factor k can be found from Equation (12.5) as presented in Section 9.6.5, repeated below for convenience:

$$F_k = \begin{cases} \left(\dfrac{F_s}{2^{k+1}}, \dfrac{F_s}{2^k}\right), k = 1, 2, \ldots, L-1 \\ \left(0, \dfrac{F_s}{2^k}\right), k = L, \end{cases} \tag{12.5}$$

where L is the maximum scaling factor in the analysis of a signal and F_s is the sampling frequency of that signal $P(t)$. Usually the sampling frequency F_s is predetermined by the dataset to be analyzed, where the lowest frequency to be analyzed follows from the purpose of analysis.

12.3 Application of Wavelets for Time-Varying Generation and Load Profiles

This section explores the applications of WT in generation and load profiles. Through the analysis of relevant signals, the variations in the signals of different frequency ranges can be characterized. This section also discusses three practical case scenarios: a system load, the power output of a wind farm and the power fluctuations in a small 10 kV distribution network (acting as a microgrid).

12.3.1 Fluctuation Analyses with FFT

The use of a FFT to determine and quantify variations in generation and load profiles was suggested in references [13,14]. By applying FFT, an impression of the variations in the

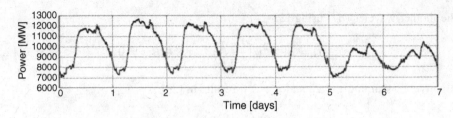

Figure 12.1 System load during one week.

frequency domain can be acquired. This can be seen in Figures 12.1 and 12.2 for a load pattern of one week, where the FFT provides the components in the frequency domain.

In Figure 12.2, different peaks in the frequency domain are indicated by gray marks corresponding to frequencies 1/(1 week) Hz and 1/(1 day) Hz. According to reference [15], large power variations occur over one-quarter of this time frame. By integrating the frequency components over different frequency ranges, the required amount of balancing capacity to balance the corresponding power variations can be determined. As FFT transforms time series into the frequency domains, any time information in the original signal is lost as periodicity is assumed in FFT. Short and rare events therefore average out if a long time series is analyzed. Shortening the time series by applying the short-time Fourier transform (STFT) with suitable time windows partially solves this issue. However, it leads to a reduction of its resolution in the frequency domain because of fixed windows in time.

12.3.2 Methodology

For a signal $P(t)$ the factors $P_{\mathrm{DWT}}(n,k)$ indicate the presence of transformed wavelet components at different scaling factors in the original signal at different time shifts. For all signals mentioned here, the signal is normalized using Equation (12.6) to be within [0,1] and the Meyer wavelet is applied for the wavelet transform:

$$P_{\mathrm{norm}}(t) = \frac{P(t) - \min[P(t)]}{\max[P(t)] - \min[P(t)]}. \tag{12.6}$$

The maximum scaling factor L for the wavelet decomposition follows from the lowest frequency to be analyzed and its sampling frequency F_{s}, which can be found from Equation (12.5). Any prior knowledge about fluctuation patterns that may perhaps be present in the signal should be considered in order to correctly define the required maximum scaling

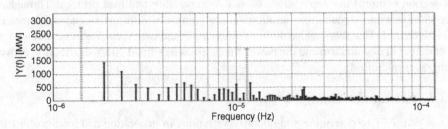

Figure 12.2 Frequency components of a system load.

factor L and the desired sampling period. As a next step, the signal is decomposed and the components $P_{DWT}(n, k)$ are those of interest. From all wavelet components $P_{DWT}(n, k)$ that represent a certain frequency range, a number of components are selected that contribute most to the original signal $P(t)$ according to Equation (12.1). The selection of these components is based on the RMS value of each individual factor.

The components considered to contribute most are those with the highest RMS values, as derived from Equation (12.7). As such, the most relevant scaling factors $P_{DWT}(n, k)$ and the original signal can be approximated by a synthetic signal. This is based on the superposition of the most relevant scaling factors as, indicated in Equation (12.1). These relevant scaling factors can also be investigated, as each component holds information about the original signal within a certain frequency bandwidth. When the individual relevant scaling factors are determined, they reveal additional information about fluctuations in the original signal at specific time periods.

From this point onwards the wavelet coefficients $P_{DWT}(n, k)$ are referred to as $A_{n,k}$ for simplicity. We therefore have

$$A_{k,\text{RMS}} = \sqrt{\frac{1}{L+1} \sum_{n=0}^{L} A_{k,n}^2}. \tag{12.7}$$

12.3.3 Load Fluctuations

To illustrate the application of wavelets in power systems, a system load was analyzed. The measurement of power data from a particular week in the year 2008 was used. A sampling period of 4 s was used, acquired from the Dutch utility. The data represent the aggregated production data from all large (i.e. > 60 MW) power plants in the Netherlands and is assumed to be equal to the aggregation of loads and losses countrywide. These data were first normalized to be within the domain [0,1] using Equation (12.6). After normalizing the data, a maximum scaling factor M of 15 was chosen using Equation (12.5) to find the components with the highest fundamental period of $1/F_M$ of 3 days. The three components $A_{n,k}$ with the highest RMS values were identified from Equation (12.7). Both the original and the synthetic load profiles are displayed in Figure 12.3.

A selection of the most relevant scaling factors as well as the moving average are displayed in Figure 12.4. A number of conclusions can be drawn from the wavelet analysis of this system load. The periods that are most relevant in the original signal are the daily and the half-daily fluctuations. It can be seen that specific daily patterns are less present during the weekend

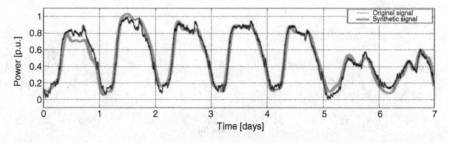

Figure 12.3 Original and synthesized load profiles based on relevant scaling factors of $A_{m,n}$.

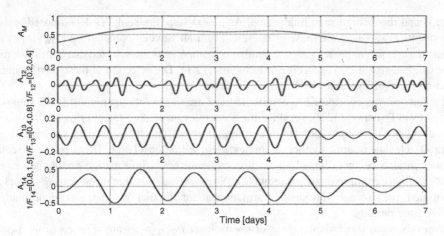

Figure 12.4 Most relevant scaling factors for a load profile during a week.

(days 6 and 7) than during weekdays (days 1 to 5). The weekly pattern can also be recognized by the moving average. It is interesting to note that these conclusions could not be drawn using conventional FFT analysis.

In addition to analysis of the system load described above, the active power load of one of the phases of a distribution transformer was monitored for a number of days. There are approximately 50 households connected to this phase. The sampling period of the measurements was of 10 min. The measured signal contained more high-frequency components than the system load due to the significantly lower aggregation level. To analyze these faster fluctuations, the data were first normalized to be within the domain [0,1] using Equation (12.6) and were subsequently analyzed through wavelets (FFT) with a maximum scaling factor of 4 to find components with a maximum fundamental period $(1/F_M)$ of 3 hours. The value of 3 hours was selected and their RMS values were chosen, based on experience. The three most relevant components of $A_{m,n}$ and $A_{M,n}$ were selected to reconstruct the signal. The original and reconstructed synthetic signals are given in Figure 12.5.

Both $A_{M,n}$ and the most relevant components were used to recreate the synthetic signal in Figure 12.5 and are displayed in Figure 12.6. From A_1 and A_2, the half-hourly and hourly fluctuating components in the load are clearly seen. It can be observed that these components are less evident during night-time (when the load is low). Again, this is a conclusion that cannot be drawn using conventional FFT analysis.

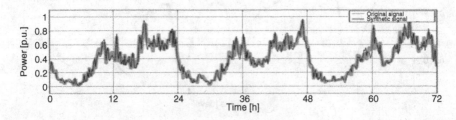

Figure 12.5 Original and synthesized load profiles during three days based on relevant scaling factors.

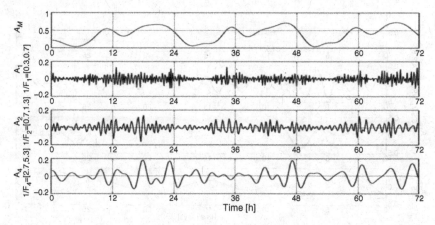

Figure 12.6 Most relevant scaling factors for a load profile during three subsequent days.

12.3.4 Wind Farm Generation Fluctuations

For this example a 25.5 MW onshore wind farm close to the city of Rotterdam in the Netherlands was selected. The wind farm consists of 17 identical wind turbines of 1.5 MW each. The turbines were connected to the 23 kV network via two connections with 8 and 9 turbines each. For both connections the aggregate current was measured during the month of May 2009. With the voltage assumed as constant, the aggregate power was calculated. The data, sampled at a period of 1 min, were available from the responsible grid operator and were normalized using Equation (12.6). A sampling period of 1 min is assumed to be short enough to recognize fluctuations in wind power for load balancing [3]. As such, wavelet analysis was performed with a maximum scaling factor M of 14 with a sampling period of 1 min; this leads to the lowest traceable period ($1/F_M$) of 3 days. Three components $A_{m,n}$ with the highest RMS value $A_{k,\mathrm{RMS}}$ were identified using Equation (12.7) and considered as the most relevant. By summing only these most relevant scaling factors and $A_{M,n}$, the original signal was synthesized. Both the original and the synthesized signals are displayed in Figure 12.7. $A_{M,n}$ and the three most relevant components ($A_{11,n}$, $A_{12,n}$ and $A_{13,n}$) are displayed in Figure 12.8.

From Figure 12.8 it can be concluded that, based on the RMS values, the three components that contribute most to the original signal have fundamental periods of 1.4–2.8, 2.8–5.7 and 5.7–11.4 days. It can therefore also be concluded that the main fluctuations of this wind farm occur on a daily scale. This also means that no electricity storage devices could be properly sized to handle these fluctuations in order to balance the power output of this wind farm.

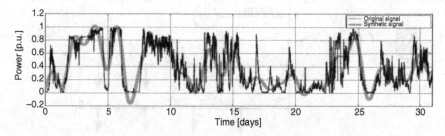

Figure 12.7 Original and synthesized wind farm generation profiles.

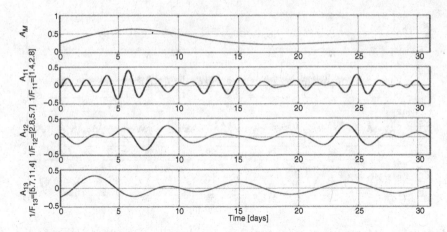

Figure 12.8 Most relevant wavelet components for a wind farm time-series.

12.3.5 Smart Microgrid

In this section, wavelet methodology is applied to to a smart microgrid to analyze power fluctuations. The microgrid in this study consists of a number of loads connected to a small 10 kV radial distribution network. A 2 MW wind turbine (W) and a conventional generator (G) are also connected to the network. The microgrid is to be operated in island mode, so the conventional generator needs to be able to deal with the aggregated fluctuations of load and wind turbine. The network topology is illustrated in Figure 12.9.

The power which is a conventional generator G needs to produce $P_G(t)$ is given by:

$$P_G(t) = \sum P_{\text{load}}(t) + \sum P_{\text{losses}}(t) - P_{\text{wind}}(t) \qquad (12.8)$$

where $P_{\text{load}}(t)$ is the power of each load, $P_{\text{losses}}(t)$ are the network losses and $P_{\text{wind}}(t)$ is the power generated by the wind turbine. The load profile in the network correlates to a month of national production data from the Dutch Tranmission Systems Operator (TSO) and is scaled to have a maximum value of 5 MW. The aggregated load was divided equally over the 8 loads in the network with a power factor of 1. The generation from the wind turbine was obtained by

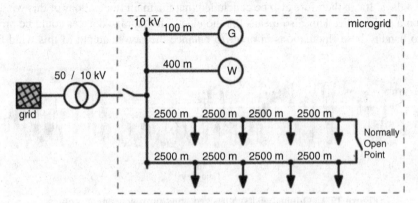

Figure 12.9 Structure of the smart microgrid.

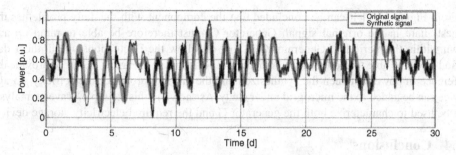

Figure 12.10 Original and synthesized power profiles of generator G during one month.

taking one month of aggregated wind data as described in Section 3.4. The wind power is scaled to have a maximum value of 2 MW and is also assumed to have a unit power factor. A month was simulated for each minute using a load flow simulation of the network in order to find the network losses $P_{losses}(t)$ and to determine the power $P_G(t)$ to be generated by the conventional generator G in order to balance the power in the network.

After completing the load flow simulations and normalizing the power profile using Equation (12.6), the power to be generated by generator G was investigated using the wavelet methodology in order to determine the characteristic fluctuations to be managed by generator G. The load flow simulations were performed for each minute during a month; the sampling period $1/F_s$ of $P_G(t)$ is 1 min. Using Equation (12.4,) the maximum scaling factor was decided to be 14 so that a maximum traceable period $1/F_M$ of 3 days occurs. As in the previous sections, the three components of $A_{m,n}$ with the highest RMS values were chosen to be most relevant. Using these components and only $A_{M,n}$ the power profile for generator G can be summarized. The original and reconstructed power profiles are given in Figure 12.10. It can be concluded from this figure that the components in the signal are equal to or smaller than 1 day and contribute most to the original signal.

Based on their RMS values calculated from Equation (12.7), the three most relevant wavelet components are $A_{10,n}$, $A_{11,n}$ and $A_{12,n}$. These, as well as the moving average $A_{M,n}$ are given in Figure 12.11.

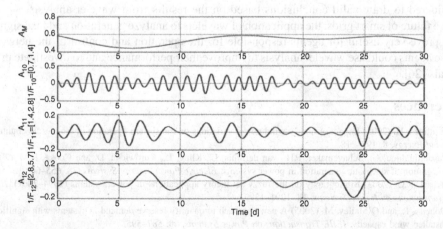

Figure 12.11 Most relevant wavelet components for the microgrid simulation during one month.

From Figure 12.11 it can be concluded that the component with the daily profile has the largest share in the original signal. Generator G must therefore be able to ramp up and down within this period. If Generator G is able to follow the fluctuations within the 1 day period it will be able to produce the synthetic profile as shown in Figure 12.10. To provide the difference in power between the synthesized and the original profiles, an electricity storage device can be added to the microgrid under study. As shown in this example, wavelet analysis can be used to characterize both the generator G and the required electricity storage device.

12.4 Conclusions

The increasing complexity of the electricity grid requires new signal processing techniques which can be used to properly and effectively analyze and diagnose the system conditions. Wavelet analysis is used more and more in power systems applications, and it is proposed that wavelet analysis is applied to determine fluctuation patterns in generation and load profiles. This is achieved by the filtering of its wavelet components based on their RMS values, and it is possible to identify the most relevant scaling factors from the analysis.

Three different case studies – a load profile, a wind farm and the operation of a microgrid – were analyzed in the context of a smart grid, demonstrating that the wavelet method can be applied to identify the most effective time range for assessment of both generation and load fluctuations. While conventional FFT algorithms only give the information as a function of frequency, STFT gives information in terms of both time and frequency. Unfortunately, it has disadvantages concerning the frequency resolution. However, since wavelet analysis yields the present frequency components as a function of time by using variable windows in the time domain, the issue of frequency resolution can be resolved.

The application of wavelet analysis as described in this chapter may prove useful both for the characterization of possible electricity storage devices and the determination of the required balancing capacities. It can also improve the bids of energy companies in energy markets by having specific information on the characteristic fluctuations of its renewable generation, providing them with the ability to counteract these by using conventional generation and electricity storage. Experience in selecting the number of scaling factors and main frequency ranges to be identified is important for accurate results. Furthermore, any prior knowledge concerning characteristic fluctuations present in the signal should be considered to draw valid conclusions based on the results from wavelet analysis.

In a future of smart grids, the application of wavelets to analyze generation and load signals may prove very useful for agents responsible for the operation and control of the network. These agents could use wavelet analysis to improve their performance and to investigate price signals [26].

References

1. Holttinen, H. (2005) Impact of hourly wind power variations on the system operation in the Nordic countries. *Wind Energy*, **8**, 197–218.
2. Papaefthymiou, G., Schavemaker, P.H., van derSluis, L., Kling, L., Kurowicka, D. and Cooke, R.M. (2006) Integration of stochastic generation in power systems. *Electric Power Energy Systems*, **28**, 655–667.
3. Bansal, R.C. (2003) Bibliography on the fuzzy set theory applications in power systems (1994–2001). *IEEE Transactions on Power Systems*, **18**, 1291–1299.
4. Doherty, R. and O'Malley, M. (2005) A new approach to quantify reserve demand in systems with significant installed wind capacity. *IEEE Transactions on Power Systems*, **20**, 587–595.

5. Allen, E.H. and Ilic, M.D. (2000) Reserve markets for power systems reliability. *IEEE Transactions on Power Systems*, **15**, 228–233.
6. Havel, P., Horácek, P., Cerný, V., and Fantík, J. (2008) Optimal planning of ancillary services for reliable power balance control. *IEEE Transactions on Power Systems*, **23**, 1375–1382.
7. Amjady, N. and Keynia, F. (2010) A new spinning reserve requirement forecast method for deregulated electricity markets. *Applied Energy*, **87**, 1870–1879.
8. Booth, R.R. (1972) Power system simulation model based on probability analysis. *IEEE Transactions on Power Systems*, **91**, 62–69.
9. Fotuhi-Firuzabad, M., Bilinton, R. and Aboreshaid, S. (1996) Spinning reserve allocation using response health analysis. *Generation, Transmission and Distribution, IEE Proceedings*, **143**, 337–343.
10. Arce, J.R., Ilic, M.D. and Garcés, F.F. (2001) Managing short-term reliability related risks. In Proceedings of Power Engineering Society Summer Meeting, Vancouver, British Columbia, Canada, July 15–19, pp. 516–522.
11. Billinton, R. and Allan, R.N. (1984) Operating reserve. In *Reliability Evaluation of Power Systems*, Pitman Publishing Limited, pp. 139–171.
12. Mazumdar, M. and Bloom, J.A. (1996) Derivation of the Balerieux formula of expected production costs based on chronological load considerations. *Electric Power Energy Systems*, **18**, 33–36.
13. Alvarado, F.L. (2002) Spectral analysis of energy-constrained reserves. In Proceedings of 35th Hawaii International Conference on Systems Science, pp. 749–756.
14. Frunt, J., Kling, W.L. and Myrzik, J.M.A. (2009) Classification of reserve capacity in future power systems. In Proceedings of 6th International Conference on European Energy Market, Leuven, Belgium, May 27–29.
15. Frunt, J., Kling, W.L. and van den Bosch, P.P.J. (2010) Classification and quantification of reserve requirements for balancing. *Electric Power Systems Research*, **80**, 1528–1534.
16. Ribeiro, P.F. (1994) Wavelet transform: an advanced tool for analyzing non-stationary distortions in power systems. ICHPS VI/94 Italy.
17. Graps, A. (1995) An introduction to wavelets. *IEEE Computing in Science and Engineering*, **2**, 1–18.
18. Galli, A.W., Heydt, G.T. and Ribeiro, P.F. (1996) Exploring the power of wavelet analysis. *IEEE Computer Applications in Power*, **9**, 37–41.
19. Dong, L., Wang, L., Liao, X., Gao, Y., Li, Y. and Wang, Z. (2009) Prediction of wind power generation based on time series wavelet transform for large wind farm. In Proceedings of 3rd International Conference on Power Electronric Systems and Applications, Hong Kong, May 20–22, pp. 1–4.
20. Lei, C. and Ran, L. (2008) Short-term wind speed forecasting model for wind farm based on wavelet decomposition. In Proceedings of 3rd International Conference on Electric Utility Deregulation and Restructuring and Power Technologies, Nanjuing, China, April 6–9, pp. 2525–2529.
21. Masson, P.J., Silveira, P.M., Duque, C. and Ribeiro, P.F. (2008) Fourier series: Visualizing Joseph Fourier's imaginative discovery via fea and time-frequency decomposition. 13th International Conference on Harmonics and Quality of Power, ICHQP. 28 Sept–1 Oct 2008, 1–5.
22. Lee, C.H., Wang, Y.J. and Huang, W.L. (2000) A literature survey of wavelets in power engineering applications. *Proceedings of the National Science Council, Republic of China(A)*, **24**, 249–258.
23. Lebedeva, E.A. and Protasov, V.Y. (2008) Meyer wavelets with least uncertainty constant. *Mathematical Notes*, **84**, 680–687.
24. Wilkinson, W.A. and Cox, M.D. (1996) Discrete wavelet analysis of power system transients. *IEEE Transactions on Power Systems*, **11**, 2038–2044.
25. Xu, J., Senroy, N., Suryanarayanan, S., and Ribeiro, P.F. (2006) Some techniques for the analysis and visualization of time-varying waveform distortions. In Proceedings of 38th North American Power Systems Symposium, pp. 257–261.
26. Frunt, J., Kling, W.L. and Ribeiro, P.F. (2011) Wavelet decomposition for power balancing analysis. *IEEE Transactions on Power Delivery*, **26** (3), 1608–1614.

13

Time-Varying Harmonic and Asymmetry Unbalances

13.1 Introduction

Traditional harmonic distortion analysis assumes balanced and steady-state conditions. However, in the real world this is far from the physical reality and experience of engineers that deal with unbalanced and time-varying components. In theory, certain harmonic components are associated with specific sequences such as positive, negative or zero. However, in real life harmonic distortions can cause a gamut of sequences for every possible time-varying frequency. The readings of these conditions complicate any analysis and need to be properly modeled. The time-varying nature of these unbalances at higher frequencies has not yet been studied in detail. This chapter introduces these unbalances with the use of the sliding-window recursive Fourier transform (SWRFT), with which the frequencies of the time-varying harmonic components are computed, calculated and then plotted for their positive, negative and zero sequences and parameters. The recorded signals used are from a real system. These time-varying parameters are used to investigate the nature of its transient phenomena and to provide information for protection and control applications.

In the past, harmonic analyses was performed taking into account only a few harmonic-producing devices such as power electronic (PE) inverters and converters, HVDC, SVC, and so on. As their supply systems are well balanced and symmetrical, most studies are carried out in the steady state and based on positive sequence network representations. However, as more and more harmonic-producing loads are connected to the grid, this 'steady-state' condition has changed significantly [1–3]. Due to the randomness of the loads and the dynamics of three-phase systems, two kinds of phenomena have become common: (1) unbalanced harmonics and (2) continuous variability of harmonics [1]. New concepts have been published, including suggestions that combine probabilistic and spectral methods (also referred as evolutionary spectrum) [4]. However, most of the techniques applied rely on Fourier transform methods that implicitly assume stationary conditions and a linearity of components.

Power Systems Signal Processing for Smart Grids, First Edition. Paulo Fernando Ribeiro, Carlos Augusto Duque, Paulo Márcio da Silveira and Augusto Santiago Cerqueira.
© 2014 John Wiley & Sons, Ltd. Published 2014 by John Wiley & Sons, Ltd.
Companion Website: http://www.wiley.com/go/signal_processing/

Utilities and industries have therefore focused their attention on methods of analysis with the ability to provide correct assessments of time-varying harmonic distortions. This issue has become crucial for control, protection, supervision and the proper diagnosis of problems.

Harmonic distortion studies in electric systems when significant variations are observed due to load or system variations have been performed using a probabilistic approach and assuming that the harmonic components vary slowly enough to affect the accuracy of the analytical and monitoring process. Another relevant issue is related to the unbalanced harmonics in three-phase systems. Finding the sequence harmonic components in balanced systems is a well-studied subject. However, when the supply voltage and loads are unbalanced, there will be large deviations from the traditional pattern. Under these circumstances, symmetrical component theory cannot be applied for accurate identification of the sequence components used for power quality assessment. The same occurs for other important applications such as protection and control of power systems [4]. Publications on time-varying harmonic unbalances is limited [5–8].

In order to analyze distorted waveforms that vary continuously in the time domain the concept of time-varying waveform distortions is discussed [9]. A sliding-window discrete Fourier transform (SWDFT) can be a useful tool as it provides the capacity to analyze and visualize voltage and current waveforms and graphically illustrate the time-varying harmonic components.

This chapter describes and illustrates how this approach can be used to determine harmonic sequence components with some practical examples. Some useful parameters in the time domain for each of these situations are discussed, knowledge of which can lead to a better understanding of unbalances and the asymmetries related to each of the harmonic frequencies.

13.2 Sequence Component Computation

Computation of the symmetrical components is depicted in Figure 13.1, based on the sliding-window recursive DFT SWRDFT presented in Chapter 9. For each phase the signal is split into its harmonics of order h using the SWDFT architecture. The magnitude and angle of each harmonic is subsequently used for the computation of its symmetric components. The result

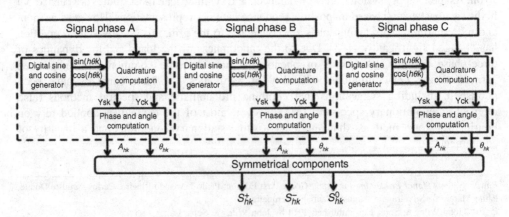

Figure 13.1 Symmetrical components computation using SWDFT.

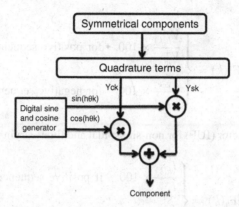

Figure 13.2 Architecture to reconstruct the components in the time domain.

are vectors of positive (S_{hk}^+), negative (S_{hk}^-) and zero (S_{hk}^0) sequences representing the phasor of harmonic h at instant k. The notation S can be used for voltage or current.

The symmetrical components are computed according to:

$$\begin{bmatrix} S_{hk}^0 \\ S_{hk}^+ \\ S_{hk}^- \end{bmatrix} = \frac{1}{3} \begin{bmatrix} 1 & 1 & 1 \\ 1 & \alpha & \alpha^2 \\ 1 & \alpha^2 & \alpha \end{bmatrix} \begin{bmatrix} S_{Ahk} \\ S_{Bhk} \\ S_{Chk} \end{bmatrix}$$ (13.1)

where S_{hk}^+, S_{hk}^- and S_{hk}^0 are the sequence components, h is the harmonic order and α can be represented by the phasor:

$$\alpha = 1\angle 120°$$ (13.2)

$$\alpha^2 = 1\angle 240°.$$ (13.3)

The symmetrical components can then be reconstructed in the time domain by using the architecture shown in Figure 13.2. The quadrature term is obtained from the symmetrical component vector. Then, using Equation (13.2) of the Fourier theory, this term is used to obtain the component in its own time domain.

13.3 Time-Varying Unbalance and Harmonic Frequencies

As defined by the European standards, the degree of unbalance or the voltage unbalance factor (VUF) is the ratio of the negative sequence voltage to the positive sequence voltage, calculated:

$$\%\text{VUF} = \frac{V^-}{V^+} \times 100$$ (13.4)

where V^+ and V^- are the positive and negative sequence voltages, respectively. This parameter is used at several harmonic frequencies for a better understanding of unbalances in non-sinusoidal and time-varying situations:

$$\%\text{VUF}_h(k) = \begin{cases} \dfrac{V^-_{h(k)}}{V^+_{h(k)}} \times 100, & \text{for positive sequence,} \\[3mm] \dfrac{V^+_{h(k)}}{V^-_{h(k)}} \times 100, & \text{for negative sequence.} \end{cases} \tag{13.5}$$

The current unbalance factor (IUF) for non-sinusoidal and time-varying conditions is defined:

$$\%\text{IUF}_h(k) = \begin{cases} \dfrac{I^-_{h(k)}}{I^+_{h(k)}} \times 100, & \text{if positive sequence,} \\[3mm] \dfrac{I^+_{h(k)}}{I^-_{h(k)}} \times 100, & \text{if negative sequence.} \end{cases} \tag{13.6}$$

13.4 Computation of Time-Varying Unbalances and Asymmetries at Harmonic Frequencies

A simulated signal was used to evaluate the proposed methodology. The main concern is to evaluate the accuracy of the signal decomposition of its harmonics and then to compute the symmetrical components and unbalances at these different harmonic frequencies.

Furthermore, the three-phase signal simulated is time-varying. The scenario below illustrates the main use of this methodology. A signal was generated according to following equations:

$$V_{A1} = \sqrt{2}M_{A1}\sin(wt + \theta_{A1}) \tag{13.7}$$

$$V_{B1} = \sqrt{2}M_{B1}\sin(wt + \varnothing_{B1}) \tag{13.8}$$

$$V_{C1} = \sqrt{2}M_{C1}\sin(wt + \alpha_{C1}) \tag{13.9}$$

$$V_{A5} = \sqrt{2}M_{A5}\sin(5\,wt + \theta_{A5}) \tag{13.10}$$

$$V_{B5} = \sqrt{2}M_{B5}B\sin(5\,wt + \varnothing_{B5}) \tag{13.11}$$

$$V_{C5} = \sqrt{2}M_{C5}\sin(5\,wt + \alpha_{C5}) \tag{13.12}$$

$$V_{A7} = \sqrt{2}M_{A7}\sin(7\,wt + \theta_{A7}) \tag{13.13}$$

$$V_{B7} = \sqrt{2}M_{B7}\sin(7\,wt + \varnothing_{B7}) \tag{13.14}$$

$$V_{C7} = \sqrt{2}M_{C7}\sin(7\,wt + \mu_{C7}) \tag{13.15}$$

where $M_{A1} = M_{B1} = M_{C1} = 1$; $M_{A5} = M_{C5} = 0.3$; $M_{A7} = M_{B7} = M_{C7} = 0.2$; $\theta_{A1} = \theta_{A5} = \theta_{A7} = 0$; $\varnothing_{B1} = \varnothing_{B5} = \alpha_{C5} = -120°$; $\alpha_{A7} = \alpha_{C7} = 120°$ and A, B and C are different phases.

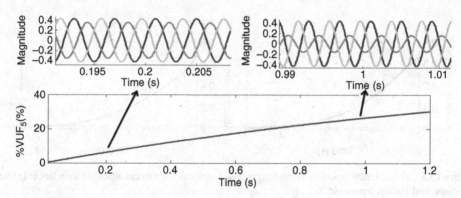

Figure 13.3 Unbalance computation at 5th harmonic for the simulated signal.

The unbalance was produced by changing the magnitude of phase B at the 5th harmonic and the phase angle of the 7th harmonic according to

$$M_{B5} = 0.3\, e^{-t} \tag{13.16}$$

$$\alpha_{B7} = 120\, e^{-t}. \tag{13.17}$$

The signal is the sum of the three components:

$$V_A = V_{A1} + V_{A5} + V_{A7} \tag{13.18}$$

$$V_B = V_{B1} + V_{B5} + V_{B7} \tag{13.19}$$

$$V_C = V_{C1} + V_{C5} + V_{C7}. \tag{13.20}$$

Figure 13.3 shows the 5th harmonic decomposed using the SWDFT and its unbalance is analyzed using the methodology above. Equation (13.19) describes the magnitude of phase B at the 5th harmonic and it decreases exponentially. Figure 13.3 shows two points in the time domain to better illustrate the unbalance calculations; two points of the phase angle at 7th harmonic changes exponentially are shown in Figure 13.4.

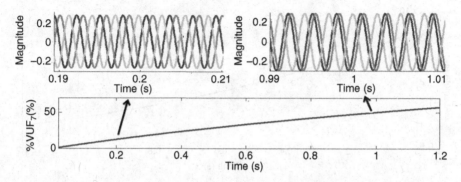

Figure 13.4 Unbalance computation at 7th harmonic for the simulated signal.

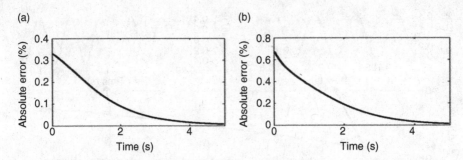

Figure 13.5 Error measurement by using the proposed methodology to calculate the unbalance: (a) 5th harmonic and (b) 7th harmonic.

The signal was decomposed according to the methodology above, and the unbalance was calculated using the symmetrical components from the signal V_A, V_B and V_C. The error was computed as the difference between the calculated theoretical value using the equations and the obtained results, and is depicted in Figure 13.5. As can be seen, the error is less significant if the unbalance is higher.

Figure 13.6 shows the sum of all zero-sequence components. The result is compared to the theoretical value, and the error shown in Figure 13.7.

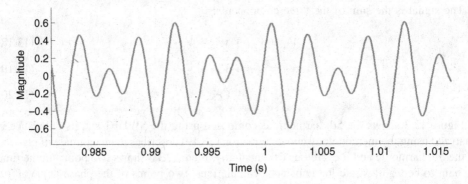

Figure 13.6 Zero-sequence component in time domain.

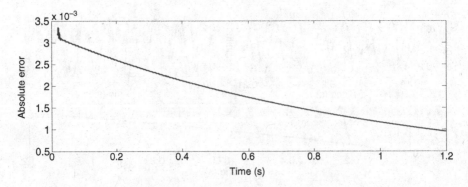

Figure 13.7 Error computing the zero-sequence component in time domain.

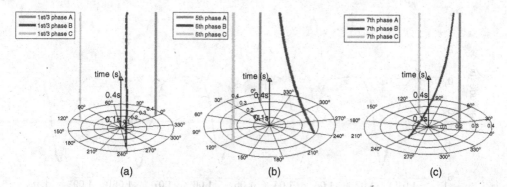

Figure 13.8 Time varying phasor of: (a) fundamental, (b) 5th harmonic and (c) 7th harmonic.

Figure 13.8 shows the phasors as a function of time. The fundamental is balanced and can easily be seen. The 5th harmonic presents magnitude variations in one of the phases and the 7th harmonic has angle variations. This graphs is useful to observe the time-varying phasors and is used in the examples described in the following section.

13.5 Examples

Examples of the methodology applied to real signals under time-varying conditions are described in the following. Sample conditions include a three-phase inrush current, a three-phase voltage during a sag and the unbalance of an electronic converter system. The computations include both the time-varying unbalance and the asymmetry during these transients. These parameters can also be applied to support control, protection, supervision and accurate diagnosis of possible network problems.

13.5.1 Inrush Current

For illustrative purposes a real three-phase inrush current was used and is shown in Figure 13.9. From the figure it can be seen that the transient behavior remains after the

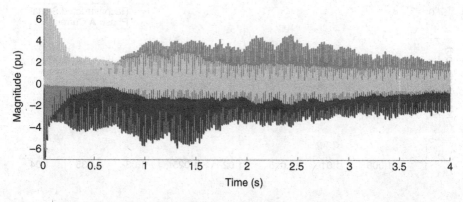

Figure 13.9 Three-phase inrush current during a transformer energization.

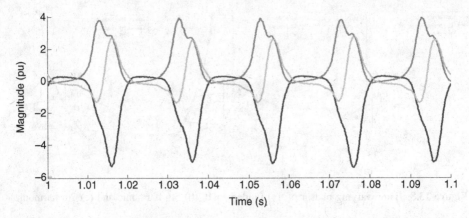

Figure 13.10 Zoom view of a three-phase inrush current during a transformer energization.

signal sample is used. The graph represents 4 s in time, so that the shape of the signal can be clearly seen and the time-varying behavior highlighted. Details can be seen in Figure 13.10.

The signal was decomposed using SWDFT. The harmonics were reconstructed and the sum of all components was computed in order to compare it to the original signal. The result of phase A is shown in Figure 13.11. A small deviation can be seen during the fast transient, but this can be disregarded. The components are subsequently used to calculate the symmetric components for each harmonic and the related unbalances and asymmetries as a function of time.

Figure 13.12 shows the sum of all zero-sequence components, representing the actual neutral to ground current which is subsequently used for the proper diagnoses of unsafe conditions. Figure 13.12b shows the asymmetry ratio (I_0/I_1) for the fundamental frequency. These parameters clarify the nature of the phenomena and can be used to set improved protection systems.

The unbalance is verified by calculating the negative-sequence component over the positive sequence, according to Equation (13.13). The results for the fundamental, second and fifth harmonics are shown in Figures 13.13–13.15 respectively. A high degree of unbalance can be

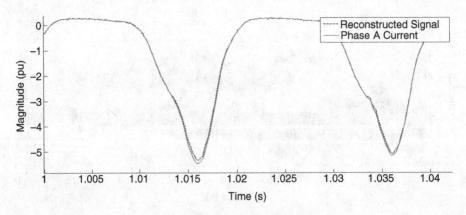

Figure 13.11 Reconstructed signal compared to the original signal.

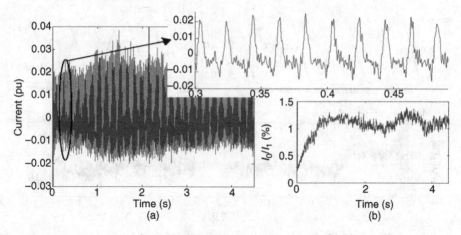

Figure 13.12 (a) Zero-sequence component in time domain and (b) asymmetry ratio.

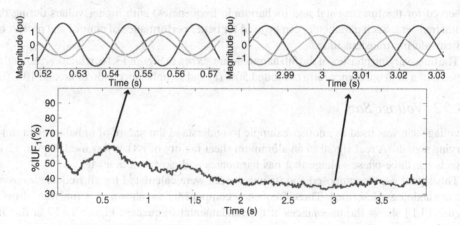

Figure 13.13 Time-varying current unbalance for fundamental component.

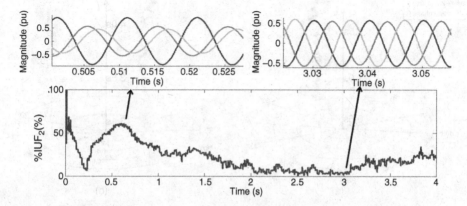

Figure 13.14 Time-varying current unbalance for 2nd harmonic.

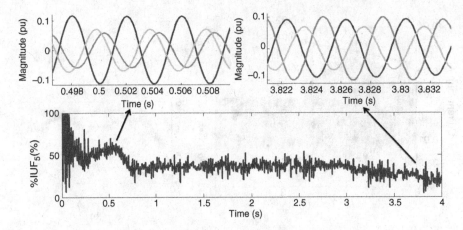

Figure 13.15 Time-varying current unbalance for 5th harmonic.

observed for the fundamental and its harmonic frequencies, with higher values during the initial transient. Such parameters can be used for better performance of control and protection systems during transient time.

The time-varying phasors of the 5th harmonic are shown in Figure 13.16, which correspond to the case of a high unbalance ratio of around 50% (a) and an unbalance ratio of around 10% (b).

13.5.2 Voltage Sag

A voltage sag was used as another example to understand the nature of unbalances in time-varying signals. A real signal in an aluminum sheet facility of 88 kV was used. Figure 13.17 depicts the three-phase voltage that has harmonics and experiences a sag.

This signal was decomposed and its unbalances were calculated for all frequencies using the methodology described. Three frequency components are shown as a function of time: Figure 13.18 shows the unbalances at the fundamental frequency; Figure 13.19 at the 5th

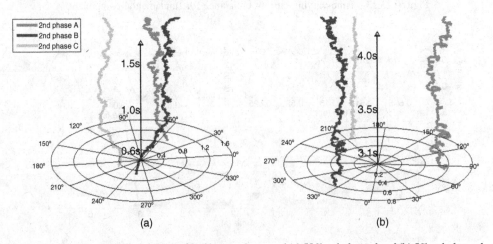

Figure 13.16 Time-varying phasors of 2nd harmonic around (a) 50% unbalanced and (b) 5% unbalanced.

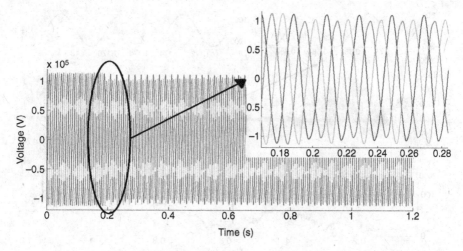

Figure 13.17   Real signal of voltage during a sag.

harmonic; and Figure 13.20 at the 7th harmonic. As expected, analysis of the sag decomposition reveals that the fundamental frequency initially has a greater unbalance, after which it slowly recovers. However, the 5th and 7th harmonics demonstrate completely different behavior during the sag with a variation range from 10% to 100%. For instance, the 5th harmonic is almost balanced at 0.6 s, although at approximately 1.1 s the unbalance is >60%.

It is important to note that the sequence changes during the sag. Initially the 5th harmonic has a negative sequence, but during the sag it becomes positive and then again negative after the event. The same analysis can be performed for the 7th harmonic: the phase sequence is

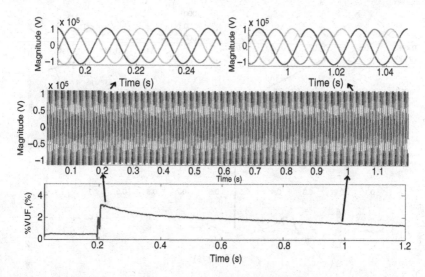

Figure 13.18   Fundamental component decomposed by SWRDFT and unbalance computation during a sag transient.

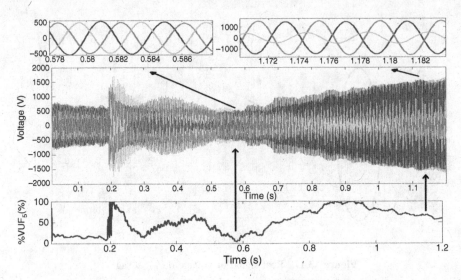

Figure 13.19 Three-phase 5th harmonic decomposed by SWRDFT and the unbalance computation during a sag transient.

switched twice during the sag and is changed again when the unbalance of the three-phase signal achieves 100%.

The phasor behavior of the 5th and 7th harmonics is depicted in Figure 13.21. As can be seen, the magnitude of 5th harmonic phases A and B increase and phase C decreases. The phase angles of phases A and B demonstrate large changes during this period. For the 7th

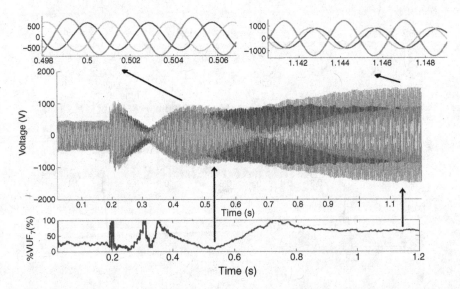

Figure 13.20 Three-phase 7th harmonic decomposed by SWRDFT and unbalance computation during a sag transient.

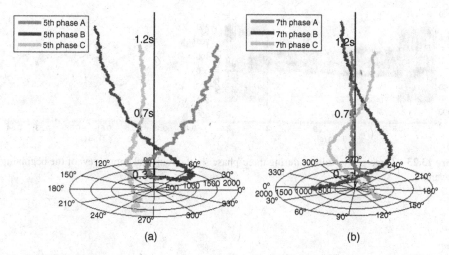

Figure 13.21   Three-phase voltage symmetrical component for (a) 5th harmonic and (b) 7th harmonic.

harmonic, notice that at 0.6 s the signal is almost balanced but, due to variations of the phase angles and magnitudes, the unbalance increases.

13.5.3 Unbalance in Converters

The method proposed in Sections 13.2–13.4 is also used to observe the unbalances during a voltage sag when converters are connected to the grid. The schematic of the three-phase system is shown in Figure 13.22.

The first trial was conducted with a three-phase short circuit of 1.5 s. The waveforms of the three-phase voltage and current are shown in Figures 13.23 and 13.24, respectively.

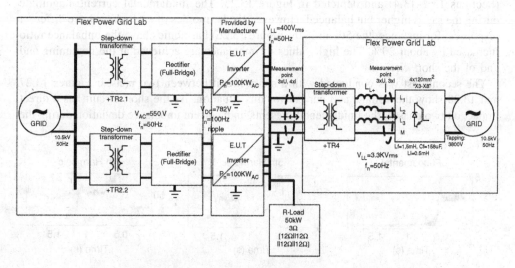

Figure 13.22   Schematic diagram of the experiment with converters.

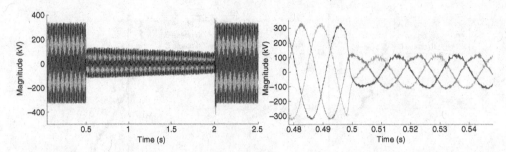

Figure 13.23 Three-phase voltage during three-phase short circuit and zoom view of the beginning of the event.

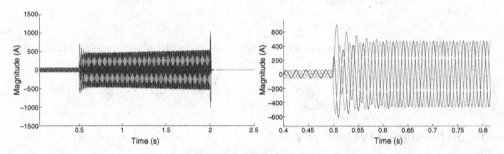

Figure 13.24 Three-phase current during three-phase short circuit and zoom view of the beginning of the event.

The behaviors of unbalances at harmonic frequencies are estimated using the method (Sections 13.2–13.4) and depicted in Figure 13.25. The fundamental current magnitude during the sag is higher but balanced. However, the harmonics of each phase have different responses; for instance, the 5th and 7th (Figure 13.26) harmonics have the unbalance ratio increased by about 20%. The high values of imbalance are achieved at the beginning and end of the short circuit.

The second test was conducted with a short circuit between two phases; Figures 13.27 and 13.28 show the voltage and current for this situation. As the short circuit has a direct impact on two phases, the fundamental presents high current unbalance deviation during the

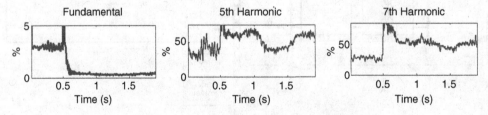

Figure 13.25 Current unbalance during voltage sag.

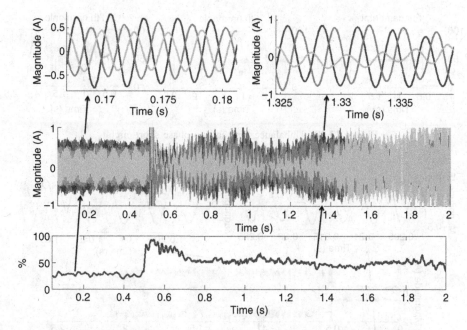

Figure 13.26   Unbalances at 7th harmonic during a three-phase short circuit.

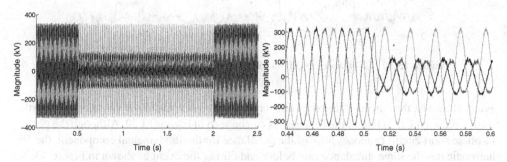

Figure 13.27   Voltage during sag for phase-to-phase short circuit.

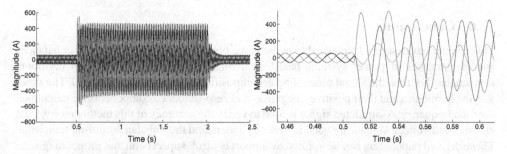

Figure 13.28   Current during voltage sag for phase-to-phase short circuit.

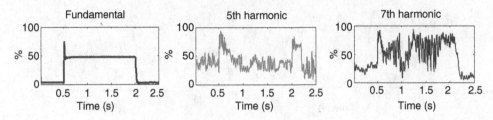

Figure 13.29   Unbalance for phase-to-phase short circuit.

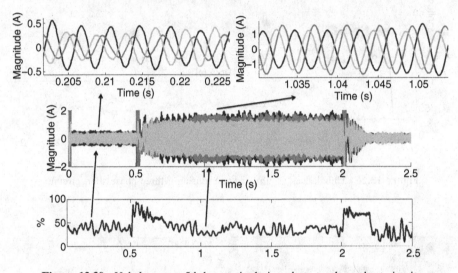

Figure 13.30   Unbalances at 5th harmonic during phase-to-phase short circuit.

event. However, the current harmonics unbalances experience particular variations during the short circuit similar to that of the first test, as can be seen in Figure 13.29. Although the phase-to-phase short circuit produces 50% of the unbalance for the fundamental component, the 5th harmonic has the same unbalance rate before and during the event as shown in Figure 13.30. The 7th harmonic has high unbalance variations due to the low magnitude of this component that experiences greater interference from harmonics and limitations of the filter.

13.6   Conclusions

This chapter has described a method to assist in the observation and evaluation of unbalances and asymmetries in power systems based on a time-varying decomposition of harmonic frequencies. For this, the signal processing decomposition method SWDFT is used. The time-varying harmonics and their positive-, negative- and zero-sequence components are calculated for each frequency. A simulated signal is used to verify the accuracy of this methodology. Real currents and voltage signals are used in order to understand the unbalances during transients. These derived parameters can be applied to support control, supervision and proper diagnosis of possible power quality and network events.

To the knowledge of the authors, this is the first time that such a time-varying decomposition has applied in this context and explicitly illustrated. This method may become a very useful tool in aiding the engineer to better understand the harmonic distortions of certain phenomena under time-varying and unbalanced system conditions.

References

1. Bonner, A., Grebe, T., Gunther, E., Hopkins, L., Marz, M.B., Mahseredjian, J., Miller, N.W. *et al.* (1996) IEEE task force on harmonics modeling and simulation: modeling and simulation of the propagation of harmonics in electric power networks. Part i: concepts, models and simulation techniques. *IEEE Transactions on Power Delivery*, **11** (1), 452–465.
2. Baghzouz, Y., Burch, R.F., Capasso, A., Cavallini, A., Emanuel, A.E., Halpin, M., Langella, R. *et al.* (Probabilistic Aspects Task Force of Harmonics Working Group) (2002) Time-varying harmonics. Part ii: harmonic summation and propagation. *IEEE Transactions on Power Delivery*, (1), 279–285.
3. Morrison, R.E. (1984) Probabilistic representation of harmonic currents in AC traction systems. *Proceedings of IEEE, Part B*, **131** (5), 181–189.
4. Ribeiro, P.F. (2003) A novel way for dealing with time-varying harmonic distortions: the concept of evolutionary spectra. Power Engineering Society General Meeting, 13–17 July, **2**, pp. 1153.
5. Marei, M.I., Saadany, E.F.E. and Salama, M.M.A. (2004) A processing unit for symmetrical components and harmonics estimation based on a new adaptive linear combiner structure. *IEEE Transactions on Power Delivery*, **19** (3), 1245–1252.
6. Henao, H., Assaf, T. and Capolino, G.A. (2003) The discrete Fourier transform for computation of symmetrical components harmonics. IEEE Bologna PowerTech Conference, June 23–26, Bologna, Italy.
7. Silveira, P.M., Duque, C.A., Baldwin, T. and Ribeiro, P.F. (2008) Time-varying power harmonic decomposition using sliding-window DFT. IEEE International Conference on Harmonics and Quality of Power, Wollongong.
8. Hartley, R. and Welles, K. (1990) Recursive computation of the Fourier transform. IEEE International Symposium on Circuits and Systems, **3**, 1792–1795.
9. Jouanne, A. and Banerjee, B. (2001) Assessment of voltage unbalance. *IEEE Transactions on Power Delivery*, **16** (4), October.

Index

Power Systems Signal Processing for Smart Grids, First Edition. Paulo Fernando Ribeiro, Carlos Augusto Duque,
Paulo Márcio da Silveira and Augusto Santiago Cerqueira.
© 2014 John Wiley & Sons, Ltd. Published 2014 by John Wiley & Sons, Ltd.
Companion Website: http://www.wiley.com/go/signal_processing/

Printed in the United States
By Bookmasters